Leibniz on Binary

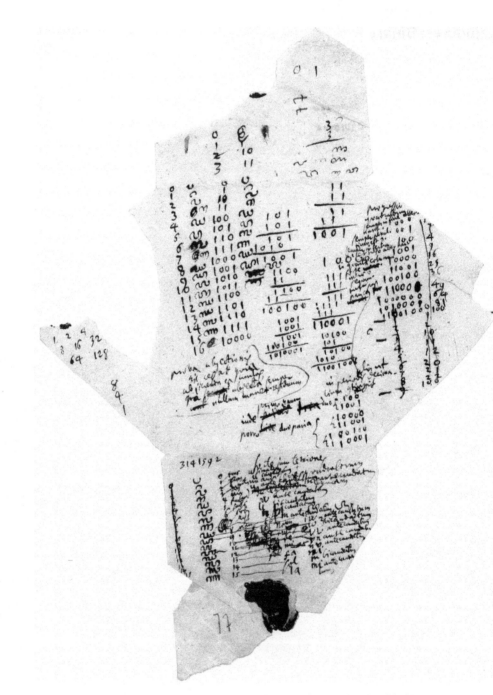

Manuscript of "Sedecimal on an Envelope" (LH 35, 3 B 5 Bl. 77).

Leibniz on Binary
The Invention of Computer Arithmetic

Lloyd Strickland and Harry Lewis

The MIT Press
Cambridge, Massachusetts
London, England

This book was set in New Times Roman by the editor using LATEX. Printed and bound in the United States of America.

Library of Congress Cataloging-in-Publication Data
Names: Leibniz, Gottfried Wilhelm, Freiherr von, 1646–1716, author. | Strickland, Lloyd, 1973– editor,
    translator, writer of added commentary. | Lewis, Harry R., editor, translator, writer of added commentary.
Title: Leibniz on binary : the invention of computer arithmetic / Lloyd Strickland and Harry Lewis.
Description: Cambridge, Massachusetts : The MIT Press, [2022] | Includes bibliographical references and index.
Identifiers: LCCN 2021059164 | ISBN 9780262544344 (paperback)
Subjects: LCSH: Leibniz, Gottfried Wilhelm, Freiherr von, 1646–1716—Translations into English | Binary
    system (Mathematics) | Mathematics, German. | Binary system (Mathematics)—History—Sources.
Classification: LCC QA141.4 .L45 2022 | DDC 513.5/2—dc23/eng20220405
LC record available at https://lccn.loc.gov/2021059164

10   9   8   7   6   5   4   3   2   1

This book is dedicated

by Lloyd to Vernon Pratt, mentor, friend, and *vir eximie*, as Leibniz would say.
   As an indefatigable source of wisdom, inspiration,
   and knowledge of where to find good desserts,
   you are too good for this world.

by Harry to the memory of his father, Emil Harold Lewis,
   who with the 16th General Hospital, from 1943 to 1945,
   helped save his father's birthland, free his grandfather's,
   and create the enduring Anglo-German-American friendship
   of which this book is fruit.
   Medio tutissimus ibis.

All programmers are optimists.
—Frederick P. Brooks Jr., *The Mythical Man-Month*

Contents

List of Figures

Abbreviations

A	Leibniz (1923–). Cited by series, volume, and page or item number (N).
DSR	Leibniz (1992).
Dutens	Leibniz (1768).
GM	Leibniz (1849–1863). Reprinted as Leibniz (1977).
Gotha	Manuscript held by Forschungsbibliothek Gotha, Erfurt, Germany.
GP	Leibniz (1875–1890).
LBr	Manuscript held by Gottfried Wilhelm Leibniz Bibliothek, Niedersächsische Landesbibliothek, Hanover, Germany. Cited by shelfmark and Blatt [sheet].
LGR	Leibniz (2016a).
LH	Manuscript held by Gottfried Wilhelm Leibniz Bibliothek, Niedersächsische Landesbibliothek, Hanover, Germany. Cited by shelfmark and Blatt [sheet].
LTS	Leibniz (2011b).
PPL	Leibniz (1969).
SLT	Leibniz (2006b).
W	Leibniz (2006a).
WOC	Leibniz (1994).
Z	Zacher (1973).

Preface

It would be pointless to say that Mr. Leibniz was a mathematician of the first rank,
[since] it is through mathematics that he is most generally known.
—Bernard le Bovier de Fontenelle (1718, 108–109)

Despite his original contributions to a host of disciplines, such as law, philosophy, politics, languages, and many areas of science, the fame and renown of the polymath Gottfried Wilhelm Leibniz (1646–1716) has always rested principally on his pioneering mathematical work, in particular his independent invention of the calculus in 1675. Another of his enduring mathematical contributions was his invention of binary arithmetic, though the utility of binary went unrecognized until it became the basis for today's world of digital computing and communications. And in a second flash of foresight, Leibniz also invented the other number system commonly used in computing, namely, base 16, or hexadecimal in modern parlance (Leibniz's own term for it was sedecimal). These two inventions are the focus of this book.

Leibniz's groundbreaking work in mathematics is all the more remarkable given that he did not begin serious study of the subject until he was in his mid-twenties. Born in Leipzig in 1646 to a professor of moral philosophy, Leibniz obtained his bachelor's and master's degrees in philosophy in 1663 and 1664, respectively, before turning to the law, in which he obtained a bachelor's degree in 1665 and a doctorate a year later. By the end of the 1660s, he was at the court of Mainz working alongside the diplomat Johann Christian von Boineburg (1622–1672) on reforming the legal code. In February 1672, Boineburg dispatched him on a diplomatic mission to Paris, where he met some of the foremost mathematicians of the day, most notably Christiaan Huygens (1629–1695). Leibniz later recalled that he had arrived in Paris with "a superb ignorance of mathematics" (GM III, 71), but this was soon rectified by Huygens's tutelage and his own insatiable thirst for inquiry. He initially focused on summing infinite series and the classical Greek problem of squaring the circle, and from his intensive work on both, he eventually arrived at the calculus.[1] In October 1675, he devised the symbols d and $\int$, which are still used today.

Despite his growing reputation as a mathematician, Leibniz was unable to secure a post in Paris. One year after his breakthrough with the calculus, and a little over four and a half years since his arrival in Paris, Leibniz departed for the northern German city of Hanover to take up the post of court counselor, a role he held for the remaining forty years of his life (though in the intervening years, his facility at moving within courtly circles enabled him to add roles in Wolfenbüttel, Berlin, and Vienna to his portfolio). Leibniz later explained that he stayed in Paris as long as he did in order "to stand out a little bit in mathematics" (A II 1, 753).

It was during his Paris years that Leibniz developed his lifelong interest in the automation of calculation. In the 1640s Blaise Pascal (1623–1662) had invented a machine capable of adding and subtracting, but Leibniz wanted to go further, proposing a machine capable of multiplying, dividing, and even (in his initial plans, in 1670, for a "new arithmetic instrument") extracting roots (LH 42, 5 Bl. 16v). His designs were refined over the next three years, but putting into practice his belief that

---

1. See Probst (2018).

wheels and cranks could perform the labors of the mind proved a challenge. A machine he demonstrated at the Royal Society in London in 1673 was, by all accounts, imperfect.[2] Further refinements were communicated to the craftsman tasked with building the device, but progress was slow. Leibniz several times reported that the machine had been completed and tested,[3] but these successes appear to have been short-lived at best, and it is doubtful that the machine ever worked reliably. Even in the last year of his life, a frustrated Leibniz was firing off instructions for fixing various faults, which he blamed on a craftsman's lack of diligence and precision.[4] Of the three versions of Leibniz's machine built in his lifetime, only the last survives. When it was rediscovered in 1879, Arthur Burkhardt (1857–1918), a leading mechanical engineer, was tasked with making it functional, but he was unsuccessful and concluded that it had probably never worked.[5] Nonetheless Burkhardt soon began manufacturing calculating machines incorporating some of Leibniz's ideas, and in the twentieth century, engineers were able to construct fully-functioning "replicas" of Leibniz's machine,[6] one of which is displayed at the G. W. Leibniz Bibliothek in Hanover.

The G. W. Leibniz Bibliotek also holds the vast majority of Leibniz's surviving writings, which total some 200,000 manuscript pages, a sizable proportion of which have yet to be published. These writings cover a wide variety of topics and disciplines, and range from completed books and journal articles to draft essays, rough jottings, and personal reading notes, as well as letters to and from more than a thousand correspondents. This material offers a window into Leibniz's prodigiously wide-ranging interests, ideas, and inventions, not to mention his various projects. Between 1680 and 1686, he sought to improve the productivity of the Harz silver mines by designing pumps and windmills to drain the floodwaters that made mining impossible after heavy rains.[7] In 1686, he took on the task of writing a history of the House of Guelph (or Welf) in order to enhance his employer's dynastic ambitions (the history was still not complete at the time of his death thirty years later, due to Leibniz's excessive thoroughness and many distractions).[8] From the early 1680s onward, he sought to facilitate a reunion between Catholics and Lutherans and, in the late 1690s, a reunion between Lutherans and Calvinists.[9] In 1700, he lobbied for the formation of the Kurfürstlich Brandenburgische Societät der Wissenschaften [Electoral Brandenburg Society of Sciences], and was immediately installed as its first president.[10] He drew up plans for a universal encyclopedia that would contain everything known, wrote Latin poetry, built a full philosophical system that would have great influence for centuries to come, and undertook pioneering studies on the origin of languages.[11]

Leibniz's scope, reach, and genius were such that he was described as "the ornament of Germany" by the Dutch scientist Herman Boerhaave (1715, 13). The last years of his life were overshadowed by the priority dispute with Sir Isaac Newton (1643–1727) over the invention of the calculus and his frantic but unsuccessful efforts to complete the history of the Guelph house. On 6 November 1716, he became bedridden by an attack of gout and arthritis, and he died on 14 November.

---

2. For further details, see Beeley (2020).

3. For example, he claimed in July 1674 that his arithmetic machine had been "successfully completed" (A III 1, 119) and in April 1695 that he had "finally had [it] completed" (LH 42, 5 Bl. 63r; English translation: Leibniz 1695a).

4. See Leibniz (1845).

5. See Morar (2015).

6. However, the "replicas" required a little technological license and modern engineering techniques to function. See Stein et al. (2006) and Kopp and Stein (2008).

7. For further information, see Wakefield (2010).

8. For further information, see Antognazza (2018).

9. For further information, see Jordan (1927).

10. For further information, see Rudolph (2018). The institution survives today as the Berlin-Brandenburg Academy of Sciences and Humanities.

11. For additional information on Leibniz's life and work, see Aiton (1985), Belaval (2005), and Antognazza (2009).

Acknowledgments

Over the course of writing this book, many debts of gratitude have been incurred, which we are pleased to be able to record here:

Julia Weckend and John Thorley kindly looked over some of the translations in this book, and their comments and suggestions led to many improvements.

Siegmund Probst checked some of our transcriptions, thereby saving us from making a number of mistakes. He also fielded a great many queries relating to dating and watermarks, answering them all with disarming alacrity.

Daniel J. Cook read drafts of chapters 22, 28, and 29 and provided a number of useful comments.

Erik Vynckier provided useful comments on a draft translation of a text that was not ultimately included in the book. Erik's comments nevertheless helped to shape our approach to other texts in the book.

Stefan Lorenz and Stephan Meier-Oeser kindly assisted with the dating of chapter 13.

Peter Bol provided valuable assistance with the transcription and translation of the Chinese in chapter 28.

Bill Gasarch and Salil Vadhan generously provided valuable mathematical advice. Matthew Lena did an extraordinarily close reading of the entire book under severe time pressure, and Owain Daniel Jones kindly proofread the front matter, introduction, and some of the headnotes.

We gratefully acknowledge the permission granted by the Gottfried Wilhelm Leibniz Bibliothek – Niedersächsische Landesbibliothek, Hanover, to reproduce scans of various manuscript pages in their holdings. Lloyd also gratefully acknowledges the financial support of the Gerda Henkel Stiftung, Düsseldorf, which awarded him a research scholarship (AZ 46/V/21) to complete this project.

Untitled manuscript from 1703, featuring a table of decimal numbers 0–40 represented in every base from 2 to 16 (LH 35, 3 B 11 Bl. 11v). Note Leibniz's use of the Roman letters a, b, c, d, e, and f for the six extra digits, anticipating the modern convention.

Introduction

Leibniz and Binary: An Overview

Behind every great human invention lies a story. The oft-told story of Leibniz's binary system, a great invention if ever there was one, goes like this:

> Leibniz invented the binary system in 1679, first outlining it in a three-page manuscript written on 15 March of that year. Binary then featured in a couple more texts written in 1679 and was next revisited almost twenty years later, in 1696, when he happened to mention the binary representation of all numbers by 1 and 0 to Duke Rudolph of Brunswick and Lüneburg, who (so the story goes) saw therein an analogy with the Christian doctrine of creation *ex nihilo*, according to which all things were created from nothing by the one God. Excited by its theological potential, in 1697, Leibniz began sending details of the binary system to Christian missionaries in China, hoping that the theological analogy would assist them in converting the Chinese. One such missionary, Joachim Bouvet, was struck by a parallel between binary notation and the hexagrams of the ancient Chinese divinatory system, the Yijing. Each of the sixty-four hexagrams consists of a stack of six horizontal lines, each either broken (a yin) or unbroken (a yang), and by equating the unbroken line with 1 and the broken line with 0, Bouvet suggested that the hexagrams might be nothing more than Leibniz's binary system in alternative notation. Leibniz required little convincing and was so excited by the idea that he had unlocked a piece of ancient Chinese wisdom that he decided to make his invention public without further delay. Within a week he had sent a short paper—"Explanation of Binary Arithmetic"—to the French periodical *Memoires de l'Académie Royale des Sciences*, which published it in 1705.

This story—developed, polished, and rehearsed often over the past fifty years—contains much that is true, at least the part from 1697 onward. The rest is what we might call a work of commentators' fiction, dominated by errors and omissions. Most notably, it says nothing about the most striking by-product of Leibniz's work on binary, namely, his invention of the base-16 number system and experimentation with a variety of symbols to represent what today are called the hexadecimal digits. And this almost 200 years before John W. Nystrom (1825–1885) developed and promoted the base-16 system in a series of articles. The reason the standard story is so distorted is that Leibniz's 200 or so writings on binary remain understudied; even 300 years after his death relatively few have even been published. The thirty-two writings collected in this volume represent the deepest dive to date into Leibniz's work on binary, and together, they tell a story quite different from that sketched above. To provide an overview of the texts in this volume and how they flesh out and correct the oft-told story of Leibniz and binary, it is best to go back to the start of the story or, rather, to what is often believed to be the start.

The earliest dated writing on the binary system in Leibniz's Nachlaß (that is, the collection of manuscripts left after his death) is "On the Binary Progression" (chapter 6), written on 15 March 1679 (that is, 25 March 1679 in the modern Gregorian calendar). The complete manuscript, a remarkably detailed and sophisticated investigation into binary, is divided into two parts. In part I,

Leibniz presents tables of binary numeration, offers algorithms for converting decimal numbers into binary, outlines the four basic arithmetic operations in binary, describes how these operations could be carried out by a calculating machine quite different in design from the decimal calculating machine on which he had been working since the early 1670s, and explores the square of a six-digit binary number. In part II, which is more experimental, Leibniz attempts to extract the square root of an eight-digit binary number and examines the repeating binary representation of fractions. The essay closes with rumination about whether arbitrary numbers are expressible using algebraic rules, or whether, to the contrary, some numbers lie beyond such finite control. Given its importance to the history of the binary system and of number systems generally, it is surprising that "On the Binary Progression" is still so little known, or at least known in its entirety.[1] The complete manuscript consists of four folio (i.e., $2°$) sheets, that is, eight pages in total, of which seven were used for the text itself.[2] Of these, only the first three are known to scholars due to an unfortunate oversight of previous editors. Hochstetter (1966, 42–47) published a facsimile of only the first three pages (i.e., $1\frac{1}{2}$ of the four folio sheets) along with a German translation. An Italian version of Hochstetter's book followed five years later (see Leibniz 1971), featuring again a facsimile of the first three pages, but this time with an Italian translation. Then, four decades later, Serra (2010b) published a facsimile of the same three pages followed by a French translation. Curiously, no transcription of the original Latin has yet been published,[3] only facsimiles of the first three pages of the manuscript along with translations of those three pages into various European languages. This has created the misleading impression that "On the Binary Progression" consists of only those three pages, which is in fact less than half of the entire text.[4] Our translation in chapter 6 thus marks the first publication of the entire text in any language.

Although "On the Binary Progression" is Leibniz's earliest *dated* writing on the binary system, is it really likely to be his *earliest writing* on the subject, as many scholars have claimed?[5] Arguably, no: given how well developed are the ideas in that text, it seems much more likely that it was preceded by other, more preliminary studies, over the course of which Leibniz developed his understanding of binary to the high standard found in "On the Binary Progression." Surprisingly, at the time of writing, only one attempt has been made to establish how the binary system developed in Leibniz's hands, almost fifty years ago by Hans Zacher, although his studies are partial and incomplete. Zacher (1973, 10–11) did, however, discover "Notes on Algebra, Arithmetic, and Geometric Series" (chapter 1), at the bottom of which Leibniz wrote this:

$$\begin{array}{r} 111 \\ \underline{101} \end{array}$$

The binary progression

0	1	2	3	4	5	6	7	8	0	0	0	0	0
0	1	10	11	100	101	110	111	1000	.		.		.

$$\begin{array}{r} 111 \\ \underline{1110.} \\ 100011 \end{array}$$

The manuscript itself has now been dated to October 1674, which places it during Leibniz's four-year stay in Paris (from March 1672 to October 1676), a time when he studied alongside leading

---

1. Scholars have only just started to explore part II of "On the Binary Progression"; see Brancato (2021).

2. The remaining page contains the wage tables for the Harz silver mines for the final quarter (quarthall luciae = quarter of St. Lucy) of 1678. Between 1680 and 1686, Leibniz attempted to build wind machines to drain the mines, but his involvement with the mines began in the fall of 1678. See Wakefield (2010).

3. One will eventually appear in series 7 (devoted to Leibniz's mathematical writings) of *Sämtliche Schriften und Briefe* (abbreviated throughout this book as A).

4. See, for example, Serra (2010a).

5. See, for example, Mungello (1977, 51–52), Pappas (1991, 125), Ryan (1991, 31), Antognazza (2009, 247 and 275n255), Bauer (2010, 14), and Ingaliso (2017, 112n12). Implicit in the claim seems to be the implausible suggestion that Leibniz thought of the binary system, and developed it to the high standard found in that text, in a single day. Despite Leibniz's undoubted genius, one can find no other example in his writings of him having an idea and developing it to a high degree of sophistication in a single day. It is therefore much more likely that Leibniz first had the idea for binary numeration before "On the Binary Progression" and then developed it.

mathematicians such as Christiaan Huygens and made a number of important mathematical advances, among them the invention of the infinitesimal calculus.[6] Could binary be another of Leibniz's mathematical inventions from his Paris years? Such a conclusion is difficult to draw with absolute certainty, since the material on binary in "Notes on Algebra, Arithmetic, and Geometric Series" was written in different ink from everything else on the sheet and therefore might have been added later, with no way to tell whether this was later the same day or days, weeks, months, or even years later. However, given that the manuscript consists of rough notes on undated scrap paper, it seems unlikely that Leibniz would return to it a long time after to add the material on binary, making it reasonable to suppose that the material does date from the latter months of 1674. (As we shall see later in this introduction, around this time, Leibniz was also working on a different number system, the base-60 or sexagesimal system; see LH 35, 3 A 26 Bl. 17 and LH 35, 8, 30 Bl. 27.)

The only other candidate text identified by Zacher as possibly written prior to "On the Binary Progression" is "Thesaurus mathematicus" [Mathematical Thesaurus], a wide-ranging piece covering various topics in arithmetic, geometry, and mechanics.[7] In a passage near the end, Leibniz first outlines how positional notation works in the decimal and duodecimal number systems before identifying binary as an alternative:

> From this outline it is clear that only these ten digits are needed: 0, 1, 2, 3, 4, 5, 6, 7, 8, 9. Those in the first position signify the equivalent number of 1s, namely no 1s, one 1, two 1s, three, four, five, six, seven, eight, nine 1s. Those in the second position signify the equivalent number of 10s, that is, 1s taken ten times; in the third position, the equivalent number of 100s, that is, 10s taken ten times, or the squares of 10; in the fourth position, the equivalent number of 1000s, that is, 100s taken ten times, or the cubes of 10, and so on. And in place of 10 one would be able to put any other number, for example, 12. For just as when the base a is 10, the square a^2 signifies 100 and the cube a^3 signifies 1000, so when a is 12, a^2 will be 12 times 12, that is, 144, and a^3 will be 12 times 144. But on this method, instead of the digits mentioned above—0, 1 etc. 9—two new digits would be needed in addition, one which would represent ten, the other which would represent eleven; but [the digits] 10 would signify twelve, and 100 would signify one hundred and forty four. And there are some who prefer to use this method of calculating over the common method, because 12 can be divided by 2, 3, 4, and 6; in addition, a calculation is completed with fewer digits. But the difference is not so great as to be worth abandoning the decimal progression. If anyone should want to use the binary progression, he would not need any digits except 0 and 1. Hence he would not need the Pythagorean table,[8] of which [we shall speak] later. But the calculation would be longer, albeit easier.[9] (LH 35, 1, 25 Bl. 3v)

On the basis of its watermark, "Thesaurus Mathematicus" was likely written in 1678 or 1679, making it one of Leibniz's earliest writings on binary numeration and possibly one that predates "On the Binary Progression."

---

6. Leibniz's later pronouncements do not enable us to date his invention of binary with any greater accuracy. In 1698, he told Johann Christian Schulenburg that he had been thinking about binary for "twenty years and more" (A II 3, 451), and in 1703, he explained to Joachim Bouvet that his ideas about binary went back "well over 20 years" (chapter 29). Both claims loosely point to Leibniz first having the idea during the latter half of his stay in Paris or shortly afterward.

7. In fact, Zacher (1973, 11, 18–19) did identify one other manuscript that he believed was written prior to 1679. The manuscript in question contains some rudimentary material on binary (namely, a table showing the values of the decimal numbers 0–8 in binary along with some binary sums), leading Zacher to claim that it was probably from 1676 and therefore from Leibniz's Paris years. However, the manuscript, a set of notes recording mathematical ideas made during a meeting with the Dutch mathematician Johann Jakob Ferguson (1630–1706), has since been dated to spring 1680 and is therefore not one of Leibniz's earliest forays into the binary system at all. See A III 3, 138.

8. The Pythagorean table is a simple multiplication table, which in Leibniz's time went up to 9 × 9.

9. In a marginal comment, Leibniz added, "We shall say more things about the binary progression below," though this promise is not fulfilled and the text says nothing further on the subject.

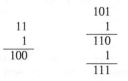

Reverse manuscript page of "The Place of Others" (LH 34 Bl. 29r).

Although not mentioned by Zacher, various other texts may have been written prior to "On the Binary Progression." For example, after having written a short text entitled "La place d'autruy" [The Place of Others] about the importance of putting ourselves in the place of others, Leibniz added in the empty margin of the reverse manuscript page a table of binary numbers from 0 to 15, while on top of his words of the text, he scrawled these two sums in binary (see LH 34 Bl. 29r; manuscript reproduced above).

$$
\begin{array}{cc}
 & 101 \\
11 & 1 \\
\underline{1} & \underline{110} \\
100 & 1 \\
 & \underline{111}
\end{array}
$$

Another text, consisting of half a page torn from a larger sheet, finds Leibniz using division to work out the binary representation of the fractions $\frac{1}{13}$ and $\frac{1}{17}$ (see LH 35, 13, 2B Bl. 155). Unfortunately, neither of these two manuscripts bears a watermark, nor do they contain any internal evidence that would enable a reliable dating. However, their embryonic treatments of binary suggest that they were written earlier than "On the Binary Progression," though how much earlier is impossible to determine. Nevertheless, they probably capture some of Leibniz's earliest explorations of binary, since one might reasonably expect his initial ventures into binary to take the form found in these manuscripts, namely, tables of binary numbers, simple binary calculations, and brief descriptions of key features of binary numeration.

Four other texts that likely capture Leibniz's early exploration of binary—or at least earlier than "On the Binary Progression"—are included in this volume. In "The Series of All Numbers, and on Binary Progression" (chapter 2), Leibniz outlines binary notation and gestures at algorithms for the basic arithmetic operations on binary numbers, while in "Binary Progression" (chapter 3), he calculates the binary representations of various fractions, such as $\frac{1}{2}$, $\frac{1}{3}$, $\frac{1}{4}$, and so on. In "Geometric Progressions and Positional Notation" (chapter 4), Leibniz notes a property of the double geometric progression (1, 2, 4, 8, 16, 32, and so on) at the heart of binary numeration, namely, that the sum of any three consecutive terms is always divisible by 7; in a passage that was subsequently deleted, he also looks to binary numeration for insight into perfect numbers (positive integers equal to the sum of their divisors other than themselves). And last, in "Binary Arithmetic Machine" (chapter 5), Leibniz sketches out a design for a binary calculating machine. Since the design was based on his preexisting decimal calculating machine, this text likely captures Leibniz's first thoughts on how to mechanize binary calculations, with the very different design outlined in "On the Binary Progression" capturing a later, more considered view. All four of these texts were undated and bear no watermarks, but they probably precede "On the Binary Progression" (15/25 March 1679), since the latter contains similar ideas in more developed form.

Other texts in this volume shed new light on Leibniz's work on binary after "On the Binary Progression." Heretofore, Leibniz was known to have touched on binary only twice between 1679 and the mid-1690s. In "On the Organon or Great Art of Thinking" (chapter 10), he fuses mathematics and philosophical theology by drawing a parallel between the binary system, in which all numbers derive from 1 and 0, and the idea that in the universe, there may be only one thing conceived through itself, namely, God, aside from whom there is nothing (privation). In his other previously known early text containing material on binary, "Summum calculi analytici fastigium" [The Highest Peak of Analytical Calculation], he observes that when an arithmetic progression is expressed in binary, the digits in any given column recur periodically. For example, the decimal sequence of odd numbers 1, 3, 5, 7, 9, 11, 13, 15, and 17 is expressed in binary thus:

$$00001$$
$$00011$$
$$00101$$
$$00111$$
$$01001$$
$$01011$$
$$01101$$
$$01111$$
$$10001$$

Leibniz notes that "while the digits of the first [column counting from the right] are formed only of 1s, the digits of the second [column] are alternately 0 and 1, of the third alternately 00 and 11, of the fourth are alternately 0000 and 1111, the digits of the fifth are 0000 0000 and 1111 1111, and so on in a geometric progression"[10] (LH 35, 13, 3 Bl. 21v; Zacher 1973, 223). The column periodicity of binary numbers would prove to be a source of fascination for Leibniz in the decades that followed, convinced as he was that the binary system, by returning to first principles (namely, 0 and 1), could reveal mathematical truths inaccessible by other means.

Such confidence did lead to some early missteps, however. For example, in "Attempted Expression of the Circle in Binary Progression" (chapter 7), Leibniz seeks to convert into binary his alternating convergent series for the quadrature, i.e., $\frac{1}{1} - \frac{1}{3} + \frac{1}{5} - \frac{1}{7} + \frac{1}{9} - \frac{1}{11}$ etc., albeit without success, while in "Binary Ancestral Calculations" (chapter 11), he uses the double geometric progression and binary notation to try to determine the number of ancestors a person alive today would have from 3000 years ago, again unsuccessfully. Even more ambitiously, in an untitled text from August 1680, Leibniz attempts to elicit an algorithm for determining prime numbers from examples of the periodicity of binary fractions, the failure of the attempt sitting uneasily with the boast found in the opening

---

10. That is, the length of each column's pattern is twice that of its neighbor to the right.

line of the text: "I am gradually obtaining access to the innermost secrets of numbers" (LH 35, 13, 3 Bl. 33).[11]

Much more fruitful was Leibniz's invention of the other number system that has come to prominence in our computer age, namely, base-16, or sedecimal in Leibniz's terminology (hexadecimal in modern parlance). Although long believed to be an invention of the engineer John W. Nystrom in the mid-nineteenth century, Leibniz's first writing on the subject dates from 1679, with others following shortly thereafter. In "Sedecimal Progression" (chapter 8), Leibniz develops an algorithm for converting decimal numbers to sedecimal and offers two distinct ways of completing the sedecimal character set by identifying the six extra digits 10–15, one using the names of musical notes—ut, re, mi, fa, sol, la—and the other using the Roman letters m, n, p, q, r, and s (today, the standard is A, B, C, D, E, and F). In the shortest text in this volume, "Binary Progression Is for Theory, Sedecimal for Practice" (chapter 9), Leibniz develops novel symbols for all sixteen sedecimal digits, and a different set of symbols in "Sedecimal on an Envelope" (chapter 12, see also frontispiece), in which he also identifies the binary equivalents of sedecimal digits, thus linking together his two chief inventions in number theory. Although between them, these three early writings on sedecimal could comfortably fit on two or three typeset pages, this should not be taken to imply a lack of interest on Leibniz's

---

11. The pieces in this volume are focused mainly on the significance of binary notation and its close relative sedecimal, patterns in sequences of binary numbers, and how to carry out arithmetic calculations in binary or sedecimal. Elsewhere, Leibniz used the simplicity of the binary system to seek out more general mathematical truths. In a 1680 manuscript (LH 35, 13, 3 Bl. 33–34), for example, he proved a restricted version of a famous theorem of Euler from the next century, that $a^{\phi(n)} \equiv 1 \pmod{n}$ whenever a and n are coprime ($\phi(n)$ being Euler's totient function, the number of positive integers less than n which are coprime with n). Leibniz's special case is that for any odd n, there is a k such that $2^k \equiv 1 \pmod{n}$. (Leibniz does not mention the obvious fact that n must be odd or that similar restrictions apply when the base is 10 or another number.)

To derive his result, Leibniz begins by reasoning that a fraction such as $\frac{1}{n}$ in any base is periodic, since when n is divided into $1.000\ldots$, the division to produce a digit of the quotient results in a remainder $< n$, which is multiplied by the base (by appending a 0) to begin the next iteration. When the value of the remainder repeats, the subsequent digits of the quotient will repeat. If the base is 2 and the denominator is $2^k - 1$ for some $k > 0$, then

$$\frac{1}{2^k - 1} = \sum_{i=1}^{\infty} 2^{-ik},$$

which has a 1 bit each k bit positions starting with the kth to the right of the binary point and 0s elsewhere.

Now (continuing with the base 2 case), let n be any odd integer. Suppose $\frac{1}{n}$ consists of an initial block of bits $c_1 \ldots c_\ell$ ($\ell \geq 0$) immediately to the right of the binary point followed by indefinite repetitions of a block $b_1 \ldots b_k$ ($k > 0$). The binary sequences $c_1 \ldots c_\ell$ and $b_1 \ldots b_k$ represent nonnegative integers $C < 2^\ell$ and $B < 2^k$. Then

$$0.\overbrace{b_1 \ldots b_k}^{k} \overbrace{b_1 \ldots b_k}^{k} \cdots = B \times 0.\,\overbrace{00\ldots01}\,\overbrace{00\ldots01} \cdots = B \times \sum_{i=1}^{\infty} 2^{-ik} = B \times \frac{1}{2^k - 1},$$

and, since multiplying by $2^{-\ell}$ shifts bits ℓ positions to the right,

$$\frac{1}{n} = 2^{-\ell} \left(C + \frac{B}{2^k - 1} \right).$$

Therefore

$$\frac{2^\ell (2^k - 1)}{n} = C(2^k - 1) + B, \text{ which is an integer.}$$

Since n, being odd, has no factor in common with 2^ℓ, n must divide $2^k - 1$. Hence, Leibniz's "wonderful theorem" that for any odd n, some multiple of n is of the form $2^k - 1$. And in fact, the nonrepeating initial segment $c_1 \ldots c_\ell$ is empty, that is, shifting the bits of $\frac{1}{n}$ left by k positions and taking the fractional part results in $\frac{1}{n}$ once again. For let A be the integer $\frac{2^k - 1}{n}$, so that $2^k = An + 1$. Multiplying $\frac{1}{n}$ by 2^k shifts the bits left k positions, but $2^k \cdot \frac{1}{n} = (An + 1) \cdot \frac{1}{n}$, of which the fractional part is $\frac{1}{n}$. (And therefore $B = A$.)

A direct way of finding a multiple of odd n of the form $2^k - 1$ considers the n remainders when $2^1, 2^2, \ldots, 2^n$ are divided by n (Astin 1984). There can be at most $n - 1$ different remainders, since a remainder of 0 is impossible when dividing an even number by an odd. Then if $1 \leq i < j \leq n$ and dividing 2^i and 2^j by n leave the same remainder, then $2^j - 2^i = 2^i (2^{j-i} - 1)$ is divisible by n, so letting $k = j - i$ yields the desired result.

part. In fact, Leibniz would return occasionally to sedecimal, developing a third form of notation for sedecimal digits in a 1701 letter to Joachim Bouvet (chapter 22), a fourth way of completing the character set using the Roman letters a, b, c, d, e, and f in an untitled text from 1703 (see LH 35, 3 B 11 Bl. 11v, reproduced facing the introducton of this book), and recommending the sedecimal system over the decimal for any future reform of common practice (see chapters 13, 22, and 30).

Contrary to the popular account that Leibniz's investigations into binary hit a long barren patch after "On the Binary Progression," the texts in this volume show that Leibniz filled many pages with such investigations between 1679 and the early 1680s. Curiously, however, he appears to have said relatively little about it to others during this time. As far as we can tell, he first mentioned the binary system to the Dutch mathematician Johann Jakob Ferguson (1630–1706) during a meeting in spring 1680; near the end of Leibniz's notes from that meeting, there is a table showing the values of the decimal numbers 0–8 in binary along with some binary sums (see A III 3, 138). In writing, Leibniz first mentioned the binary system in a letter to the German mathematician Detlev Clüver (1645–1708) of 31 August/10 September 1680: "I have quite often fruitfully used binary logistics of progression in which there occur no other digits besides 0 and 1, nor is that for common use but for the sake of theory, and so it does not seem the common method of counting should be disturbed without good reason" (A III 3, 263). Less than two years later, Leibniz mentioned the binary system to another mathematician acquaintance of his, Ehrenfried Walter von Tschirnhaus (1651–1708), in a letter written at the end of June 1682: "The binary progression would be particularly useful for expressions of quantity in numbers, because it is basic and simplest, and I do not doubt that there would be many harmonies to be found therein, not likewise to be found in other progressions" (A III 3, 655–656). Although Leibniz failed to stimulate any interest in binary in any of these interlocutors, this would eventually change in 1696.

In May of that year, Leibniz explained the binary system in person to Rudolph August, Duke of Brunswick and Lüneburg (1627–1704), or at least explained enough of it to draw an analogy between the representation of all numbers by 1 and 0 and the theologically orthodox doctrine of creation *ex nihilo*, that is, the creation of all things out of nothing by (one) God. While it is often claimed that the binary-creation analogy was the duke's idea,[12] the analogy appears in "Remarks on Weigel" (chapter 13), a text Leibniz wrote at least a year before his discussion with the duke. In that text, Leibniz makes a series of critical observations on the work of his old mathematics tutor, Erhard Weigel (1625–1699), who had advocated the adoption of a quaternary, or base-4, system some years before. After noting Weigel's preference for base 4, Leibniz insists that binary is superior for matters of theory and sedecimal for matters of practice, before outlining the binary-creation analogy he would later relate to Duke Rudolph August.

By all accounts, the duke himself was delighted with the analogy and this prompted Leibniz to draft a short paper outlining the basics of the binary system (chapters 15 and 16), along with a short covering letter (chapter 14). At the start of 1697, Leibniz wrote again to the duke, this time suggesting that the binary-creation analogy be commemorated in the form of a medal, to be struck at the duke's command (chapter 17). Although no such medal was ever struck, Leibniz later learned that the duke had commemorated the binary-creation analogy in a different way, by having a set of wax seals designed (chapter 18). Now convinced of the theological value of the binary-creation analogy, particularly its potential to illuminate the doctrine of creation *ex nihilo*, in early 1697 Leibniz wrote to a Jesuit missionary in China, Claudio Filippo Grimaldi, presenting the analogy and some of the mathematics underpinning the binary system (chapter 19). Undeterred by the lack of reply, in 1701, he repeated the attempt with another Jesuit missionary in China, Joachim Bouvet (chapter 22).

The discovery of the binary-creation analogy, and the duke's royal approval, prompted Leibniz to communicate the binary system to a series of mathematical acquaintances from the late 1690s onward. A recurring theme in this correspondence was the column-periodicity of number sequences expressed in binary. In truth, this had exercised Leibniz for some time, though with little progress to show for the many manuscript pages he had devoted to it. However, there were some occasional

---

12. See Zacher (1973, 36–37), Aiton (1985, 206), and Antognazza (2009, 357).

advances, such as his discerning in "Periods" (chapter 20) the "general rule" that the first half of any sequence of digits of a period is the complement (i.e., the opposite) of the second half. Realizing that he could devote insufficient time to determining other such rules, Leibniz attempted to enlist other mathematicians, such as Philippe Naudé (1654–1729), by outlining what he had achieved thus far (chapter 21). His only success in this regard was with Pierre Dangicourt (1664–1727), who composed an essay on the topic in 1701, "On the Periodic Intervals of Binaries in the Columns of Natural Numbers" (A III 8, 534–545).

Eventually, Leibniz cast his net wider than his own mathematical acquaintances. In early 1701, he composed a treatise entitled "Essay on a New Science of Numbers" for the Académie Royale des Sciences [Royal Academy of Sciences] in Paris, which had elected him as a foreign member the previous year (chapter 23). By his own admission, the mathematically rich treatise, which captured the key advances he had made thus far, was designed to inspire other members of the Académie to develop binary even further. In this it did not succeed, and a disappointed Leibniz eventually declined the Académie's offer to publish the paper in its annual proceedings, the *Histoire de l'Académie Royale des Sciences* [*History of the Royal Academy of Sciences*]. However, later in 1701, he did find the time to make some further advances himself. In "Binary Addition" (chapter 24), he developed a simplified method for adding together many binary numbers. His investigations into the column periodicity of binary numbers led him to experiment with simplified notation for long strings of digits in "Periods and Powers" (chapter 26), though ultimately the notation was abandoned. In "Periods in Binary" (chapter 25), he determined that a "summatrix" column, formed by summing initial terms from another column, is itself periodic. He formalized this finding in "Demonstration That Columns of Sequences Exhibiting Powers of Arithmetic Progressions, or Numbers Composed from These, Are Periodic" (chapter 27), which marked the last of Leibniz's mathematical advances in binary.

Perhaps the most intriguing chapter in the Leibniz-binary story began on 1 April 1703, when Leibniz received a response to the letter he had sent the Jesuit missionary Joachim Bouvet almost two years before. Bouvet was a long-time student of the ancient Chinese divination system, the Yijing, or Book of Changes, and upon reading Leibniz's account of binary arithmetic, he was struck by a correlation between Leibniz's binary numeration and the sixty-four hexagrams of the Yijing. Each hexagram consists of six stacked horizontal lines, each line being either unbroken (representing yin) or broken (representing yang), and by construing the unbroken line as 1 and the broken line as 0, Bouvet concluded that the hexagrams were a form of binary numeration (chapter 28). Thrilled by the idea that binary was the key to deciphering an ancient Chinese enigma that had baffled scholars for millennia, Leibniz wrote an enthusiastic reply (chapter 29) in which he floated the suggestion that the hexagrams, understood as an encoded binary system, may also have been intended to represent the creation of all things from nothing by God, giving a new twist to an idea he had entertained since the mid-1690s. Buoyed by Bouvet's hypothesis that binary was the key to deciphering the secrets of the hexagrams of the Yijing, Leibniz wrote a new paper for the Académie Royale des Sciences to replace the one he had withdrawn. This new paper, "Explanation of Binary Arithmetic" (chapter 30), opens with an introduction to the binary system, with Leibniz explaining binary notation, giving examples of all four standard arithmetic operations, and presenting the periodicity in the columns of natural numbers expressed in binary as evidence of the wonderful order to be found in his new calculus. The paper then turns, about halfway through, to outlining Bouvet's hypothesis that the binary system was foreshadowed in the sixty-four hexagrams of the Yijing.

"Explanation of Binary Arithmetic" was eventually published by the Académie Royale in 1705, but despite Leibniz's efforts to determine the fate of his paper, he never did learn of its publication. Despite or perhaps because of this, he continued to promote the binary system whenever the opportunity presented itself. Sometimes this was with the long-held aim of encouraging others to investigate it further; such was the purpose of "On Binary" (chapter 32), Leibniz's last systematic presentation of the binary system, intended for the mathematician Jakob Hermann (1678–1733). His overtures to other mathematicians met with a range of responses, usually measured enthusiasm or polite indifference, but sometimes the response was a little more challenging. César Caze (1641–1719), for example, took it upon himself to throw doubt on the suggestion of a link between binary and the hexagrams of the Yijing and also put it to Leibniz that other mathematicians had invented binary before him, forcing him on to the defensive in his reply (chapter 31).

The last decade of Leibniz's life was marked by much less activity in both investigating and promoting his binary system. Perhaps most significantly, he helped Dangicourt rework the essay he had initially composed back in 1701, with it ultimately being published in 1710 as "On the Periods of Columns in a Binary-Expressed Sequence of Numbers of an Arithmetic Progression" (see Dangicourt 1710). Leibniz also managed to induce another correspondent, Theobald Overbeck (1672–1719), to investigate column periodicity of binary numbers (see LH 35, 12, 1 Bl. 190–191). But much of Leibniz's attention, so far as binary was concerned, was focused on celebrating the supposed correlation between binary numeration and the hexagrams of the Yijing. In "Discourse on the Natural Theology of the Chinese," a lengthy essay written in the last year of his life, Leibniz noted previous attempts to explain the hexagrams mystically or philosophically, before claiming that "actually, the 64 figures represent a Binary Arithmetic which apparently this great legislator [Fu Xi] possessed, and which I have rediscovered some thousands of years later" (WOC 132).

While Bouvet's hypothesis of a correlation between binary notation and the hexagrams of the Yijing ultimately proved untenable,[13] Leibniz's suggestion here that he had not so much discovered binary arithmetic as rediscovered it turns out to have been prescient after all, since as we shall see, the history of binary extends back far beyond the seventeenth century, to ancient Egypt.

A Brief History of Binary and Other Nondecimal Systems

Certain reckoning methods used in antiquity are best, if anachronistically, explained in terms of the binary system. For example, the Egyptian method of multiplication, illustrated in the Rhind papyrus of c. 1700 BCE, computes a product $m \times n$ by repeatedly doubling n—a process known as *duplation*—and then adding up certain of the resulting values $2^0 n, 2^1 n, 2^2 n, \ldots$. Alongside the duplation of n, a parallel duplation process is begun starting with 1, yielding $2^0, 2^1, 2^2, \ldots$, stopping before the result exceeds m. A unique subset of these powers of 2 sums to m—namely, 2^i for those i such that the ith bit of the binary representation of m is 1—and $m \times n$ is then the sum of $2^i n$ for those values of i. That is, if the binary representation of m is $a_k \ldots a_0$, where each a_i is 0 or 1, then $m \times n$ is computed as

$$\left(\sum_{i=0}^{k} a_i 2^i \right) \times n = \sum_{i=0}^{k} \left(a_i 2^i n \right).$$

For example, 23×27 is computed by repeatedly doubling 1 (resulting in 1, 2, 4, 8, 16 and stopping there since $32 > 23$), while repeatedly doubling 27 (27, 54, 108, 216, 432) and then adding up $27 + 54 + 108 + 432$ to get the result 621, those terms having been selected since $1 + 2 + 4 + 16 = 23$ (Newman 1956, I: 172). So this method works because every positive integer has a unique binary representation ($23 = 10111_2$ in this case).[14]

The so-called Russian peasant algorithm for multiplication (Gimmestad 1991) works similarly, except that in place of the duplation of 1, the multiplier m is repeatedly halved—a process called *mediation*—until it is reduced to 1. The terms $2^i n$ that are summed up to yield the product are those for which the mediated multiplier is odd. So to continue with the same example, the mediated multipliers are 23, 11, 5, 2, 1, so the product is the sum of the duplated terms $2^i n$ corresponding to the odd numbers in this sequence, namely (as before), $27 + 54 + 108 + 432$.

Use of pan balances led naturally to systems of measurement with a binary structure. Old English and Scottish units, some still in use in the United States, were in powers-of-2 ratios to one another. Swinton (1789, 29), for example, includes a table of Scottish measures for liquids then in use (4 gills to the mutchkin, 2 mutchkins to the chopin, 2 chopins to the pint, 2 pints to the quart, 4 quarts to the gallon, 8 gallons to the barrel), complete with all the powers of 2 needed to convert any unit to any other (for example, 1024 gills to the barrel). In the same spirit, weighing an item on a pan balance

---

13. For details on why Bouvet's hypothesis is incorrect, see Sypniewski (2005) and Zhonglian (2000).

14. This method should not be confused with binary multiplication—the multiplier and multiplicand themselves are represented in a nonpositional form of decimal notation (Peet 1923, 11–13).

using weights in the proportions 1:2:4:8 ... amounts to determining the binary representation of the item's weight. For example, if the available weights are 1, 2, 4, 8, and 16 ounces, an item weighing 23 ounces in one pan will be counterbalanced by placing the 1-, 2-, 4-, and 16-ounce weights (but not the 8-ounce) in the other pan (23 being 10111_2).

Such methods of reckoning and measuring, while dependent on the principle of unique binary representation of numbers, lack any suggestion of a binary positional notation. But nondecimal positional systems do have a long and independent history, dating back more than four millennia. Ifrah (1998, chap. 12) shows how Sumerian children learned to do sexagesimal arithmetic (base 60), not just sums but division (the exercise $1,152,000 \div 7 = 164,571$ with a remainder of 3 is worked out on a tablet from c. 2650 BCE). Base 60 may have been derived from duodecimal: Ifrah (1998, chap. 9) conjectures that duodecimal counting started with one-handed counting, using the thumb to count on the twelve phalanges of the other four fingers.[15] Duodecimal survives in the merchandizing of goods by the dozen and gross and in hours on the dial; sexagesimal in minutes, seconds, and degrees of the circle.

And yet the principle that numbers can be represented positionally using arbitrary bases seems to have remained unnoticed by mathematicians until Blaise Pascal (1665, 42) mentioned it quite offhandedly.[16] His initial subject was the rule for "casting out 9s" to determine whether an integer represented in decimal notation was divisible by 9, but he generalizes the rule to other divisors and also to other bases. The method works, he says, "not just in our decimal system of numeration (which has been established not as a result of natural necessity, as the common man thinks, but as a result of human custom, and quite foolishly, to be sure), but in a system of numeration based on any progression whatsoever." The only nondecimal example he presents is duodecimal, showing (Pascal 1665, 47–48) that casting out nines would then become casting out elevens. Having spent enough time on this "novelty," he ends with "I'll stop here lest I become tiresome to the reader by going through too many details."

But by this time, a form of binary reckoning, without binary notation, had been detailed by John Napier (1617) in his *Rabdologiæ* (English translation: Napier 2017, 649–749). In the final part of this work, "Location Arithmetic," Napier details a chessboard-style instrument for numerical calculations, cautioning in the Preface: "There is one small difficulty in working with it, and that is that the numbers it uses differ from ordinary numbers, so that one must begin by expressing ordinary numbers in the new form and end by reducing them to common form" (Napier 2017, 727). In Napier's representation, each power of 2 is associated with a Roman letter—*a* for 1, *b* for 2, *c* for 4, *d* for 8, *e* for 16, and so on—and numbers are written in a form reminiscent of Roman numerals by concatenating the letters corresponding to what we would now call the 1 bits in its binary representation. So $23 = 10111_2$, for example, would be written as *abce*. Or to be precise, that would be the canonical representation of 23; any other sequence of letters that sum up to 23 would be equivalent—*abcdd*, for example, or *aaace*—and these various forms could be derived from one another by replacing a *c* by two *b*s or two *b*s by a *c*, for example. Napier's multiplication method involves starting with a checkerboard (of any size) labeled along both its bottom edge (right to left) and right edge (bottom to top) *a*, *b*, *c*, ...; putting tokens on those two margins to represent what are essentially the binary representations of the two operands; putting a token at the intersection of any row and column that both have tokens at the margin (the intersection between the row representing 2^i and the column representing 2^j being the square representing 2^{i+j}); removing from the margins the tokens representing the operands; and finally accumulating at the bottom margin the tokens on the various northeast-to-southwest diagonals (as all interior squares on any one such diagonal have the same value of $i + j$ and

---

15. His conjecture that duodecimal began with one-handed counting (a technique he describes as still in use in the Middle East) would imply that the factorization of 12 as 4×3 was part of the origin of duodecimal, with the factor of 5 to reach 60 having a more obvious anatomical interpretation. He cites linguistic evidence for the existence of a pre-Sumerian base-5 system, which he conjectures became base 60 after contact with a base-12 culture.

16. This publication is a posthumous collection, Pascal having died in 1662. Adamson (1995, 280) gives the date of the relevant essay, *De numeris multiplicibus*, as 1654.

hence all represent the same power of 2). All that is left is to reduce the piles of tokens on the bottom margin to canonical form via replacements of the kind already mentioned—and then translate them into the final decimal result.[17]

Napier went on to devise similar methods for division and even the extraction of square roots, a quite substantial labor. As a calculating aid, his assembly of rods and boards and tokens might best be compared to an abacus, a way that human calculators could substitute the movement of pieces for "the tedium of calculation," as he called it. It did not catch on; Napier died soon after this publication, and the work excited negligible interest. And perhaps that was as Napier intended and expected. He describes it as "more of a lark than a labor"; it was something of a game, a term he proudly used in an epigram at the beginning of the book to describe his ambitions for arithmetic more generally.[18]

Leibniz cited Napier's *Rabdologiæ* in his habilitation thesis of March 1666, written when he was nineteen years old (A VI 1, 203), and recalled reading Napier's book in his 1705 letter to César Caze ("I once saw Napier's book in Latin without considering then its relation to binary arithmetic, which is also noticeable in the divisions of weights among assayers"; see chapter 31). So it may well be that Leibniz's thinking about binary was influenced by Napier's devices, as well as by the use of weights in geometric progression and traditional arithmetic tricks involving duplation and mediation. But as Leibniz's first scribblings about binary were about binary notation, it is important to explore where, if anywhere, the idea might have come from to use positional notation with only two digit values.

At about the same time Napier was working on *Rabdologiæ*, proper binary notation and binary arithmetic made a brief and unheralded appearance in England and then in a flash disappeared. Thomas Hariot (1560–1621, sometimes "Harriot") was a mathematician and astronomer of great ability who published little; one of his few works is an account of his trip to the Virginia colony (Harriot 1590). In the mid-twentieth century, John Shirley (1951) discovered previously unpublished Hariot manuscripts in the British Museum, on the basis of which he and others (e.g., Glaser 1981, 11–14; Wolf 2000, 31; Ineichen 2008, 14) have claimed that Hariot developed the binary system first. There is no doubt that the manuscripts show calculations in binary—$11\,1011 + 111\,0111 = 1011\,0010$ and another addition, two subtractions, two multiplications, and conversion back and forth to decimal, but no division. There is no indication of Hariot's motivation and no evidence that anyone except Hariot himself saw these manuscript pages in the 330 years or more that elapsed between their writing and their publication.[19] Certainly Leibniz never saw them.

Nor, as we shall see, was Leibniz aware of Juan Caramuel's *Mathesis biceps* of 1670, which contains a short article entitled "On Binary Arithmetic." In this article, Caramuel presents a table of binary numbers up to 32, using 0 and the letter *a* to form the sequence of binary numbers (for example, in his notation, 25 is *aa*00*a*). He then notes a convergence between binary arithmetic and music, "for each observes numbers, the former abstracted from material things, the latter discovered

---

17. Kolpas (2019) has a nicely illustrated explanation of Napier's "chessboard calculator." Here is a simple example: to compute 3×3, Napier would multiply $ab \times ab = abbc$ as shown below and then transform $abbc \to acc \to ad = 1 + 8 = 9$.

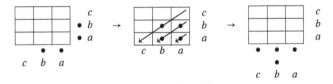

18. Arithmetic has now become a game.
   Gone is the tedium of calculation.
   And *Logarithms, Chess-board, Rod, and Strip,*
   Confirm once more great the Napier's reputation.
   (Napier 2017, 655)

19. Bauer (2010, 20) states that Napier published "a few years before Harriot," but as Hariot died four years after Napier's publication of *Rabdologiæ* and Shirley (1951) does not provide any information that could be used to date Hariot's manuscripts, no basis is evident for knowing which happened first.

in sounds," in particular the intervals in music, since a double geometric progression models the succession of octaves (Caramuel 1670, XXVII). Caramuel does not accord any further significance to the binary system, however, and in subsequent sections proceeds to identify and discuss a variety of different number systems, all of which are presented in positional notation. Aside from base 2, he discusses bases 3, 4, 5, 6, 7, 8, 9, 10, 12, and 60 (Caramuel 1670, XLV–LXI). For each of these, Caramuel provides a table of numbers, identifies nations or thinkers that had favored that system, and gives a series of examples of how each particular base number is embedded or reflected in the natural or supernatural world.[20] For example, when discussing "ternary arithmetic," Caramuel notes the theological importance of the doctrine "of the mystery of the most holy trinity," that angels are divided into three classes, and that the heavens are threefold, consisting of the Aëreum (in which the birds fly), the Aethereum (location of the stars), and the Empyreum (the throne of God). Similarly, in his discussion of "quaternary arithmetic," Caramuel (1670, L) points out that there are four elements and four cardinal directions, in "septenary arithmetic," seven principal planets, and so on (Caramuel 1670, LIV). His principal aim in treating the various bases is to show that they are all natural, being present in mundane and spiritual things without our thinking of them as bases. Accordingly, in his treatment of the different number systems, Caramuel often seems more concerned with a kind of number mysticism than he is with mathematics per se (in the words of Zacher 1973, 30, Caramuel "treats his systems without any mathematical or practical interest"). Even leaving this aside, his discussion of binary is disappointing because of what it doesn't offer: although the heading "binary arithmetic" would lead the reader to suppose that he will offer a full-blown binary arithmetic, Caramuel makes no attempt to show how arithmetic would work in that system, contenting himself with presenting a table of numbers up to 32.

Much more ambitious as an engagement with a nondecimal number system was the *Tetractys*, a short work published in 1673 by the mathematician and philosopher Erhard Weigel (1625–1699) to recommend conversion to the quaternary or base-4 system. Weigel favored the quaternary partly because it was easier to work with than the decimal, having fewer distinct digit values, and partly because he thought 4 to be the more natural unit of measure, being reflected everywhere in nature. To secure this latter point, Weigel (1673, 37–40) argues that the world, understood metaphysically, spiritually, and physically, consistently exemplifies the number 4; for example, there are 4 modes of being (necessary, contingent, possible, and impossible), 4 spirits (God, angel, devil, and rational soul), 4 types of parallelogram (square, rhombus, rhomboid, and rectangle), 4 islands of East Asia (Maluku, Philippines, Formosa, and Japan), and so on. To undercut potential resistance to his proposal, Weigel (1673, 9) notes that both the learned and unlearned already use quaternary measures routinely, such that "they make one digit from four barleycorns [i.e., from the breadth of four barleycorns], one palm from four digits, one foot from four palms, one pace from four feet, one perch from four paces ... and also separate day and night into four parts, the hour into four quarters," and so forth.[21] To further smooth the introduction of the new system, at least in Germany, Weigel (1673, 13) devises new German terms for key values: *Secht* for four fours and *Schock* for four *Secht* (thereby revaluing the term *Schock*, traditionally used to refer to the decimal number 60). More important from a mathematical perspective, Weigel (1673, 16–23) outlines methods of performing all four arithmetic operations in base 4, including the use of a dot to indicate carries when adding or multiplying, as well as methods for the extraction of square and cube roots.

---

20. The last of these features is likely derived from Schwenter (1636, 117–122), who provides a lengthy list of ways that the numbers 2–30 are embedded or reflected in the physical and spiritual realms. In a later work, Harsdörffer (1653, 80–128) examines the philosophical, theological, and scientific significance of an even wider range of numbers, namely, 1–24, 30, 40, 50, 60, 70, 80, 90, 100, and 1000. Caramuel uses many of Schwenter's examples.

21. Note that not all Weigel's measurements coincide with what Clavius (1570, 263) had earlier claimed was a system of measurements laid down by mathematicians as a universal standard, in an effort to avoid regional or national variations. For example, Clavius (1570, 264) describes a perch as equivalent to 10 feet (or 40 palms), not 16 feet (or 64 palms), as Weigel states.

Leibniz's Influences

The deep and varied interest in nondecimal number systems during the seventeenth century raises the question of who or what may have influenced Leibniz in his development of binary. This usually takes the form of trying to identify a single thinker from whom Leibniz may have drawn inspiration. Hence, Couturat (1901, 473)[22] suggested that Leibniz was influenced by Weigel, a view later endorsed by Ingaliso (2017, 111–112), while Tropfke (1980, 12) wondered whether Leibniz had been influenced by Weigel or Caramuel.[23]

Of these suggested influences or inspirations, that of Caramuel can be swiftly ruled out by the following remark Leibniz made in a letter to Friedrich Simon Löffler of 11 January 1711:

> Regarding Caramuel's *Mathesis biceps [Old] and New*, for which he asks ten thalers, I am unable to judge well because I have not yet seen it, and I fear it may contain vain subtleties, which is not unusual for Caramuel. (Dutens V, 418)

As Leibniz had not even seen a copy of Caramuel's book more than three decades after he had developed the binary system, clearly it could not have inspired or influenced him in this regard.[24]

As for the potential influence of Weigel's quaternary system on Leibniz's development of binary, this was first suggested in 1701 by Johann Bernoulli, who, upon learning of Leibniz's "new kind of arithmetic," informed him that "Weigel once devised something not unlike this …, the only difference being that in place of your binary sequence he adopts the quaternary" (A III 8, 614). In his response to Bernoulli, Leibniz flatly denied Weigel's influence, insisting that he had "reflected upon this [i.e., the binary system] for many years before any mention was made of the *Tetractys* being revived" (A III 8, 639). Unfortunately, Leibniz's implicit assertion that he developed the binary system before 1673 (the year Weigel's *Tetractys* was published) cannot be verified, as there is no evidence of his writings on binary going back that far. But the textual evidence we do have at least supports his claim that he had not been influenced by Weigel's book: the earliest evidence we have of his having read it is his brief reading notes on it,[25] and from the watermark, we know that they were written in the first half of 1683, by which point Leibniz had already written many manuscripts on binary. Prior to that, Leibniz appears to have had no awareness of the quaternary system or Weigel's advocacy of it, since there is no mention of either in any of his pre-1683 writings. In fact, in his pre-1683 writings on binary, Leibniz mentions only one other nondecimal number system, namely, the duodecimal, which he believed some had preferred to the decimal (see, for example, LH 35, 1, 25 Bl. 3v; LH 35,

---

22. "[The binary system] probably had been suggested to him by the *Tetractys* of his old master Weigel, published in 1673."

23. "Perhaps Leibniz was inspired to create the dyadic by Weigel's four-system or the work of Caramuel y Lobkowitz."

24. This fact has not stopped Ares et al. (2018) from claiming that Leibniz not only knew of Caramuel's work on binary but deliberately plagiarized it. Oddly, they cite the same passage from Leibniz's 1711 letter to Löffler in order to support this claim, apparently not realizing that it does the very opposite. The other textual "evidence" Ares et al. provide is no better. First, they point the reader to Leibniz's dissertation for his Master of Philosophy degree. While Leibniz does mention Caramuel in this text (see A VI 1, 88), he does not mention the *Mathesis biceps*, nor indeed could he, since Leibniz's dissertation was written in 1664, six years *before* Caramuel published the *Mathesis biceps*! The final piece of "evidence" of Leibniz's supposed plagiarism offered by Ares et al. is a letter to Michael Gottlieb Hansch of 13 December 1714, in which Leibniz encourages his correspondent to undertake further investigation of the number periods he has found in the binary system and the rules underlying them (Dutens V, 170). However, there is no mention of Caramuel or the *Mathesis biceps* in this letter. The charge of plagiarism leveled by Ares et al. is therefore not merely baseless but manufactured and need not be taken seriously. More sober is the assessment of Eberhard Knobloch (2018, 242) that "there is not the least evidence that this book [Caramuel's *Mathesis biceps*] influenced Leibniz's invention of the dyadic."

25. See A VI 4, 1162–1163. The notes are not particularly interesting in themselves, as Leibniz mainly restricts himself to summarizing some of Weigel's claims. However, at the end, Leibniz does briefly offer his own thoughts, writing, "I think the binary is best absolutely, the ternary in planes, and the octal in solids" (A VI 4, 1163). Around the same time, Leibniz also made notes on the *Tetractyn*, the companion work by the Pythagorean Society (1672); see A VI 4, 1163–1164.

13, 3 Bl. 33).[26] If, as seems likely, the quaternary system appeared on Leibniz's radar only in 1683, when he read Weigel's book on the subject, the case for thinking Weigel influenced Leibniz in the development of binary must be considered a lost cause.[27]

On the question of Leibniz's influences, then, aside from ruling out Caramuel as a candidate, and all but doing the same for Weigel, nothing definitive can be said except that Leibniz's idea for binary did not arise in a vacuum. He intimates as much himself on the occasions he offers his own narrative regarding his invention of binary, though the story he tells is not consistent and likely incomplete. In his "Essay on a New Science of Numbers" from February 1701, he identifies his inspiration as the use other people had made of nondecimal systems:

> Everyone agrees that the decimal progression is arbitrary, so others have sometimes been used. That made me think of the binary, or double geometric progression, which is the simplest and most natural. (p. 138 in chapter 23)

A similar claim was made in his letter to Bouvet of 15 February 1701:

> As we are usually accustomed to make use of progression by ten, and as some have used other progressions, I wanted to consider what would be the simplest progression possible, which is the binary progression or the double geometric progression. (pp. 127–128 in chapter 22)

While these writings do not specify the bases of the number systems that others had used, in a letter to Grimaldi written in early 1697, Leibniz identifies them as the duodecimal and quaternary:

> It is apparent that some considered the duodecimal to be more useful while others took pleasure in the Pythagorean tetractys. At some point it occurred to me to consider what would ultimately be revealed if we used the simplest of all [progressions], namely the dyadic or binary. (p. 110 in chapter 19)

The story Leibniz would tell decades after his invention of the binary system, then, was that he had hit upon binary as the simplest number system in conscious reflection upon other systems, sometimes unspecified but occasionally identified as the duodecimal and quaternary. Scholars, of course, must be wary of taking such *ex post facto* narratives on trust and will want textual evidence that confirms them before considering them reliable. And in this case, the textual evidence only partly supports Leibniz's story, for while we know that he was aware of the duodecimal system as least as early

---

26. "It is well known that all fractions can be expressed by an infinite sequence of integers of a certain progression, for example, the decimal, or even the duodecimal, or the one I prefer overall, the binary" (LH 35, 13, 3 Bl. 33, from August 1680). The relevant passage from LH 35, 1, 25 Bl. 3v is quoted above. In a letter written in 1712, Leibniz (mistakenly) identifies the German mathematician Daniel Schwenter (1585–1636) as a proponent of the duodecimal system: "Some report in the German *Deliciae mathematicae* has given preference to the duodecimal progression, in which eleven digits will be needed, namely 0, 1, 2, 3, 4, 5, 6, 7, 8, 9, δ, ε, where δ is 10 and ε is 11" (LBr 705 Bl. 93r). The reference is to Schwenter (1636) or perhaps Harsdörffer (1651) or Harsdörffer (1653). However, in none of those works does Schwenter mention, let alone endorse, the duodecimal system. Leibniz had cited Schwenter (1636) in a dissertation written in 1666 (see A VI 1, 173; English translation: PPL 78); it is not surprising that almost fifty years later, he had only an imperfect recollection of a work he had read in his youth.

27. It is worth noting that Leibniz had studied under Weigel at the University of Jena for the summer term of 1663. Whether Weigel was advocating the quaternary system at this time—a decade before he published the *Tetractys*—is unknown. However, it is at least possible that he was. In the preface to the counterpart publication to Weigel's *Tetractys*—namely, the *Tetractyn*, written by "The Pythagorean Society" of the University of Jena—it is noted that when the Society was formed, it had scarcely started to discuss Pythagoras's famous tetrad when Weigel was appointed director (Pythagorean Society 1672, 4). That appointment was made in 1662, though how soon after that Weigel developed the quaternary system—which is quite different from the Pythagorean tetrad—is unknown. There is also no evidence that Leibniz was a member of the Pythagorean Society during his term in Jena, though of course this cannot be ruled out. As Zacher (1973, 33) notes, one could certainly form and entertain the hypothesis that Leibniz had learned about the quaternary system through Weigel in 1663, but as there is no evidence to support the hypothesis, it should be treated as nothing more than speculation.

as 1670,[28] there is, as noted above, no evidence in his writings that he had any awareness of the quaternary system prior to 1683. Moreover, as also noted above, Leibniz elsewhere explicitly denied that he knew of the quaternary system prior to his invention of binary.

In truth, one thing that beleaguers most attempts to determine influence, whether on oneself or others, is the natural impulse to restrict oneself to identifying a single influential figure or single influential idea. Given the complexity of the human mind and the range of different factors impinging thereon, this impulse is probably best resisted as a general rule. Certainly, in the case of Leibniz's invention of binary, yielding to it (as he himself did and as some of his commentators have done) is likely to make us miss much of what is important. For as we have seen, the idea of exploring nondecimal number bases was not uncommon in the seventeenth century, and Leibniz himself was aware of some of this work. He may not have been privy to Hariot's experiments with binary arithmetic, Caramuel's flirtation with binary notation, or perhaps even Weigel's work on the quaternary system, but he did know Napier's *Rabdologiæ* as well as Pascal's offhand remark about nondecimal bases being possible (alluding to it occasionally; see, for example, chapter 19). He was also aware—as indeed was everyone else!—of the wide use of nondecimal number bases for measurements and coinage, and he even explored some of these himself prior to his invention of binary. Most notably, in the middle of 1674, he drafted two manuscripts on the sexagesimal system. In these two manuscripts, Leibniz concerns himself with the mathematical foundations of the sexagesimal system, working out procedures for converting sexagesimal numbers to decimal and for multiplying a three-digit sexagesimal number by an improper fraction (see LH 35, 3 A 26 Bl. 17 and LH 35, 8, 30 Bl. 27). No doubt it could be claimed that Leibniz's investigation into the mathematical foundations of the long-known sexagesimal system directly or indirectly spurred him to consider other number bases, but such a claim would be undermotivated and in any case repeats the mistake of supposing that a single figure or idea was his sole inspiration. A better conclusion to draw from Leibniz's work on sexagesimal is that it would have made it easier for him to adapt the algorithms for multiplication and division to other bases, as he did with binary and sedecimal shortly thereafter. Other than that, perhaps all that can be said about who or what inspired Leibniz to invent binary is that no simple or easy answer presents itself, and all that can be done by way of trying to answer it is what we already have done, namely sketching what we might call the mathscape of his time, noting the cognate ideas with which he would been familiar and treating the complex of these as the catalyst for his invention of binary, while declining to privilege any one of them in particular.

Leibniz's Legacy: The Birth of Computer Arithmetic

The use of duplation and mediation in arithmetic, as well as the use of positional notation with nondecimal bases, runs deep in human history and mathematical experimentation. By the seventeenth century, the overlapping waves of influence on individual scholars had become so turbulent that it is impossible to give a simple answer to a big question such as, "Who gave us the binary system?" Narrower questions are more approachable. If the question is, "Who wrote the first binary sum?" the answer seems to be Thomas Hariot, from what we know today, but it would surprise no one if an earlier and unrelated example turned up in some manuscript collection in the future. If the question is, "Who first realized that any integer can be expressed as a sum of powers of 2?" the answer could be the scribe of the Rhind papyrus (though the record includes only worked examples) or some Egyptian arithmetician who figured out the trick a millennium earlier and passed it down through a now-lost chain of cultural transmission.[29] Or perhaps credit for binary decomposition should go to the forgotten grain merchant who first used a pan balance with weights in geometric progression.

But if the question is, "Who gave binary computer arithmetic to the modern world?" the answer is Leibniz. We will make the case for forward influence below, but we pause here to reflect on what change of direction in intellectual history is presented by the disorderly collection of manuscripts

---

28. Early in 1670, Leibniz wrote, "In arithmetic, and in other disciplines as well, truths remain the same even if notations are changed, and it does not matter whether a decimal or a duodecimal number system is used" (A VI 2, 429; English translation: PPL 128).

29. Separately, Eglash (2005, chap. 7) proposes that binary began in sub-Saharan Africa.

and publications presented in these pages. Leibniz's work combines three elements that had not been seen together before and have never been completely disjoined since. First, he (like Hariot) made the leap from positional notation in the abstract, and multiple specific examples of positional number systems, to the special case of base 2. Second, Leibniz developed a complete suite of paper-and-pen algorithms for binary arithmetic, including not just the four standard arithmetic operations, but also extraction of square roots, the use of two's complement arithmetic, and the use of (as we would call it today) boolean algebra to describe the results of binary operations in terms of their operands. And third, he developed all this with an agenda to mechanize thought and actually designed machines (not just calculating aids, like Napier's chessboard) that could, in principle, carry out binary arithmetic calculations with minimal human intervention.

To say a few more words about the third point: discrete mathematics has always had game-like qualities. When only whole numbers are allowed in the solution to a problem, what otherwise looks amenable to well-understood methods can turn into an intractable puzzle. The "Cattle Problem" of Archimedes (Bell 1895) resisted solution for two millennia. With its stipulations that there are $\frac{1}{3} + \frac{1}{4}$ as many white cows as black cattle and so on, finding the size of the herd looks like a straightforward exercise in solving linear equations—except that certain of the bulls are to be arranged in a perfect square and others in an equilateral triangle. (The smallest solution turns out to be on the order of 10^{206545}.) Such puzzles have always brought mathematicians joy, because they have the appearance of reality (we can all picture arranging bulls in rows and columns) without any of the true utility that motivated the development of linear algebra or the infinitesimal calculus, for example.

So when Napier said that his location arithmetic was a "lark" and he had reduced calculation to a "game," he meant it. But when Leibniz said that binary was for theory, he was only half serious. He surely recognized that no one would want to carry out arithmetic with numbers twenty or thirty bits long (though his manuscripts show his own attempts) and that mathematically, binary notation was a theoretical extremum on a spectrum where decimal stood at a more useful midpoint. But he immediately followed that thought with a sedecimal notation "for practice," recognizing that binary-sedecimal conversion was trivial. And he designed not one but two different calculating engines based on binary principles. Perhaps binary is for theory, and perhaps he knew he would have trouble building the binary machines he sketched (he had enough trouble with his decimal machine), but he also had supreme confidence that thought could be mechanized and that the day would come when human disputes would be settled by calculation. His binary explorations were, from the beginning, part of that large agenda for him and became part of the development of the decision-making computers that would be built two and a half centuries later.

In his ambivalence about the utility of his discrete mathematical explorations, Leibniz foreshadows the spirit of computer science and diverges from the spirit of mathematics. When G. H. Hardy famously characterized his favorite branch of mathematics, number theory, as "useless," he was speaking as a premier mathematician of his time. When Alan Turing a few years later, and Rivest, Shamir, and Adleman (1978) some decades after that, used that theory for powerful cryptography and cryptanalysis, they were birthing computer science. In proposing to mechanize binary arithmetic, Leibniz was their ultimate progenitor.

A line can be drawn between Leibniz's writings on binary and its adoption in the first modern computers designed by Konrad Zuse (1910–1995), the pioneering computer scientist who in 1934 made the decision to use binary numbers and code in his Z1 and subsequent machines. As we shall see, this line passes through the various books, articles, and encyclopedia and dictionary entries that promulgated Leibniz's work on binary from his own time to the 1930s. To trace this line, let us begin with published treatments during Leibniz's own lifetime.

Aside from Leibniz's own "Explanation of Binary Arithmetic," published in 1705, two further treatments of his binary system were published during his lifetime, both of which he had a hand in. One of these was the aforementioned essay by Dangicourt on column periodicity of binary numbers, which was written at Leibniz's behest and published in *Miscellanea Berolinensia*, a journal of which Leibniz was the editor. The other, an essay by Leibniz's friend and long-time correspondent, Wilhelm Tentzel (1659–1707), was published in 1705 under the title: "Explanation of Binary Arithmetic, Which Was Used by the Chinese 3000 Years Ago and Lost by Them but Recently Found Again by Us." Unlike Dangicourt's essay, which does not mention Leibniz's name at all, let alone credit him with the invention of the binary system (and this presumably with Leibniz's blessing, since he commissioned and published the essay), the essay by Tentzel (1705, 81) opens with an acknowledgment of his debt to Leibniz: "To curious people in general, especially mathematicians and arithmeticians, I offer something quite new for the New Year, which Privy Counselor Leibniz of Hanover discovered and recently communicated to me in accordance with his usual kindness." He then offers brief details of binary numeration, examples of the four standard arithmetic operations in binary, and the periodicity of columns of binary numbers, all within the space of four pages. The remaining twenty-eight pages are devoted to an outline and detailed exploration of Bouvet's hypothesis of a correlation between binary numeration and the hexagrams of the Yijing. Tentzel (1705, 112) concludes his study by describing the table of sixty-four hexagrams as a "labyrinth" from which not even Confucius could escape, but that "Leibniz alone has discovered the thread of Ariadne," echoing Leibniz's claim that an ancient Chinese enigma had been solved by a European. While Tentzel's essay, essentially an expanded version of—and apology for—Leibniz's "Explanation of Binary Arithmetic," went largely unnoticed, its nonmathematical focus would be mirrored in many of the treatments of the binary system during the eighteenth and (to a lesser extent) the nineteenth centuries, as we shall see.

In the decades after Leibniz's death, references to his work on binary inevitably drew upon the few of his writings on the subject that had been published at the time, in particular the 2/12 January 1697 letter to Duke Rudolph, with its focus on the binary-creation analogy and proposal for a commemorative medal, "Explanation of Binary Arithmetic," which proposed a parallel between binary and the hexagrams of the Yijing, and "Discourse on the Natural Theology of the Chinese," which likewise explored the link between binary and the hexagrams.[30] Leibniz's 1697 letter to Duke Rudolph appeared in six different books published between 1720 and 1768,[31] and was discussed in various others, such as Wideburg (1718). While interest in the binary-creation analogy waned over the course of the eighteenth century, it was occasionally resuscitated by those keen to deride it, such as Laplace (1825, 211), who wrote,

> Leibniz thought he saw the image of creation in his binary arithmetic, in which he used only the two characters, 0 and 1. He imagined that God could be represented by 1, and nothingness by 0; the Supreme Being had drawn all beings from nothing, like 1 with 0 expresses all numbers in this system of arithmetic. This idea pleased Leibniz so much that he communicated it to the Jesuit Grimaldi, president of the tribunal of Mathematics in China, in the hope that this symbol of creation would convert the emperor of the time, who was particularly fond of science, to Christianity. I report this feature only to show how far the prejudices of infancy can lead the greatest men astray.

The vast majority of eighteenth-century treatments of Leibniz and binary followed the thread of "Explanation of Binary Arithmetic," by outlining both binary notation and its apparent correlation with the hexagrams of the Yijing. Such an account was added to the biography of Leibniz by Lamprecht (1740, 74–75), as well as to some editions of Leibniz's works, such as the German edition of the *Theodicy* (Leibniz 1744, 824–828). An article on "binary arithmetic," written by Samuel Formey (1711–1797), was also included in the monumental enterprise of Diderot and D'Alembert (1751, 257–258), the *Encyclopédie*. Opening with "Binary arithmetic is a new kind of arithmetic that Mr. Leibniz based on the shortest and simplest progression," the article focuses on binary notation and conversion of decimal to binary, though a few lines are devoted to the claimed connection between binary and the hexagrams of the Yijing, about which no judgment is offered. Other encyclopedia editors opted not to include a dedicated entry on binary but rather discussed the possible binary–Yijing connection in other articles; Croker, Williams, and Clark (1766, n.p.), for example,

---

30. Other writings of Leibniz on binary that were published in the eighteenth century often had the same focus, for example, his letter to Ferdinand Orban of 27 August 1705, which was first published in Will (1778, 57–65); English translation: Leibniz (1705b).

31. Namely, Leibniz (1720, 103–112), Nolte (1734, 1–16), Ludovici (1737, 132–138), Leibniz (1739, I: 234–239), Leibniz (1740, 92–100), and Leibniz (1768, III, 346–348).

did so in an article on "Chinese philosophy," in which they asserted that "Leibnitz deciphered the aenigma, and demonstrated that Fohi's [Fuxi's] two lines were only the elements of binary arithmetic." Striking a more cautious note was Georg Bernhard Bilfinger (1693–1750), who dealt with the binary and Yijing question at the end of a lengthy book on Chinese moral and political philosophy. While accepting that Leibniz and Bouvet were probably right in thinking that binary was the key to deciphering the hexagrams of the Yijing, Bilfinger (1724, 359–360) pointed out that the Chinese usually arrange the hexagrams in a different way from the ordering sent to Leibniz by Bouvet (i.e., the so-called Fuxi ordering), which led him to insist that the Chinese, whose mystery it was, ought to have the final say on whether Bouvet and Leibniz were right to treat the hexagrams as a form of binary numeration.

While the possible link between binary and the hexagrams of the Yijing dominated the eighteenth-century treatments of Leibniz's binary system in the secondary and tertiary literature, there were some serious mathematical treatments also. Wenceslaus Josephus Pelicanus (1712) published a short book whose title translates as "The Perfect Arithmetic for Anyone Who Does Not Know How to Count to Three," in which he offered a detailed exposition and analysis of the binary number system, albeit under his own preferred term "dual arithmetic." After describing binary notation, Pelicanus describes how the four standard arithmetic operations work in binary before considering how to perform the same operations with binary fractions. He also investigates the extraction of square and cube roots and provides tables for easy conversion between binary and decimal. Curiously, Pelicanus says nothing about the inventor of the "dual arithmetic" and mentions Leibniz not once, which suggests that Pelicanus was either yet another independent inventor of binary or not very good at acknowledging intellectual debts. Although Leibniz was made aware that Pelicanus was working on binary, he appears not to have seen Pelicanus's book;[32] had he done so, he might have realized that it was precisely the sort of work that he could and should have written thirty years earlier. Twenty-five years after Leibniz's death, Buffon (1741, 219–221) published a short paper outlining a method for the conversion of numbers between different bases, including binary, using logarithms; the paper's summary, written by the perpetual secretary of the Académie Royale des Sciences, Bernard le Bovier de Fontenelle (1657–1757), reminds the reader that Leibniz had published on the binary system back at the turn of the eighteenth century. Three decades after Buffon's paper, Brander (1775, 3) published a short pamphlet on binary arithmetic (which "owes its invention . . . to the excellent Mr. Leibniz," he noted at the outset), detailing algorithms for the four base-2 arithmetic operations, which he appears to have worked out independently of both Leibniz and Pelicanus.

During the eighteenth century, outlines of the binary system also started to appear in mathematics textbooks, often followed by an account of Bouvet and Leibniz's hypothesis that binary was the key to deciphering the Yijing. In some cases this hypothesis was merely described without assessment (for example, see Vieth 1796, 13), while in others, it was evaluated, not always positively (for example, see Ozanam 1790, 5, in which the hypothesis is described as "more imaginative than sound").[33] But authors of textbooks, encyclopedias, and other reference works varied in how much space they were prepared to devote to the Yijing during their treatment of binary. Andrés (1790, 71–73) mentions the hypothesis of the Yijing almost in passing in his outline of binary notation, while Hutton (1795, I: 144, 206) mentions it not at all.

In this matter, little changed in the first two-thirds or so of the nineteenth century. Encyclopedias and other reference works from this time treated Leibniz's binary system—and it was almost always seen as Leibniz's[34]—as a curious, if impractical, form of numeration, often devoting as much space to binary notation as to the purported parallel with the hexagrams of the Yijing, with an occasional

---

32. See Michael Gottlieb Hansch's letter to Leibniz of 4 October 1714 (LBr 361 Bl. 49–50), in which he outlines Pelicanus's method of determining the decimal equivalent of a long binary string but does not mention Pelicanus's book.

33. The same verdict is also to be found, verbatim, in Montucla (1799, 457).

34. Even brief references to binary notation in encyclopedias and dictionaries were typically accompanied by an assertion that Leibniz was the author thereof; see, for example, Lieber, Wigglesworth, and Bradford (1854, 335) and Heyse (1859, 289).

mention of the binary-creation analogy thrown in for good measure (see, for example, Barlow 1814, n.p., art: "binary arithmetic"; Brewster 1832, II: 382–384; Anon 1857, III: 591; Schlömilch and Witzschel 1858, 338). Even serious mathematical works tended to follow this (by then rather tired) formula; for example, in an essay on the history of different number systems, Krist (1859, 49–50) wrote,

> For $g = 2$ one arrives at the so-called dyadic or binary system, which needs only the digits 1 and 0 for number figures, according to which our 2 is represented by 10, our 3 by 11, our 4 by 100, our 5 by 101, and so on. This system is of historical interest insofar as it was established around 1697 by the great German philosopher Leibniz, after it had already been found in 1670 by the bishop of Campagna and Satriano, Johann Caramuel. Leibniz, however, came up with it independently of the latter, and dealt with it merely in the hope of being able to examine the properties of numbers more easily with its help. Leibniz also showed, and after him Brander, how the dyadic system can be used for calculation, but without the intention of thereby displacing the already prevailing decimal system. Interesting is the fact that Leibniz used the dyadic system to symbolize the origin of the universe. In a letter to the Duke of Brunswick from 1697, he added a medallion on which he sketched light and darkness behind a table filled with dyadic numbers, and to which he wrote the following sentence: "In order to bring forth everything out of nothing, unity is sufficient" in Latin as a circumscription. At the same time, Leibniz also expressed the intention of sending his number system to the missionary Grimaldi in China, in the hope that the Chinese emperor might be converted to Christianity through its profound meaning. Although this hope was not realized, the missionary Bouvet was informed of the dyadic and was thereby able to unravel many ancient Chinese manuscripts, which at the same time revealed that the dyadic system was already known to the Chinese and that among them the invention of the same is attributed to their emperor Fuxi, the founder of the Chinese empire and Chinese scholarship. But Fuxi is said to have lived around 2200 BC.

It is a notable that, until deep into the nineteenth century, virtually all treatments and discussions of binary and the hexagrams of the Yijing merely repeated the claims made by Bouvet and Leibniz at the very start of the eighteenth century. Moreover, those who sought to assess these claims did so either with a partial understanding of the Yijing itself or in complete ignorance of it, as it was not available to European scholars in its entirety until the first complete translation into a European language—Latin—appeared in the 1830s (see Mohl 1834, 1839), with various others following thereafter (for example, McClatchie 1876 and Legge 1882). The availability of the full text of the Yijing engendered an explosion of scholarship, and eventually the deeper understanding of the Yijing by European scholars began to put strain on the hypothesis that its hexagrams were a form of binary notation. For example, Cantor (1863, 48) considered the evidence that the Chinese had used a base-2 system in the hexagrams of the Yijing and dismissed it, claiming that binary numeration was an original invention of Leibniz. Thanks to such efforts, as the nineteenth century wore on, it became increasingly common for binary to be treated more as a system of numeration and arithmetic than as an intercultural curiosity; see, for example, Privat-Deschanel and Focillon (1864, 264), Lübsen (1869, 232–234), Figuier (1870, 73), and Collignon (1897). And by the end of the nineteenth century, the idea of mechanizing calculations in binary arose for the first time since Leibniz first entertained it in 1679, with Lucas (1891, 148) stating that "this system [i.e., binary] would lend itself more naturally than any other to the manufacture of arithmetic machines." Inspired by this suggestion, Peano (1898, 10) designed and built a stenograph based on "binary writing," where groups of binary digits corresponded to the common sounds of various languages, making it possible to input shorthand transcriptions faster than in a conventional stenograph.

This decoupling of binary and the Yijing, along with the accompanying re-mathematization of Leibniz's binary system, was driven not just by a greater understanding of the Yijing but also by increasingly greater availability of Leibniz's writings on binary. In the mid-nineteenth century, the German mathematician Karl Immanuel Gerhardt (1816–1899) published a seven-volume series of Leibniz's mathematical writings, *Mathematische Schriften*, containing many previously unpublished pieces. These included important works on binary such as "Demonstration That Columns of Sequences Exhibiting Powers of Arithmetic Progressions, or Numbers Composed from These, Are Periodic" and "On Binary," neither of which had been published before (see GM VII 228–238, corresponding to chapters 27 and 32 of the present book). Gerhardt's volumes also brought together items from Leibniz's correspondence in which he discussed binary, including letters to Johann Christian Schulenburg of 1698 (GM VII 238–243) and Johann Bernoulli of 1701 (GM III 2, 656–657, 660–662, 669–670). All this enabled nineteenth-century readers to get a better—albeit still very incomplete—understanding of Leibniz's studies on the subject.

Tracing the references to Leibniz and binary across the eighteenth and nineteenth centuries is not merely an exercise in *Rezeptionsgeschichte* but also makes it possible to show that Leibniz's work on binary appeared in the sort of mathematics textbooks that would likely have been used by Konrad Zuse during school or at the Technical University of Berlin in the late 1920s and 1930s. Indeed, there is no shortage of German-language textbooks from the first quarter of the twentieth century that (a) mentioned the binary system and (b) indicated Leibniz's authorship thereof. For example:

- Tropfke (1902, 4n3): "in *Histoire de l'acad. d. Paris* 1703 (printed 1705), pp. 85–89 (*Gesammelte Werke*, ed. Gerhardt, third series, volume 7, Halle 1863, pp. 223–227) he [Leibniz] discusses the advantages and disadvantages of the binary system."
- Wieleitner and Braunmühl (1911, 74) notes Leibniz's development of binary and refers the reader to Leibniz's "Explanation of Binary Arithmetic" as well as letters to Schulenburg, Bernoulli and so on (all of which were available in Gerhardt's *Mathematische Schriften* by that point).
- Weber (1922, 30): "With the exception of 1, any number g can be taken as the base number of a number system with g digits 0, 1, 2, ... $(g-1)$. The theoretically simplest is the binary system with the base number 2, which manages with only two digits, 0 and 1 (Leibniz, *Math. Schriften*, ed. C. J. Gerhardt 7, 223)."

The first two of these works were and still are held by the library at the Technical University of Berlin,[35] where Zuse studied between 1927 and 1935. The same library also holds original copies of Leibniz's *Mathematische Schriften*. Whether in the aforementioned works or others, it would not have been difficult for Zuse to come across references to Leibniz and binary, making it likely that he would have read about them firsthand prior to beginning work on his first computer in 1934. That computer, the Z1, was organized on the principle of binary representation of numbers—several years before John von Neumann (1993, written in 1945) adopted binary for the EDVAC and roughly simultaneously with the adoption by Howard Hathaway Aiken of decimal for the "Automatic Sequence Controlled Calculator" he was designing at Harvard (Aiken, Oettinger, and Bartee 1964, written in 1937). And indeed, Zuse emphasized that he adopted the binary system in the early 1930s and that the "birth of modern computer science" was with his realization that "all data could be represented through bit patterns" (Zuse 1987) and "data processing begins with the bit" (Zuse 1980, 621). Zuse singled out Leibniz as the forefather of the modern computer, in no small part because of his work on the binary system:

> The story [of the computer's prehistory] begins in the seventeenth century with Leibniz, who, together with Schickard and Pascal, was one of the pioneers of computing machine construction. He developed the mathematics of the binary number system and made one of the first formulations of symbolic logic—what we today call propositional calculus or boolean algebra. It was precisely this symbolic logic that later proved extraordinarily useful in the construction of computing machines and led to the generalization of the concept of computation. (Zuse 1993, 33)

In bestowing such an honor upon Leibniz, Zuse can be plausibly read as acknowledging an intellectual debt and, just as plausibly, influence. Petrocelli (2019) suggests that Zuse's use of the term

---

35. The search function for the catalog is available online at https://www.tu.berlin/en/ub/.

"Dyadik" is a direct borrowing from Leibniz.[36] While others such as Hariot and Caramuel may have claims to have entertained the idea of binary numeration before Leibniz, their work played no part in the birth of the computer age. Rather, it was Leibniz's independent invention and development of binary that proved to have the only impact in this regard—an assessment shared by other scholars, for example, Garfinkel and Grunspan (2018, 32) ("All modern computers use binary notation and perform arithmetic using the same laws that Leibniz first devised"), Stein (2006, 57) ("Leibniz's Machina Arithmeticae Dyadicae is the forefather of today's computers"), and Wiener (1961, 19) ("If I were to choose a patron saint for cybernetics out of the history of science, I should have to choose Leibniz").[37]

---

36. Leibniz seems to have been responsible for introducing *Dyadik* into the German vocabulary. This German word is still used exclusively to refer to "the binary system of arithmetic." The German word *binär* was not used to refer to the binary system until the twentieth century and even now is not the usual term, which is *Dyadik* or sometimes *Zweiersystem*, or *Dualsystem*.

37. An alternative view, that Leibniz's work was largely irrelevant to the development of the modern binary computer, has been asserted forcefully by Bernhard Dotzler:

- "Structurally, the back-projection of computer binarism onto the dyadic is almost the same story as the former identification of the system of binary numbers with the Yijing. Since its hexagrams are made up of only two elements—the whole and the broken line, they can formally be described as a binary system. However, this former interpretation is as wrong in terms of content as the updated one is in functional terms." (Dotzler 2010, 29)

- "So one could say that with dyadics, esotericism was once set against esotericism: the esotericism of the dyadic penetration of creation against the esotericism of the Yijing interpretation, which had been declared false. To remember this, of course, cannot aim at bringing the associated metaphysics back into play. It is only a matter of keeping in focus this formerly different purpose of the binary number system: namely esoteric, and not cybernetics. Before it was seized by the binarism of information technology, the dyadic was an ontological instrument of understanding. Accordingly, it characterizes a functionality that may differ only minimally from the binary of the computer, but one which is fundamentally different." (Dotzler 2010, 31)

- "With this, however, the dyadic stands for a paradox, which then counteracts the myth of its anticipatory conspiracy with the binary of the computer, with cybernetics and digital arithmetic. The formal does not correspond to a functional correspondence, and that means: The equation of the binarism of today with the dyadic of yore is actually—fiction." (Dotzler 2010, 33)

However, it is clear from the timeline of Leibniz's extensive algorithmic writings on binary arithmetic and his design sketches for two kinds of binary calculator that he saw the binary system as far more than an "ontological instrument" and that he did so well before he was made aware of the Yijing hexagrams. The assertion by Merzbach and Boyer (2011, 388) that Leibniz's "noting of the binary system of numeration" was one of his "relatively minor contributions" can perhaps best be reconciled with its influence as an acknowledgment of the extraordinary breadth and range of Leibniz's other work.

About the Texts, Translations, and Apparatus

The Texts

At the time of writing, almost 350 years have elapsed since Leibniz devised his binary system, and despite its importance, remarkably few of his writings on the subject have been published, and fewer still are available in English translation. In Leibniz's Nachlaß, there exist around 100 manuscripts devoted to binary and a similar number of letters in which binary is discussed (usually alongside other things). While a complete English edition of this corpus would be desirable, space constraints forced us to be selective. In narrowing our selection to thirty-two texts, we have focused on those writings which, taken together, tell the story of binary as it developed in Leibniz's hands, from first thoughts through to dissemination and eventual publication. It is, as we shall see, if not quite an epic then at least a protracted and sprawling story, featuring mathematical, philosophical, theological, and cultural twists and turns and such disparate artifacts as medals, wax seals, and the hexagrams of the Yijing.

Among the writings that we chose to omit were ones devoted to sequences of binary numbers along with examples of addition,[38] workings out of various binary sums,[39] investigations into the regularities in columns of binary numbers,[40] binary fractions,[41] the periodicity of binary fractions,[42] and binary algebra.[43] However, many of the ideas found in these omitted writings can also be found in the texts we have included. Twelve of our thirty-two texts have yet to be published in their original language (namely, chapters 2–4, 6–12, 20, 24, and 25) and all but three are appearing in English for the first time.[44]

Leibniz saved virtually everything he wrote, so the vast majority of his extant writings on binary were never intended for publication or dissemination in any form.[45] Many were simply rough notes or exploratory drafts in which he thinks as he writes, seeing where the ideas lead. Despite their rawness and sporadic clumsiness, these pieces afford a valuable insight into Leibniz's working methods, complete with their wrong turns, dead ends, and occasional calculation errors or lapses in reasoning as he charts new territory. Thus when reading this book, it is good to keep in mind that Leibniz was writing mostly for himself.

The Translations

As so many of Leibniz's writings on binary have yet to be published, including most of the important early ones that show Leibniz developing and experimenting with the system, we could not rely on published transcriptions and so made our own transcriptions from the original manuscripts, including of those writings for which a published transcription is available. Transcribing Leibniz's manuscript writings on binary proved to be a challenge, in no small part because many of them are hastily scribbled drafts often featuring dense tables, cramped scratchwork, and frequent deletions and revisions. We have sought to replicate the manuscript material as faithfully as the limitations of the printed page allow. We should note, however, that we have corrected Leibniz's errors (where we have spotted them!), indicating this in the footnotes, and have provided his deletions only where we consider

---

38. See LH 35, 3 B 5 Bl. 58.
39. See LH 35, 3 B 5 Bl. 28 and 62.
40. See LH 35, 3 B 5 Bl. 15–18, 25–26, 57, and 94–95.
41. See LH 35, 3 B 5 Bl. 47, 51, 52, and 89–90.
42. See LH 35, 3 B 5 Bl. 11–12.
43. See LH 35, 3 B 5 Bl. 78.
44. The three that have appeared in English before are "On the Organon or Great Art of Thinking" (chapter 10) in Leibniz (1973, 10–17); Leibniz's letter to Duke Rudolph August of 2/12 January 1697 (chapter 17) in Ching and Oxtoby (1992, 70–76) and partially in Cajori (1916, 560–563) and Glaser (1981, 31–36); and "Explanation of Binary Arithmetic" (chapter 30) in Glaser (1981, 39–43).
45. As Knobloch (2004, 75) observes, "Leibniz's texts reflect his thinking as thinking while writing. Every idea, every question, every doubt, every access and optimism, every provisional result, every plan or intention is written down."

them to have some significance or interest for contemporary scholars working in one or other of the fields of mathematics, computer science, philosophy, or theology (or the histories thereof).

Translating this material also proved quite challenging at times, usually due to Leibniz's terminology; the mathematics of his time had few agreed terms and little standardized notation. While we have tried to stay as close to the source language as possible, a literal translation of some of Leibniz's terms and symbols would have made for difficult reading today. We note below some of our more important translation choices.

Binary   When referring to his notation of 0 and 1, Leibniz interchangeably used the borrowed Greek term *dyadica* and the Latin *binarius*, both of which mean "pertaining to two." Where possible, we have translated both terms as "binary," since this has greater currency in English than does the equivalent "dyadic," at least as regards the base-2 number system. Occasionally, Leibniz uses both *dyadica* and *binarius* separated by the "or" of equivalence, to indicate that they are alternative terms, and in such cases, we have translated the former as "dyadic" in order to avoid the obvious problem with "binary or binary." On the German word *Dyadik*, see note 36 above.

Carry   Leibniz used the Latin term *residuus* (remainder, the remaining, left over) to refer to a quantity being transferred from one column to another. But "remainder" is a term now used in mathematics only to refer to an amount left over after a division operation; hence, when Leibniz uses *residuus* to refer to the transfer of quantities between columns, we translate it by the noun "carry," in line with modern usage. Leibniz also uses phrases such as the Latin *transmittit aliquid* (literally: transmits something) when the addition of two digits in one column causes a carry to the next; in such a case, we have translated the phrase as "causes a carry."

Digit   In line with its etymology and usage at the time, Leibniz reserved the term "digit" (Latin: *digitus*) for the decimal numerals 0–9. When referring to binary digits, whether represented by 0 and 1 or by letters, he used the Latin *notae* (marks, signs) or the French *notes* (notes, marks). Since mathematicians and computer scientists now routinely refer to "binary digits," and since referring instead to marks, signs, or notes risks confusion, we have translated *notae* and *notes* as "digits" except for a few cases in which Leibniz used those terms to refer to the trigrams of the Yijing or their component lines, where we have opted for "trigrams" or "glyphs," respectively. (In our commentaries but not in the translation, we also use the conventional "bit" for "binary digit.")

Dot   When indicating carries in a column of binary digits, Leibniz used the Latin *punctum*, the French *point*, and the German *Punct* (*Punkt* in modern German), all of which are most naturally translated as "point." However, in the context of a mark indicating a carry, we have opted to translate these terms as "dot," the term commonly used for a diacritic mark over a letter or numeral, as opposed to the decimal or binary "point" introducing a number's fractional part.

Progression   Leibniz commonly referred to a sequence of binary digits by the Latin *progressio*, the French *progression*, and the German *Fortschreitung*, all of which are commonly rendered into English as "progression." Despite some reservations, we have decided to translate these terms by "progression" rather than "sequence," which is the English word we have used to translate the Latin *series* and French *série*. Others sometimes translate *progressio dyadica* as "binary system"; we have refrained from rendering *progressio* in this way to give readers the discretion to decide for themselves when Leibniz is referring simply to a geometric series and when to an associated system of notation and arithmetic.

Double geometric progression   Leibniz uses the expression *progressio geometrica dupla*, literally "double geometric progression," to refer to the geometric progression in which the ratio of successive elements is 2, that is, 1, 2, 4, 8, 16, ...[46] (Or, sometimes, the progression in which the ratio is $\frac{1}{2}$, that is: 1, $\frac{1}{2}$, $\frac{1}{4}$, $\frac{1}{8}$, $\frac{1}{16}$, ...) We use the literal English translation of this phrase since it is so much less

---

46. Leibniz is inconsistent as to whether $1 = 2^0$ is a member of the double geometric progression. In chapter 4 he explicitly includes it, but the deleted text in footnote 8 of chapter 23 requires that 2^1 be the first member of the progression, and the lower-left medal design near the end of chapter 17 affixes a star to 2, 4, 8, and 16, but not to 1.

wordy than a more precise phrasing, even though it may unintentionally suggest that the progression itself is "double" somehow.

The Apparatus

For each text, we have provided full details of the manuscript sources and available transcriptions, as well as a "headnote" in which we explain the context and content of the text as well as give reasons for our dating of the manuscript source in the many cases where Leibniz did not date the text himself. It should be noted that those texts written before March 1700 are given two dates: the first follows the Julian calendar, the second the Gregorian calendar, which was finally adopted in Protestant Germany on 1 March 1700. Until the switch, Leibniz dated his writings according to the Julian calendar, which was ten days behind the Gregorian. Although we have not amended any of the dates Leibniz supplied on his manuscripts, we have followed the convention of using a dual-dating system, consisting of old- and new-style dates, when referring to dates falling before the switch to the Gregorian calendar, for example, writing a date as 15/25 March, where the first date is that of the Julian calendar (old style) and the second that of the modern Gregorian calendar (new style).

In addition to the headnotes and footnotes, we have also supplied commentaries to explain Leibniz's thinking and/or to identify his errors and how these affect his calculations. These commentaries are displayed against a gray background and are positioned alongside or underneath the relevant part of a text; the gray box indicates that the comments within are not part of the text proper but are an editorial comment thereon.

The layout of the manuscript material requires some comment. While it might seem most fitting to reproduce the layout of Leibniz's manuscripts exactly on the printed page, this was not always possible or desirable in practice. Many of the manuscripts we have used are cluttered and untidy, with Leibniz adding new material wherever he could find space, sometimes indicating by a line where in the text it should fit, sometimes not. Some feature passages written in the margin or vertically down the side of the page. Many of the manuscripts contain lengthy tables and/or rows of sums, sometimes with no workings out, and many also contain material that Leibniz chose to delete. Representing all of this faithfully on the printed page would have made many of the texts unreadable. Hence, we have sought to order and arrange the material as best we can, seeking to remain faithful to Leibniz's intentions where these could be discerned. When Leibniz indicates where a marginal addition should go, we put it there, likewise with tables and sums, so far as typesetting limitations permit. Material whose placement could not be discerned is recorded in footnotes. In a direct quotation from Leibniz, a period or comma positioned outside the closing double quotation mark, rather than inside as would be conventional, indicates that the text quoted does not include that punctuation mark.

We have also used the following conventions:

- [ ]   In our editorial matter, square brackets are used to indicate English translations of titles or organizations given in their original language. In the translations (including translated material in the headnotes or footnotes), square brackets denote editorial interpolations, usually to indicate words omitted by the original author.
- *italics*   In the translations, italics are used to emphasize words or passages that were underlined in the manuscript, as well as to indicate book titles and words or expressions left in other languages.
- grayed-out text   In chapter 1, we have used grayed-out text to indicate manuscript material that was not in Leibniz's hand.

Typographical Conventions   Each chapter begins with a chapter number, title, and date, for example:

5   Binary Arithmetic Machine (before 15/25 March 1679)

If Leibniz supplied a title, we use it as the chapter title and also present it at the beginning of the text itself; if he did not include a title, we have devised one, but have not put this at the beginning of the text itself as it is not part of the manuscript.

Next comes information about the manuscript and available transcriptions, for example:

Manuscript:
M: LH 42, 5 Bl. 61. Draft. Latin.

Transcription:
LVM: Mackensen (1974, 256, 259).

Where there exists only a single manuscript source, it is designated M, but where there are multiple manuscript sources, they are labeled M1, M2, and so on, and in such cases, we indicate at the start of the translation which source(s) we have used. We have likewise provided an abbreviation for available transcriptions. Where there are discrepancies between our reading of a manuscript source and that found in a published transcription, we record it in a footnote, referring to each by the abbreviations indicated at the start of the chapter. Similarly, where there are variations between different manuscript sources, this is recorded in a footnote. Abbreviations such as "LH" are explained in the "Abbreviations" section. "Bl" stands for "Blatt," German for "sheet." "Mackensen (1974)" can be identified in the bibliography at the end of this book.

Our headnote follows and is ended by the following ornament:

⊸∘◦⧉∘◦

Leibniz's text follows and continues to the end of the chapter.

> Material that appears in a box on a gray background, like this, is our explanatory gloss. Sometimes it appears alongside Leibniz's text; sometimes, as here, as a paragraph interpolated between sections of Leibniz's text.

To improve readability, long binary numerals are sometimes presented as blocks of four bits separated by narrow spaces, for example, "10100001" as "1010 0001."

A numeral with slashes through digits, "1̸0̸1̸0̸ 0̸001", for example, transcribes a numeral that Leibniz has crossed out digit by digit, in whole or in part (in this case, Leibniz wrote "1010 0001" and crossed out each of the first five digits). This is not an instance of Leibniz correcting a mistake; rather, he is checking off digits in the course of carrying out an arithmetic algorithm, for long division perhaps. Depending on typographical convenience, deleted passages are noted either in footnotes or by ~~crossing them~~ out.

When captured in a footnote, deleted material is presented between ▷ and ◁ (the former symbol indicating where a deleted passage begins, the latter where it ends). Variants between different manuscripts of the same text are recorded in a similar way. After ◁ we indicate whether the material is a deletion or a variant in a different source from that used for the translation of a chapter. In cases where a deleted passage includes a restart which is also deleted, we indicate this using numbers in the following way:

{1} the deleted passage starts; {2} Leibniz strikes out everything in the whole of {1} and starts again.

1 Notes on Algebra, Arithmetic, and Geometric Series (October 1674)

Manuscript:
M: LH 35, 3A, 22 Bl. 1. Draft. French and Latin.

Transcription:
A: A VII 1, 882–883.

In March 1672, Leibniz was sent to Paris on a diplomatic errand by his patron, Johann Christian von Boineburg (1622–1672).[1] Leibniz would remain in Paris until October 1676, when he departed for Hanover to take up a new post there, and in much of the intervening time was devoted to studying under the tutelage of some of the best mathematicians of the age, including Christiaan Huygens (1629–1695) and Edme Mariotte (1620–1684). Leibniz's immersion in mathematics yielded thousands of pages of material on various branches of the subject, including geometry, algebra, and number theory. This material ranges from rough notes to draft pieces in which Leibniz works through his ideas on paper, as well as to letters and polished papers intended for publication. Although Leibniz's most famous mathematical invention during his so-called Paris period was the infinitesimal calculus, the binary system was also invented during this time, with the following text, evidently just a page of rough notes on algebra, geometry, and arithmetic, containing Leibniz's first foray into the area. The grayed-out text—most of the top half of the page—is in Mariotte's hand; everything underneath (starting with the obscure "3* equals 2" in the oval) is in Leibniz's. The infinite sum $\frac{1}{1} - \frac{1}{3} + \frac{1}{5} - \frac{1}{7} + \frac{1}{9} - \frac{1}{11} + \ldots$, which has the value $\frac{\pi}{4}$, is also in Leibniz's hand and forms part of a formula for the quadrature of the circle he had determined in the first half of 1674.[2]

The sequence of formulas involving four variables a, b, c, d is hard to interpret.[3] The geometric diagrams are also of uncertain significance. One figure shows a small square, bisected, with an inscribed circle and a second square erected on a diagonal of the first. The ratio of the area of the circle to the area of the smaller square is the same fraction $\frac{\pi}{4}$. It is also true that the smallest of angles from the center of the circle, 45°, is $\frac{\pi}{4}$ radians. And finally, the ratio of the side of the larger square to the side of the smaller is $\sqrt{2}$, an expression appearing elsewhere on the page.

The section in Leibniz's hand includes several recognizable calculations.

- $\frac{1}{2}$ and $\frac{1}{3}$ expressed as fractions of a power of ten (so that the digits of the numerator are those of a decimal fraction, e.g., $\frac{1}{3} = .3333\ldots$).

---

1. For details, see Strickland (2023).

2. For details, see Crippa (2019). Although Leibniz circulated details of this formula among friends as early as October 1674 (see, e.g., A III 1, 154), he did not publish it until 1682; see Leibniz (1682); reprinted in Leibniz (2011a, 7); English translation Leibniz (1742).

3. In the manuscript, Mariotte uses the sign ∞ for =.

- To the right, a calculation of the digits of $\frac{1}{7} = .142857\ldots$. The quotient is to the right of the \ sign, the dividend is $1000\ldots$, and the digits above the dividend are successive remainders as intermediate dividends are divided by 7.
- A table showing the values of the decimal numbers 0–8 in binary.
- What seems to be the decimal sum $8 + 2 + 1 = 11$, without any evident appearance of the corresponding binary numeral 1011.
- A multiplication of the multiplier $5 = 101_2$ by the multiplicand $7 = 111_2$ to yield the product $100011_2 = 35$. The method is the standard grade-school multiplication algorithm adapted to binary, but Leibniz shortens the work by writing intermediate products only for the nonzero bits of the multiplier, one row containing 111 and the other row containing 111 padded on the right with a 0 and shifted one place to the left (so the 111 is aligned two positions to the left, corresponding to the existence of a 1 in the third-from-the-right position of the multiplier).
- At the bottom left the binary sum $10 + 10 + 10 = 110$, that is, $2 + 2 + 2 = 6$.
- At the bottom right a calculation of $\frac{7}{6} = 1.1666\ldots$, which may be related to the comment several lines above about 7 and 6.
- A few other formulas of unclear significance. It is tempting to read the "10" in the statement "xx equals 10" as binary and relate it to the $\sqrt{2}$ appearing elsewhere on the page, but there is simply not enough coherence to these various expressions to draw a connection.

The watermark is documented to October 1674, but the material on binary, although unquestionably in Leibniz's hand, appears to have been written in different ink from the rest of his material and so may have been added later.[4]

---

4. For this reason, the editors of Leibniz (1923–) state that the dating of Leibniz's deliberations on binary is uncertain. See A VII I, 882.

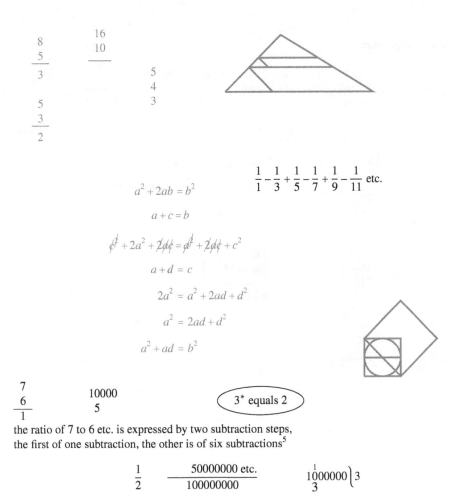

$$\frac{1}{1} - \frac{1}{3} + \frac{1}{5} - \frac{1}{7} + \frac{1}{9} - \frac{1}{11} \text{ etc.}$$

$$a^2 + 2ab = b^2$$

$$a + c = b$$

$$a^{2} + 2a^2 + 2ac = a^{2} + 2ac + c^2$$

$$a + d = c$$

$$2a^2 = a^2 + 2ad + d^2$$

$$a^2 = 2ad + d^2$$

$$a^2 + ad = b^2$$

3* equals 2

the ratio of 7 to 6 etc. is expressed by two subtraction steps,
the first of one subtraction, the other is of six subtractions[5]

$$\frac{1}{2} \qquad \frac{50000000 \text{ etc.}}{100000000} \qquad 1\overset{1}{0}00000 \big| 3$$

$$\frac{1}{3} \quad \text{equals} \quad \frac{3333333}{10000000} \qquad 10000000000 \big| 14282$$

$$\frac{7}{6} \qquad \sqrt{2}$$

11

2000000000   8+2+1

The binary progression

0	1	2	3	4	5	6	7	8	0	0	0	0	0
0	1	10	11	100	101	110	111	1000	.	.	.		

$$
\begin{array}{r}
111 \\
101 \\
\hline
111 \\
1110. \\
\hline
100011
\end{array}
$$

$$\frac{200}{100}$$

$$\frac{20000}{10000}$$

100 $\qquad$ $x^2 + 2x$ equals 10 $\qquad$ xx equals 10

$xx + 2x$ equals 10

$a^2 + a + 1$

$$
\begin{array}{r}
10 \\
10 \\
10 \\
\hline
110
\end{array}
$$

$$
\begin{array}{r}
100 \\
10
\end{array}
$$

$$
\begin{array}{r}
\cancel{1444} \\
\cancel{7000}0000 \\
\hline
\cancel{6666}
\end{array}
\left.\rule{0pt}{2.5em}\right\}
\quad
\begin{array}{c}
1.166 \text{ etc.} \\
\hline
10000
\end{array}
$$

2 The Series of All Numbers, and on Binary Progression (before 15/25 March 1679)

Manuscript:
M: LH 35, 3 B 2 Bl. 6. Draft. Latin.

After his initial foray into the binary system during his Paris years (see chapter 1), the manuscript evidence suggests that Leibniz developed the idea across a series of short draft pieces written as and when inspiration struck. For example, having completed a short text entitled "La place d'autruy" [The Place of Others] about the importance of putting ourselves in the place of others, he added in the margin the numbers from 0 to 15 in binary, while on top of the text itself he scrawled the calculations $11 + 1 = 100$ and $101 + 1 = 110 + 1 = 111$ (see LH 34 Bl. 29r; reproduced in the introduction, this volume).[1] Another text, on half a page torn from a larger sheet, finds Leibniz using division to work out the binary representation of the fractions $\frac{1}{13}$ and $\frac{1}{17}$ (see LH 35, 13, 2 B Bl. 155). Lack of watermarks makes dating such pieces difficult, but their rudimentary content suggests that they are early explorations. The same is true of the texts in this and the next two chapters, in which Leibniz experiments with binary notation and arithmetic, and notes patterns in various binary sequences. Since such ideas appear in more developed form in Leibniz's first dated manuscript on binary, "On the Binary Progression" from 15/25 March 1679 (see chapter 6), these writings were likely made earlier than that,[2] though there is nothing to indicate whether this was days, weeks, or even months earlier, nor is there sufficient evidence for arranging them in a particular order. Accordingly, we date them all as simply before 15/25 March 1679.

The following piece develops the basics—that binary notation can express all rational numbers, and that standard arithmetic operations can be adapted to it; decimal-to-binary conversion is sketched. The exposition also incorporates flashes of much more sophisticated suggestions, though these are neither worked out here nor connected in obvious ways to binary notation.

The Series of All Numbers, and on Binary Progression

Because all numbers proceed in order from these: 1, 2, 4, 8, 16, it will be useful to investigate how the properties of numbers can be deduced therefrom.[3] The demonstration of one [property] is to be sought, namely that all numbers can be produced by the mere addition of these, but this is

---

1. "La place d'autruy" is published in A IV 3, 903–904, though the material on binary is omitted. An English translation can be found in Dascal (2006, 164–165), again without the material on binary.

2. The claim that Leibniz's first writing on binary was "On the Binary Progression" has been commonly made in recent decades, sometimes implicitly, usually explicitly; see, for example, Mungello (1977, 51–52), Pappas (1991, 125), Ryan (1991, 31), Antognazza (2009, 247 and 275n255), and Bauer (2010, 14). In each case, the claim is apparently made without awareness of the manuscripts described above or those translated in chapters 2–4.

3. Written vertically down the left-hand margin of this paragraph is this: "It can also be written thus: $y * 4y^2$, likewise as 10101 etc."

demonstrated in this way, because if any number is expressed by a binary progression,[4] there arises from that no other way of expressing than this: 1001010001 etc., for a two cannot be placed in such an expression just as a ten is not placed in the decimal. Wonderful things arise from this expression. For it alone is determined by nature, all the others are arbitrary. Addition, multiplication, and division are very easy in this system. [It is] remarkable for periodic progressions in expressible quantities which are not whole or rational.[5] In a nutshell, every secret of arithmetic and geometry lies hidden in this system. Let us see how numbers are expressed in it.

1	2	3	4	5	6	7	8	9
1	10	11	100	101	110	111	1000	1001

10	11	12	13	14	15	16	17	18
1010	1011	1100	1101	1110	1111	10000	10001	10010

19	20	21	22	23	24	25
10011	10100	10101	10110	10111	11000	11001

From this, both the mode of varying and the rule of progression are evident, for—to give an example—to discover all [numbers] between 10000 and 100000, i.e. between 16 and 32, first add 1 to 10000. From this arises 10 and all its variations (10, 11), and from this 100 and all its variations (and all variations of 100 itself arise from variations of 1 and 10 added together), from this 1000 and all of its variations (which are made from the variations of 1, 10, and 100 added together). NB, from this continuous replication is also evident the true demonstration of that wonderful proposition, why the sums of replicated numbers known as figurate numbers are numbers of a geometric progression diminished by one.

The figurate numbers are the triangular numbers 1, 3, 6, 10, ..., the tetrahedral (or pyramidal) numbers 1, 4, 10, 20, 35, ..., and so on. The nth triangular number is the number of dots in a triangular array with one dot in the first row, two in the second, and so on, and the nth tetrahedral number is the number of dots in a tetrahedron with such triangles of increasing size in successive layers. Leibniz refers to these as "replicated" numbers because the nth element of each series is the sum of the first n elements of the previous. The nth element of the kth series is the binomial coefficient $\binom{n}{k}$ (where the triangular numbers are the series for $k = 2$ and the tetrahedral numbers are those for $k = 3$). The figurate numbers for a fixed dimension k are familiar as the diagonals of Pascal's triangle, the kth element of the nth row being the number of different paths from the apex down to that element (there being $\binom{n}{k}$ possible ways of choosing k southwest steps, the rest being southeast moves).

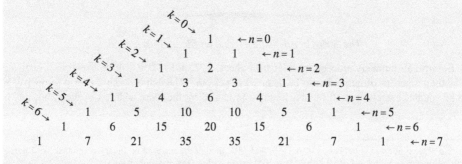

---

4. Leibniz in fact wrote "decadicam" [decimal] here, but this is clearly a mistake.

5. Periodicities in binary numerals, and in tables of numerals, come up for analysis repeatedly in this collection. Only here does Leibniz oddly suggest periodicity in irrationals, and given the extravagance of the next sentence, the statement need not be closely parsed.

Leibniz had studied properties of these numbers in the context of his earliest original mathematical work, having been introduced to them by Huygens in 1672 (see Rabouin 2015, 57). The sum of each row of Pascal's triangle—that is, holding n constant and summing over different values of k—is $\sum_{k=0}^n \binom{n}{k} = 2^n$ (set $x = y = 1$ in the binomial formula $(x+y)^n = \sum_{k=0}^n \binom{n}{k} x^k y^{n-k}$). Or, regarding the entries as figurate numbers and omitting the degenerate $k = 0$ case, the sum of all figurate numbers for which the side of the figure is of length n is 1 less than the elements of the geometric series $2^1, 2^2, 2^3, \ldots, 2^n, \ldots$. Leibniz later returns to the properties of figurate numbers expressed in binary (for example, in chapters 23, 27, and 30).

You may change a given number of the decimal progression into binary. One needs to change only the numbers 10, 100, 1000 into binary and the numbers 1, 2, 3, 4, 5, 6, 7, 8 and 9. From which the powers of ten itself may be multiplied by single numbers expressed in binary. But if a larger table is given, one need only divide a given number into two parts, like 13265 into 13, 1000, and 265.[6] Nothing but wonders are observed in this progression. I think that multiplying can be done immediately by addition, division without much effort, and the logarithms of given numbers can be found easily. It is to be demonstrated whether tetragonism is possible, etc.[7]

---

6. Leibniz seems to suggest that converting 13265_{10} to binary will be simplified by expressing $13265 = 13 \times 1000 + 265$ and working with a known binary representation of 1000_{10}.

7. Tetragonism is a term used for the quadrature (or squaring) of the circle, with Leibniz often referring to his solution of the problem as "arithmetical tetragonism" (see, e.g., A II 1, 632; A VI 6, 85). Leibniz later attempted to express his formula for squaring the circle in binary, without success. See chapter 7.

First manuscript page of "Binary Progression" (LH 35, 3 B 2 Bl. 7r).

3 Binary Progression (before 15/25 March 1679)

Manuscript:
M: LH 35, 3 B 2 Bl. 7. Draft. Latin.

In this manuscript, Leibniz calculates the binary representation of certain simple fractions, such as $\frac{1}{3}$, by adapting a long division algorithm to binary. This binary algorithm is simpler than the decimal process because at each stage, the divisor (appropriately shifted) divides the current dividend either just once or not at all, yielding either a 1 or a 0 in the quotient. The general workflow is to put the dividend to the left of the $\backslash$ sign and the quotient to the right, and to write the remainder above the dividend, crossing out digits of the dividend from left to right so that the new dividend is the concatenation of the current remainder with the remaining digits of the original dividend; meanwhile, digits of the old remainder are themselves crossed out to be replaced by digits of the new remainder, and so on. Below the original dividend, the digits of the divisor are written, crossed out, and shifted over. So in both the divisor and the dividend, the eye sometimes has to cobble together a multidigit numeral by zig-zagging up and down among rows of digits while moving left to right.

Some of the fractions are computed in more than one way, for example, $\frac{2}{3}$ by division, by adding the binary for $\frac{1}{3}$ to itself, and by adding the binaries for $\frac{1}{2}$ and $\frac{1}{6}$. Finally the binary for $\frac{1}{3}$ is multiplied by 3 to get the infinite binary $0.1111\ldots$, which is equal to $\sum_{i=1}^{\infty} 2^{-i} = 1$, much as in decimal, $0.999\ldots = 1$.

The manuscript is unusually abstruse because the work is experimental, sometimes incorrect, and composed almost entirely of 0s and 1s, making it hard to be confident of the sequence in which digits were written and crossed out. Moreover, almost the entire first page is crossed out, sometimes with more than one kind of mark. We also find Leibniz using his own symbol for equality, namely, Π, which he developed in June 1674 and used at least until early 1680, although throughout this time, he also used other symbols for equality, including =. In line with modern usage, we have rendered Leibniz's Π as =.

For the dating of this manuscript, see chapter 2. Chapter 6 includes cleaner examples of binary long division, for which this manuscript seems to have been practice. The examples of sedecimal division in chapter 8 show Leibniz's intermediate calculations step by step.

Binary Progression

(1) 1   (2) 10   (3) 111   (4) 100   (5) 101
(6) 110   (7) 111   (8) 1000   (9) 1001   (10) 1010   (11) 1011   (12) 1100
(13) 1101   (14) 1110   (15) 1111   (16) 10000

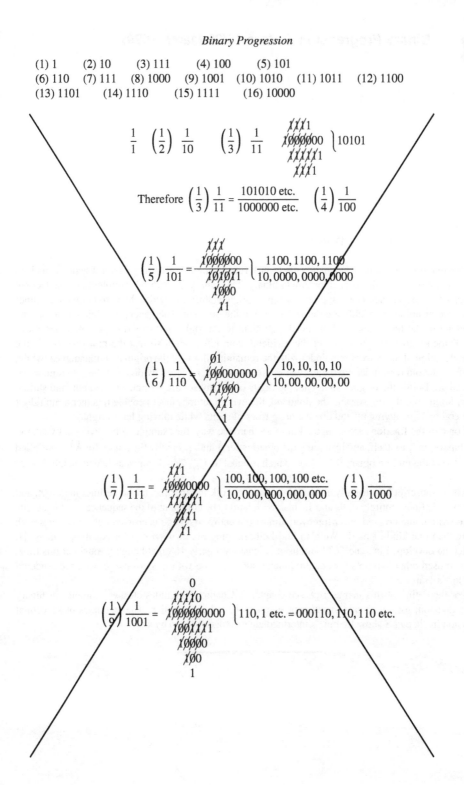

$\dfrac{1}{1}$ $\left(\dfrac{1}{2}\right)\dfrac{1}{10}$ $\left(\dfrac{1}{3}\right)\dfrac{1}{11}$

Therefore $\left(\dfrac{1}{3}\right)\dfrac{1}{11} = \dfrac{101010 \text{ etc.}}{1000000 \text{ etc.}}$ $\left(\dfrac{1}{4}\right)\dfrac{1}{100}$

$\left(\dfrac{1}{5}\right)\dfrac{1}{101}$

$\left(\dfrac{1}{6}\right)\dfrac{1}{110}$

$\left(\dfrac{1}{7}\right)\dfrac{1}{111}$ $\left(\dfrac{1}{8}\right)\dfrac{1}{1000}$

$\left(\dfrac{1}{9}\right)\dfrac{1}{1001}$

In the sum beginning $\frac{1}{2} = \ldots 0100\,0000$ below, the value of $\frac{1}{6}$ is incorrect, and therefore the result is incorrect. Leibniz implicitly places a binary point after the initial 0 of both addends, and since $\frac{1}{6} = 0.0010101\ldots$, the sum should be $0100\,0000 + 0001\,0101 = 0101\,0101$.

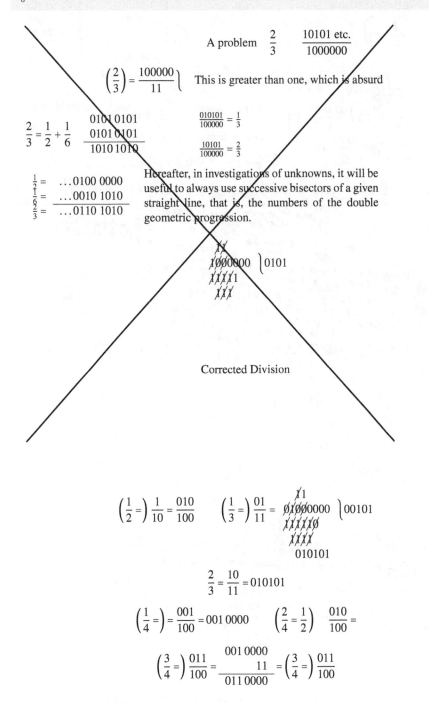

A problem $\quad \dfrac{2}{3} \qquad \dfrac{10101 \text{ etc.}}{1000000}$

$\left(\dfrac{2}{3}\right) = \dfrac{100000}{11} \;\Big|$ This is greater than one, which is absurd

$\dfrac{2}{3} = \dfrac{1}{2} + \dfrac{1}{6}$ $\quad$
$\begin{array}{r} 0101\,0101 \\ 0101\,0101 \\ \hline 1010\,1010 \end{array}$
$\qquad \dfrac{010101}{100000} = \dfrac{1}{3}$

$\qquad \dfrac{10101}{100000} = \dfrac{2}{3}$

$\begin{array}{l} \frac{1}{2} = \;\ldots 0100\,0000 \\ \frac{1}{6} = \;\ldots 0010\,1010 \\ \hline \frac{2}{3} = \;\ldots 0110\,1010 \end{array}$

Hereafter, in investigations of unknowns, it will be useful to always use successive bisectors of a given straight line, that is, the numbers of the double geometric progression.

1000000 $\big| 0101$
11111
111

Corrected Division

$\left(\dfrac{1}{2} =\right) \dfrac{1}{10} = \dfrac{010}{100} \qquad \left(\dfrac{1}{3} =\right) \dfrac{01}{11} = \begin{array}{c} 1 \\ 01000000 \;\big| 00101 \\ 111110 \\ 111 \\ \hline 010101 \end{array}$

$\dfrac{2}{3} = \dfrac{10}{11} = 010101$

$\left(\dfrac{1}{4} =\right) = \dfrac{001}{100} = 001\,0000 \qquad \left(\dfrac{2}{4} = \dfrac{1}{2}\right) \quad \dfrac{010}{100} =$

$\left(\dfrac{3}{4} =\right) \dfrac{011}{100} = \dfrac{\begin{array}{c}001\,0000\\11\end{array}}{011\,0000} = \left(\dfrac{3}{4} =\right) \dfrac{011}{100}$

Reverse manuscript page of "Binary Progression" (LH 35, 3 B 2 Bl. 7v).

$$\left(\frac{1}{5}\right)\frac{001}{101} = \begin{array}{l}\cancel{1111}\,\cancel{11} \\ 00\cancel{100}\,\cancel{0000} \\ 1\cancel{0}\,\cancel{1111}\,\cancel{1111} \\ 1\cancel{000}\,\cancel{0000} \\ \cancel{111}\,\cancel{1111} \end{array} \left| \begin{array}{l} 0\,0011\,0011 \\ 0\,0011\,0011 \\ \hline 001100110 \end{array} \right. \quad \begin{array}{l}\left(=\tfrac{1}{5}\right)=\frac{001}{101} \\[4pt] \left(=\tfrac{2}{5}\right)=\frac{010}{101}\end{array}$$

$$\begin{array}{l} 0\,0011\,0011\,0011\,0011 \\ \hspace{4.2cm} 11 \\ \hline 0\,0011\,0011\,0011\,0011 \\ 00\,0110\,0110\,0110\,011 \\ \hline 00\,1001\,1001\,1001\,1001 \end{array} \qquad \begin{array}{l} 0\,0011\,0011 \\ \hline 0\,1001\,1001 \\[6pt] =\tfrac{2}{5}=\frac{010}{101} \end{array}$$

To multiply by 11 is [to multiply] first by 1 and then by 10. To multiply by 10 is to chop off a 0 at the left.[1]

$\frac{1}{8}+\frac{1}{32}+\frac{1}{128}+\frac{1}{512}$ etc. $=\frac{1}{3}$ All fractions are composed of infinite terms of the fractions of any geometric progression.

$$\begin{array}{ll} 0\,0011\,0011\,0011\,0011 & \left(=\tfrac{1}{5}\right)=\frac{001}{101} \\ 0\,0011\,0011\,0011\,0011 \\ \hline 0011001100110110 & \left(=\tfrac{2}{5}\right)=\frac{010}{101} \\ 0\,0011\,0011\,0011\,0011 \\ \hline 0\,1001\,1001\,1001\,1001 & \left(=\tfrac{3}{5}\right)=\frac{011}{101} \\ 0\,0011\,0011\,0011\,0011 \\ \hline 0\,1100\,1100\,1100\,1100 & \left(=\tfrac{4}{5}\right)=\frac{100}{101} \end{array} \qquad \begin{array}{r} 256 \\ 2 \\ \hline 512 \end{array}$$

$$\left(\frac{1}{6}=\right)\frac{001}{110} = \begin{array}{l}\cancel{111}\,\cancel{1} \\ 00\cancel{1}\cancel{00000} \\ \cancel{11}\cancel{000000} \\ \cancel{111111} \\ \cancel{1111} \end{array} \left| \begin{array}{l} 0\,0010\,1010\,1 \\ 0\,1010\,1010\,1 \\ \hline 01\,1010\,1010 \end{array} \right. \quad \begin{array}{l}\left(=\tfrac{1}{6}\right) \\[8pt] \left(=\tfrac{5}{6}\right)\end{array}$$

$$\frac{1}{2}=\frac{01}{10}=01]0000$$

$$\frac{1}{3}=\frac{01}{11}=0]01,01,01$$

$$\frac{1}{4}=\frac{001}{100}=001]0,0,0,00$$

$$\frac{1}{5}=\frac{001}{101}=0]0011,0011$$

$$\frac{1}{6}=\frac{001}{110}=00]01,01,01,$$

$$\frac{1}{7}=\frac{001}{111}=0]001,001$$

$$\frac{1}{8}=\frac{0001}{1000}=0001]00000$$

$$\frac{1}{9}=\frac{0001}{1001}=0]000111000111$$

$$\frac{1}{11}=\frac{0001}{1011}=0]000101101,000101101$$

---

1. That is, to multiply by 10_2 a binary fraction with a 0 immediately after the binary point, remove that first 0.

$$
\begin{array}{r}
0010101010 \\
11 \\
\hline
0010101010 \\
0101010101 \\
\hline
0111111111 \quad \text{etc.}
\end{array}
$$

Proof

And 0111111111 etc. to infinity is the same as $1 = \frac{1}{2} + \frac{1}{4} + \frac{1}{8} + \frac{1}{16}$ etc.[2]

---

2. Leibniz perhaps first captured this observation in a text scrap (a small strip torn from a larger page), which reads in full: "In binary, 0.111111111 and so on to infinity is the same as 1.0000000000 and so on to infinity, that is, $\frac{1}{2} + \frac{1}{4} + \frac{1}{8} + \frac{1}{16}$ and so on to infinity is the same as 1." (LH 35, 3 B 11 Bl. 24)

In a much later manuscript, without a watermark but filed with other mathematical writings from 1703, Leibniz exploits the equality $1 = 0.111\ldots$ to develop an algebraic method for determining the binary expansions of reciprocals. To see how it works, suppose we are looking for the binary expansion of $\frac{1}{7}$. Since the length of the period is unknown, Leibniz first denotes the expansion by a long string of letters: $xwvtsrqpnml$, then multiplies this by 7, that is, 111 in binary, and sets the result equal to 1, that is, 0.1111111 in binary. To multiply the unknown string by 111 is to multiply it by $1 + 10 + 100$. This is done by adding together three copies of the string: one without a shift; one shifted left once; and the third shifted left twice:

	...	x	w	v	t	s	r	q	p	n	m	ℓ
+		x	w	v	t	s	r	q	p	n	m	ℓ	
+	x	w	v	t	s	r	q	p	n	m	ℓ		
=	...	1	1	1	1	1	1	1	1	1	1	1

He then infers the values of the variables, right to left: in the rightmost position, since the sum is 1, $\ell = 1$; in the second from right, $m + \ell = 1$, so $m = 0$; in the third from right, $n + m + \ell = 1$, so $n = 0$; in the fourth from right, $p + n + m = 1$, so $p = 1$. And so on: $q = 0$, $r = 0$, $s = 1$, $t = 0$, $v = 0$ etc. This is sufficient to show that the binary expansion of $\frac{1}{7}$ is $0.001001001\ldots$, and hence the repeating block is $\overline{001}$.

The same method is used to find the binary representation of $\frac{1}{9}$ (that is, $\frac{1}{1001}$ in binary), except that rows corresponding to multiplication by 0 are omitted:

	...	z	y	x	w	v	t	s	r	q	p	n	m	ℓ
+	x	w	v	t	s	r	q	p	n	m	ℓ	0	0	
	...	1	1	1	1	1	1	1	1	1	1	1	1	1

Now $\ell = 1$ once again, but also $m = 1$ and $n = 1$. As $p + \ell = 1$, p must be 0, and similarly q and r must be 0. As $t + q = 1$, t must be 1, and the pattern repeats; the repeating block is $\overline{000111}$, as Leibniz verifies by supplying the intermediate rows:

	...	z	y	x	w	v	t	s	r	q	p	n	m	ℓ
+	x	w	v	t	s	r	q	p	n	m	ℓ	0	0	
	...	1	0	0	0	1	1	1	0	0	0	1	1	1
	...	0	1	1	1	0	0	0	1	1	1	0	0	
	...	1	1	1	1	1	1	1	1	1	1	1	1	1

Longer periods require longer strings of letters, and when working out the period of $\frac{1}{11}$, Leibniz expands the letter string to $\gamma\beta\alpha zyxwvtsrqpnml$, though a calculation error gives him the wrong result. See LH 35, 3 B 5 Bl. 11–12.

4 Geometric Progressions and Positional Notation
(before 15/25 March 1679)

Manuscript:
M: LH 35, 3 B 5 Bl. 46. Draft. Latin.

Leibniz here experiments with the representation of positive integers as the sum of selected powers of an integer $b \geq 2$ (the $b = 2$ case being the basis for binary notation). He also considers certain divisibility relationships related to such representations, and touches on perfect numbers. For the dating of this text, see chapter 2.

———————⊷◦○◦⊷◦○———————

Numbers of the double geometric progression: 1, 2, 4, 8, 16, 32, 64 etc. Because $1 + 2 + 4$ makes 7, hence any three consecutive terms of the double geometric progression are divisible by 7. For $2 + 4 + 8$ makes 14, and $4 + 8 + 16$ makes 28, and $8 + 16 + 32$ makes 56, and so on. This is demonstrated at first glance from my manner of expression. For according to the binary expression, 111, that is, $4 + 2 + 1$, is 7, and 1110, that is, $8 + 4 + 2$ is twice 7,[1] and 11100 is four times 7, and so on.

Hence, generally, every sum of numbers of any consecutive geometric progression can be divided by the sum of the same number of initial numbers, that is, of the numbers of the same progression taken from 1. And hence, two sums of the same number of numbers of the same consecutive geometric progression have a common divisor.

Leibniz first observes that the sum of any three consecutive members of the double geometric progression is divisible by 7, because for any $k = 0, 1, 2 \ldots ,$

$$\sum_{i=k}^{k+2} 2^i = 2^k \sum_{i=0}^{2} 2^i = 7 \cdot 2^k.$$

He then generalizes that proposition to an arbitrary base, not just 2, and a consecutive sequence of any length, not just 3. That is,

$$\sum_{i=k}^{k+\ell} b^i = \left(\sum_{i=0}^{\ell} b^i\right) \cdot b^k,$$

for any $b \geq 2$, any $k \geq 0$, and any $\ell \geq 0$. And therefore, $\sum_{i=j}^{j+\ell} b^i$ and $\sum_{i=k}^{k+\ell} b^i$ have the common factor $\sum_{i=0}^{\ell} b^i$, for any $j, k, \ell \geq 0$.

---

1. Leibniz erroneously wrote "$8 + 4 + 1$ is twice 7". (The placement of the period outside the quotation marks here and in a few other contexts is explained in the introduction, on p. 24.)

Indeed, the sum of the numbers of a consecutive geometric progression can be divided by the sum of the numbers of the same progression starting from 1, and any sum of component dividend numbers from a geometric progression is exactly divisible by the sum of its component divisors.

A number composed by adding together numbers of a consecutive geometric progression has no divisors other than a power of the base of the progression (which can be known from itself) and the number itself composed from [adding] the same number of initial [numbers] of the same progression, and its divisors.

Indeed, further still: any number composed by adding the numbers of some geometric progression can be divided by a number [composed] of the initial numbers, that is, [of the numbers] beginning from 1 of the same geometric progression, in the same way as can numbers that contain each other or are proportional to each other, and in addition [it can be divided] by any power of the progression.

Every number can be composed by adding numbers of any lower geometric progression whatsoever.

What we have said about numbers composed by mere addition should be said about numbers composed by a mixture of addition and subtraction.

Leibniz first repeats the immediately preceding proposition. then generalizes it by not requiring that the selected elements of the geometric progression be in a consecutive sequence. The sum of any such subset is divisible by the sum of the corresponding members of the geometric progression, starting from its first element. That is, if $S = \{i_1, \ldots, i_s\}$ is any set of indices, $0 \leq i_1 < i_2 < \cdots < i_s$, then $\sum_{i \in S} b^i$ is divisible by $\sum_{i \in S} b^{i - i_1}$, the other divisor being b^{i_1}. The first part of the proposition is the special case of the second part in which $i_1, \ldots, i_s$ are consecutive.

The second paragraph states that the divisors of $\sum_{i=k}^{k+\ell} b^i$ are the divisors of b^k and of $\sum_{i=0}^{\ell} b^i$.

The rest develops the general principle on which positional notation works (binary or decimal, for example). For any $b \geq 2$ and any integer n, it is possible to write n as $n = \sum_{i \in S} b^i$, for some set of indices $S = \{i_1, \ldots, i_s\}$. The base b will be smaller than n, unless the base-b representation of n is a single digit. For $b > 2$, we would ordinarily think of the powers b^i as having coefficients $< b$, but if the i_j need not be distinct for different j, then Leibniz's statement is true as written.

What follows appears on the back of the manuscript page. It has all been crossed out, except for the small display at the top, which may have been meant as a sketch of the product $11 \times 10 = 110$.

A Euclidean perfect number is an even number that is equal to the sum of its divisors (including 1 but excluding the number itself), for example, $6 = 1 + 2 + 3$ and $28 = 1 + 2 + 4 + 7 + 14$. Euclid proved (Proposition IX.36) that $2^{p-1}(2^p - 1)$ is perfect whenever $2^p - 1$ is prime. (6 and 28 are the $p = 2$ and $p = 3$ cases. Almost seventy years after Leibniz penned this manuscript, Leonhard Euler proved that the *only* even perfect numbers are the ones generated by Euclid's method. It is still unknown whether any odd perfect numbers exist.) Here Leibniz seems to be looking for insight into these perfect numbers by expressing them and their factors in binary.

Leibniz's text breaks off after the expression "$7 + 4 + 2-$". Probably the final statement was meant to be "28 equals $14 + 7 + 4 + 2 + 1$," showing 28 to be perfect.

$$
\begin{array}{r}
11 \\
1 \\
\hline
11 \; 0
\end{array}
$$

Construction of perfect Euclidean number

11 times 10 gives $100 + 10$, likewise 111 times 110 gives $111 + 110 + 10$

11 times 10 gives $100 + 10$, that is, gives $11 + 10 + 1$, which are all divisors of 110 itself, if 11 is prime

111 times 100 gives $10000 + 100 + 100$, that is, gives $1110 + 1000 + 101 + 1$, which are all divisors of 1110 itself, if 111 is prime. Of course, 28 equals $7 + 4 + 2-$

Machina Arithmeticae Dyadicae

Numeros dyadicos per machinam addere, subtrahere, multiplicare et dividere facillimum est. Nam ad quemvis binarium unitatem transferre in sequens facile est.

Et autem commode et rotas fieri posse quatuor dentium, et cylindro inscribi 0, 1, 01, ita circuitus unus binas transportationes habebit.

... facilis est ... in 0 et 1 ... in ipso multiplicante vel dividente, idy sine rotis mea ... quod in ... non procedit, sed tam multiplicans qua multiplicatori initio designentur in machina, et unde propellatur tantum regula, ita multiplicator surrexit loco multiplicandi designato, per 0 totis deprimetur ... alio ubi est 1, rursus ... et remisuo ... et durante intervallo fient transportationes.

... pro divisione vel subtractione, nisi quod res initio non fieri designari ... sed inspiciend... durante propulsione

sed maxima

First manuscript page of "Binary Arithmetic Machine" (LH 42, 5 Bl. 61r).

5 Binary Arithmetic Machine (before 15/25 March 1679)

Manuscript:
M: LH 42, 5 Bl. 61. Draft. Latin.

Transcription:
LVM: Mackensen (1974, 256, 259).

By 1672, Leibniz had begun serious work on the design and construction of a decimal calculating machine capable of carrying out the four basic arithmetic operations (Jones 2018). The design was far from complete when, at the end of the decade, he jotted down this sketchy description of a binary machine founded upon the same basic ideas. (The manuscript is undated and bears no watermark; our reasoning for this date is given below.) Moreover, the mechanical-engineering challenges of building the decimal machine had not been overcome, and some never would be, despite huge investments of time and money. Summing multidigit numbers entailed propagating carries through long chains of mechanical linkages, such as cams or gears. Longer chains required more force, with its greater attendant chance of breakage; longer chains also added imprecision to the positioning of parts at the end of the chain, such that the device might not clearly indicate the correct result—a problem that plagued Charles Babbage also two centuries later.

These problems would be severely exacerbated in a binary device, whose chains of linkages would be about three times as long as those in a decimal device. And as Leibniz himself observed, "the greatest difficulty is to convert a binary number into an ordinary one or vice versa." He seems not to have refined the design beyond that presented here. The balls-in-channels machine he describes in "On the Binary Progression" (chapter 6) is based on an entirely different design; in a deleted passage from that text, he even describes it as "without wheels," as though to distinguish it from this design. For these reasons, we place this manuscript earlier than the one describing the machine based on cascading balls; after all, having conceived the idea of a binary calculating machine, it is likely Leibniz first thought of adapting the design for his decimal machine, as he does here, before considering an entirely different design, as he does in "On the Binary Progression" of 15/25 March 1679.

The figure probably represents a wheel with two teeth, but Leibniz's description of the mechanism is not easy to interpret. The design was even less complete than that of his decimal machine, which did not properly handle the basic problem of propagating carries.

Binary Arithmetic Machine

It is very easy to add, subtract, multiply, and divide binary numbers by machine, because for any binary number it is easy to transfer the units to the next position.

And I think wheels can be conveniently made with four teeth, and 0, 1, 0, 1 inscribed on the cylinder, so one revolution will cause two carries.

Even the variation in 0 and 1 is easy to do in the multiplicand or dividend itself, and that without wheels.

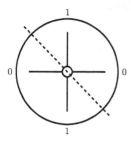

What does not happen in my ordinary machine could be accomplished [here], [namely] that both the multiplicand and the multiplier are designated initially in the machine and then only the rod[1] is driven forward; consequently, the multiplier, as it approaches a position of the multiplicand denoted by 0, will be pushed down completely or rather moved aside so that it does nothing, but when approaching another position where there is 1, it will rise up again and do its work, and during the interval the carries will occur.

As for division or subtraction, except that the matter cannot be designated from the beginning, but must be considered during propulsion; but the greatest difficulty is to convert a binary number into an ordinary one or vice versa.

From an ordinary number, the numbers of a double geometric progression should be subtracted in order, starting with the largest. These are designated by the number of teeth on cylinders or wheels, so there will be 12 subtracting cylinders for 12 numbers of a double progression.[2]

Rods protrude at the very bottom of the front of the wheels, where teeth are in the same plane like rakes. But the rod of that number which cannot be subtracted prescribes by being pushed to the right, and will be brought back into its own position where they will pass over.

The rods seem to take up a little more space than the cylinders, but otherwise the operation [proceeds] easily because there is no need for rotation.

The number will appear on the side as soon as the subtraction has been done or not, that is, the number expressed in binary, 0 and 1, will be obtained.

---

1. We have translated the Latin *regula* as "rod" throughout this chapter, following Mackensen (1974), though he notes that in certain contexts here, it may refer to a cam or a tooth of a wheel.

2. progression. ▷ Technically, it can be done that the cylinder passes by the one whose number is larger than the one to be subtracted. ◁ *deleted*.

6 On the Binary Progression (15/25 March 1679)

Manuscript:
M: LH 35, 3 B 2 Bl. 1–4. Draft. Latin.

Leibniz's early experiments with the binary system came to fruition in the following text, his first dated text on the subject (the fact that he dated it indicates that he considered it to be important). The manuscript consists of two parts, both dated the same day. Part I includes Leibniz's first systematic articulation of the algorithms for conversion between decimal and binary notation and for addition, subtraction, multiplication, and division in binary numerals. He also starts to explore patterns in the results of such manipulations. Leibniz shows, for example, that the penultimate bit of a perfect square is always 0—akin to the fact that in decimal, perfect squares never end in 2, 3, 7, or 8. He then squares a 6-bit number *abcdef*, analyzing the bits of the square in terms of the bits of the base (Leibniz uses the term *latus*, literally "side"—the side of a geometric square—but we have instead used "base," which is closer to normal English usage in this context). The last bit of the square is $f^2 = f$, the penultimate bit is $ef + fe = 2ef = 0$ mod 2, and so on. As all variables are binary, Leibniz's various equations and shortcuts are essentially exercises in boolean logic, though it would be another century and a half before George Boole would formulate laws such as $x^2 = x$ (Boole 1854, §2.9).

In part II, Leibniz tries to extract square roots by inverting the squaring algebra, breaking an 8-bit radicand into blocks of two bits (just as a standard decimal square root algorithm breaks the radicand into blocks of two decimal digits). The first bit of the root is inevitably 1; Leibniz derives algebraic expressions for the next two bits but gets bogged down in the algebra and switches to carrying out the algorithm to calculate the specific number $\sqrt{2}$, representing 2 as 0 10 00 00. . . . He also determines the repetitive patterns in the binary representations of $\frac{1}{k}$ for $k = 2, \ldots, 9$.

Leibniz's language is at times challenging and his intent hard to follow. His hope, however, seems to be that the periodicity of fractions when expressed in decimal or binary notation—that is, as the sum of terms of an infinite geometric progression—can be generalized to other kinds of "periodic expressions," so that arbitrary irrationals will also be expressible by finite means (as some certainly are, for example $\sqrt{2}$ as the solution of the polynomial equation $x^2 = 2$). That thought leads to probes toward known methods of solving polynomials (using the now obsolete notions of exaltation and pure and affected powers) and the question of whether any transcendental number exists.[1]

---

1. In 1673, Leibniz started to use the term "transcendent" for figures, curves, and problems (see, for example, A VII 3, 266–267), later defining transcendental quantities as those that "transcend all algebraic equations" (Leibniz 1686b, 294 and 295). A transcendental number is now defined as one that is not the root of any polynomial equation with rational coefficients. Leibniz's foray here notwithstanding, the existence of transcendentals was not properly demonstrated until the mid-nineteenth century.

15 March 1679

On the Binary Progression—Part I

Numeration

1	2	3	4	5	6	7	8	9	10	11	12	13	14	15	16
1	10	11	100	101	110	111	1000	1001	1010	1011	1100	1101	1110	1111	10000

17	18	19	20	21	22	23	24	25	26	27	28	29	30	31	32
10001	10010	10011	10100	10101	10110	10111	11000	11001	11010	11011	11100	11101	11110	11111	100000

```
32  100000
33  100001
34  ||||10
35  ||||11
36  |||100
37  |||||1
38  ||||10
39  |||||1
40  |1000
41  ||||1
42  |||10
43  |||||1      011111
44  ||100       100000
45  |||||1      100001
46  ||||10      100010
47  |||||1      100011
48  10000       100100
49  ||||1       100101
50  |||10       100110
51  |||||1      100111
52  ||100       101000
53  |||||1      etc.
54  ||||10
55  |||||1      From this it is
56  |1000       evident that
57  ||||1       the 0 and 1
58  |||10       alternate in
59  |||||1      position, first
60  ||100       on every line,
61  |||||1      then on every
62  ||||10      four, then on
63  |||||1      every eight,
64  1000000     and so on, in a
65  ||||1       geometric
66  |||10       progression. In
67  |||||1      the same way,
68  ||100       the
69  |||||1      progression
70  ||||10      according to
71  |||||1      twos, threes
72  |1000       etc. should be
73  ||||1       sought to make
74  |||10       clear which
75  |||||1      [numbers] are
76  ||100       prime and
77  |||||1      which are
78  ||||10      more
79  |||||1      derivative.
80  10000       Likewise the
81  ||||1       progression of
82  |||10       powers.
83  |||||1
84  ||100
85  |||||1
86  ||||10
87  |||||1
88  |1000
89  ||||1
90  |||10
91  |||||1
92  ||100
93  |||||1
94  ||||10
95  |||||1
96  10000
97  ||||1
98  |||10
99  |||||1
100 ||100
```

The progression written [on the left] can easily be continued by going from[2] right to left and writing a 0 when the number above it is a 1 until there is also a 0 in the number above, under which a 1 is written. Nor is there need to go any further, for the other digits stay the same as in the one above. Thus from

$$1010111 = 87$$

we get $1011000 = 88$.

It's the same as saying that

$$1011000 \text{ is } 2^6 + * + 2^4 + 2^3 + * * * \qquad 64$$
$$64 \quad + 16 + 8 \qquad\qquad 16$$
$$\underline{ 8}$$
$$88$$

This is because a 1 in the fourth position, i.e. 1000, signifies the cube of the base of the progression, for just as in the ordinary progression it signifies the cube of ten, i.e. a thousand,[3] so in our progression it signifies the cube of two, namely eight. Likewise, a 1 in the fifth position signifies the fourth power, i.e. 16, and in the sixth position it signifies the fifth power, i.e. 32 and then in the seventh position the sixth power, i.e. 64.

It should be noted that, if you go from top to bottom in the progression, from the digit of the number above to the digits of the numbers directly underneath, 1s always return at regular intervals, and in the final position 1 returns after a single 0. But in the second position, i.e. the square, the 1 returns after three [increments], in the third position, i.e. the cube, after seven, in the fourth position, i.e. the fourth power, after fifteen. From this we now see a convenient way to convert the decimal expression of a given number to a binary expression. Take the number 365:[4] we successively take half, then half of half, and when the remainder is written next to these half values and the digits written in order so that the bottom one is furthest left, and so on, they will yield the binary expression we were looking for.

---

2. In the right margin, Leibniz wrote, "Multiplication is perfect in the binary system because it happens through mere addition without (as in the other number systems) the Pythagorean table being presupposed." (A "Pythagorean table" shows operands and results for an operator, here a simple multiplication table.)

3. Leibniz mistakenly wrote "centum" [a hundred] in place of "mille" [a thousand].

4. Leibniz actually wrote "36" here, but his example is of 365.

15 martij 1679.　　　De Progressione Dyadica ~　　Pars I.

Numeratio

1	2	3	4	5	6	7	8	9	10	11	12	13	14	15	16
1	10	11	100	101	110	111	1000	1001	1010	1011	1100	1101	1110	1111	10000

17	18	19	20	21	22	23	24	25	26	27	28	29	30	31	32
10001	10010	10011	10100	10101	10110	10111	11000	11001	11010	11011	11100	11101	11110	11111	100000

[The remainder of the page consists of Leibniz's densely written manuscript notes in Latin, largely illegible, including the expression 1011000 esse $2^6 + * + 2^4 + 2^3 + * *$ and further calculations.]

First manuscript page of "On the Binary Progression" (LH 35, 3 B 2 Bl. 1v).

365
182 1
 91 0
 45 1
 22 1
 11 0
  5 1
  2 1
  1 0
    1

> Leibniz generates the bits from low-order to high as follows: 365 divided by 2 is 182 with remainder 1 (so the 1's bit is 1); 182 divided by 2 is 91 with remainder 0 (so the 2's bit is 0); 91 divided by 2 is 45 with remainder 1 (so the 4's bit is 1); and so on. The process inevitably arrives at 1, which when divided by 2 yields a quotient of 0 with remainder 1—the final bit generated is always a 1. The result is then read bottom up: 101101101 (which, unhelpfully for the illustrative purpose, happens to be a palindrome).

The same method can be used to convert any progression to another. So 365 will be 1 0110 1101, namely:

$$
\begin{array}{rr}
1\ 0000\ 0000 & 256 \\
100\ 0000 & 64 \\
10\ 0000 & 32 \\
1000 & 8 \\
100 & 4 \\
1 & 1 \\
\hline
& 365 \\
\end{array}
$$

In a similar way, a binary number will be converted into decimal either by adding up all of its powers of two expressed in decimal, as here 256, 64 etc., or by dividing the given binary number by ten expressed in binary,[5] with the remainder written in the order I have already mentioned: Dividing by

1010) 101101101
       100100 (first quotient),   the remainder is 101, which equals 5
           11 (second quotient), the remainder is 110, which equals 6
               11, which equals 3

Adding numbers is so easy with this method that they cannot be dictated faster than they can be added, so that it is not necessary to write down the [addend] numbers, since the sums can be written down immediately. For example, I am first told

the number 10110,
then 11011
and I immediately write 110001.[6]

But if several columns are to be added, such as[7]

$$
\begin{array}{r}
\cdot\ \cdot\ \vdots\ \cdots\ \vdots\ \cdot \\
\cdot\cdot\ \vdots\ \cdots\ \vdots\cdot \\
1011\ 0110 \\
1110\ 0101 \\
100\ 1100 \\
101\ 0101 \\
1101\ 1011 \\
\hline
11\ 0001\ 1001 \\
\end{array}
$$

---

5. Leibniz mistakenly wrote "decadice" ["decimally"] here in place of "dyadice" ["in binary"].

6. Leibniz mistakenly wrote the sum as 1000001.

7. The addition has an error. The second column from the right totals 3, that is, 1 with a carry of 1 into the third column, rather than 0 with a carry of 2 as incorrectly shown. As a result, the sum is off by 2: $182 + 229 + 76 + 85 + 219 = 791$, rather than 793 as shown. (Also, the leftmost column should have one carry rather than two, though the sum is recorded as though it had but a single carry.)

Then count the 1s in a column; if their number is odd,[8] write a 1 below it; and transfer half of the number of 1s to the next column, either in dots or in ordinary numbers. Accordingly, to do addition in this number system you need only to be able to count the 1s, whereas in other ones you need to be able at least to add simple numbers, e.g. 8 + 5 = 13. Anyone who does not know how to do this cannot easily add up in the decimal system.

With the same method, subtraction proceeds very easily, and even the mixing of subtraction and addition, for only the 1s have to be brought together. But when subtraction is mixed with addition, a 1 must be added that corresponds to each 1 to be subtracted, because these cancel each other out.

For example:[9]

	A			A
L	1· +		+	1011 0111
M	1· +		+	100 1101
N	1· +		+	101 0101
P	1· −		−	111 0011
Q	0 −		−	0
R	1 +		+	1
S	1· −		−	1
T	1 +		+	1
V	1 +		+	1
W	1· −		−	1

In column A start from L, where there is a 1 with a plus sign. Therefore this 1 is marked with a dot at the same time as is the other 1 with a minus sign in P. Similarly, there is a +1 in M. It is marked with a dot along with the −1 in S. Lastly, there is a +1 in N. It is marked with a dot together with the −1 in W. All the −1s are marked with a dot and there remains the +1s not marked with a dot, which are counted, and if they are without a companion [then] they together with their signs are transferred to the next column, in this case with +, otherwise with − if there were more −1s.

Leibniz slips up here—intuitively, a carry transfers only *half* of a given column's excess to its neighbor. Suppose c is the net of the current column's +1s and −1s (so c may be positive, negative, or zero). If c is even, then the result digit for the current column is 0, and a carry (or borrow) of $\frac{c}{2}$ is added to the next column; but if c is odd, then the current column's result digit is 1, and a carry (or borrow) of $\frac{c-1}{2}$ is added to the next column. Note that if $c = +1$, there is no carry or borrow; but if $c = -1$, a borrow of $\frac{-1-1}{2} = -1$ is added to the next column. That is, in the positive case no carry is needed unless $c \geq +2$, but in the negative case a borrow is needed even if $c = -1$.

It should also be noted that for the subtrahend number, its complement [up] to 100000 etc. can be placed, and then subtraction is not needed but only addition, according to the method I have also explained for the decimal system elsewhere. As, for example:

The subtrahend is	11 0101
Supposing it should be subtracted from	100 0000
It is evident that this gives	00 1011
For if you add this:	11 0101
It gives us	100 0000

Leibniz here anticipates the technique used in computers today to find $x - y$ in binary: form what is now called the two's complement of y, then add that to x. The two's complement of y is the

---

8. Leibniz mistakenly wrote "par" ["even"] in place of "impar" ["odd"].

9. In the manuscript, the sum on the right is part of the main text; the explanatory table on the left is written in the left margin.

number that must be added to y to yield 0 (with a carryout at the left, which is discarded) or, equivalently, the result of subtracting y from $100\ldots0$, with as many 0s as y has bits.

And in this way you can write down the complements immediately if you write the same in the first digit on the right and then the opposite for the others, namely changing 0 into 1 and 1 into 0. Finally, it should be noted at the end that you must subtract 1 from the next column, to which no digit of the subtrahend extends, if you want to complete the subtraction by mere addition using complement.

This way of calculating the two's complement works only for numbers ending in 1 (such as Leibniz's example, 110101). The correct rule is to complement each bit and then add 1, discarding any carryout from the leftmost bit position. Leibniz doesn't quite say that the two's complement of a positive number can be thought of as a negative number, but that is what this passage boils down to.

I now turn to multiplication. Here again it is evident that nothing can more easily be imagined, for you do not need a Pythagorean table,[10] and this multiplication is the only one that does not require you to know any other multiplication. For you write only the number or 0 in its place.

$$
\begin{array}{r}
1011\ 101 \\
1110 \\
\hline
\ddot{1}0\ddot{1}\ddot{1}\ 1010 \\
1\ 0111\ 01 \\
10\ 1110\ 1 \\
\hline
101\ 0001\ 0110
\end{array}
$$

This type of calculation could be carried out by a machine[11] quite easily and without effort in the following manner: Suppose a box, which is perforated in such a way that the holes can be opened and closed, is open in the places corresponding to 1 while the places corresponding to 0 remain closed. Let small cubes or balls fall into channels through the open spaces and nothing through the others, and the box is moved and transported from column to column in such a way as the multiplication requires. The channels represent the columns, and a ball cannot pass from one channel to another unless the machine is moved. Then all the balls flow into the next channel, with the one that remains in a hole always removed, if indeed it wants to pass through the gate alone. For the thing can be built so that two always necessarily flow out together and otherwise will not flow out.

Leibniz seems to be imagining a contrivance such as the following for computing the product of an n-bit multiplier and n-bit multiplicand. A row of $2n$ "channels"—one for each bit position in the product—constitute an accumulator; initially all channels are empty. A box containing balls has n shutters spaced at the same intervals as are the channels, and these shutters are set open or shut according to the 1s and 0s of the multiplicand.

The box is positioned so that its rightmost shutter (whether open or closed) is above the rightmost channel, that is, above the 1's channel of the accumulator. Then, if the multiplier has a 1 in its low-order (1's) bit, some mechanism is actuated which causes each open shutter to drop one ball into the channel beneath it, so that each channel corresponding to a 1 in the multiplicand receives a ball; but if the multiplier has a 0 in its low-order bit, then the dropping mechanism is not actuated. The box is then shifted left so that that the rightmost shutter is above the accumulator's 2's channel, and the dropping mechanism is again activated or not activated according to whether the 2's bit of the multiplier is 1 or 0. The process is repeated for each bit of the multiplier. When a channel has accumulated two balls, a carry mechanism somehow passes one ball to the column immediately to the left while disposing of the other.

---

10. That is, the multiplication table.
11. machine ▷ without wheels ◁ *deleted*.

In this calculus, division is done without a Pythagorean table and likewise without trial and error. Let's see how quickly it can be done:[12]

Leibniz works out two examples of division: $365 \div 10 = 36$, remainder 5 (in binary, $1\,0110\,1101 \div 1010 = 10\,0100$, remainder 101), and $36 \div 10 = 3$, remainder 6 (in binary, $10\,0100 \div 1010 = 11$, remainder 110). He is using a binary version of what has been called the "French method" of long division, in which when a digit of the quotient is written down, the product of that digit and the divisor is not written down, only the remainder left when that product is subtracted from the appropriate digits of the dividend. See Morrison (1952). Leibniz starts the second example several times, crossing out flawed attempts before producing a version he declares to be "best." The crossed-out numerals reproduced here are, however, from his final and successful try. As high-order digits of the dividend are no longer needed, those digits are crossed out and digits of the new, reduced dividend are written above it. The final dividend is the remainder, 110 in the example, reading from top left to bottom right of the top three rows of the display that begins "Commonly thus." Likewise the divisor is rewritten shifted over one digit on successive rows, with digits crossed out as digits of the quotient are produced.

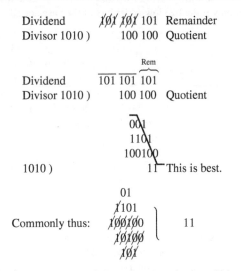

It should be noted that if you have to remove 1 from 0, you write it as if 1 were in place of 0. But then 1 has to become 0 in the following 1. But if what follows is not 1, but 0, then every 0 that follows is converted to 1, and the first 1 which follows these 0s (going from right to left) is converted to 0, just like in decimal notation, except that there 0s are changed to 9s.

Thus far we have considered the algorithm of this progression. I come now to its algebra, which I understand in a completely different way from the common method, for I shall take variables not as unknown quantities but as symbols designating sought-for numbers expressed in binary. Till now, no one has done this. And all things are wonderfully brought together by this method.

For an unknown is almost known to the extent that this one thing alone remains to be asked about it, namely whether it is 1 or 0. Then, and this is remarkable, the squares or cubes or any other powers of unknowns disturb nothing, as in ordinary algebra, for because the unknown signifies 1 or

---

12. done. ▷ There is no need to cross out; it is sufficient, indeed better, to write it like this:

        101101101
    1010)   1001

You write the dividend better this way: 101101101
The divisor is written below, in this way: 1010
The quotient can be written below. ◁ *deleted*.

0, therefore its square signifies the same as itself, for the square, cube etc. of 1 itself is 1, just as the power of 0 itself is 0. And so, if ever all things are reduced to the simplest expressions, surely it is here, for we need only 1 and 0. In addition, nothing more universal can be imagined. For I shall show that fractions and irrational quantities, indeed even transcendental quantities, like logarithms and angles, or arcs, can be determined as far as is possible in this way.

Therefore, let there be any binary integer, finite (or for finite fractions and irrational and transcendental numbers) infinite.

Leibniz squares an arbitrary binary number ending with the bits $abcdef$. To make his tableau cover the case of "infinite" numbers—fractions with nonterminating binary representations—envision $abcdef$ as the first terms of an infinite sum $\sum a_i x^i$, where $a_5, \ldots, a_0 = a, b, c, d, e, f$, and x is either 2 (for integers, with only finitely many of the a_i nonzero) or 2^{-1} (for fractions). In the simple case of a 6-bit integer, the value to be squared (the "base") is $2^5 a + 2^4 b + 2^3 c + 2^2 d + 2^1 e + 2^0 f$, and the column headers are 1 more than the exponent of 2 for that bit position. Where a term has a coefficient of 2, he moves the same term with a coefficient of 1 to the next position to the left. Leibniz then proves algebraically that the low-order bit of the square is the same as the low-order bit of the base, the penultimate bit of the square is always 0, and the antepenultimate bit of the square is 1 if and only if the two low-order bits of the base are 10.

$$
\begin{array}{ccccccccccc}
 & & & & 5 & 4 & 3 & 2 & 1 & & \\
 & & & a & + b & + c & + d & + e & + f & & \\
 & & & a & + b & + c & + d & + e & + f & & \\
\hline
 & & & & + 2af & + 2bf & + 2cf & + 2df & + 2ef & + f^2 & \\
 & & & + 2ae & + 2be & + 2ce & + 2de & + e^2 & & & \\
 & & + 2ad & + 2bd & + 2cd & + d^2 & & & & & \\
 & + 2ac & + 2bc & + c^2 & & & & & & & \\
+ a^2 & + 2ab & + b^2 & & & & & & & & \\
\hline
ab & ac & ad & ae & af & bf & cf & df & ef & 0 & f \\
a & & bc & bd & be & ce & de & & e & & \\
 & & b & & cd & & d & & & & \\
 & & & & c & & & & & &
\end{array}
$$

Its square, which may be easily formed, can be estimated from its dimensions, namely a^2 or a is 12 dimensions, ab is 11 dimensions but makes 12 because it has been doubled, ac is 10 dimensions but makes 11, ad is 10 dimensions, likewise b, likewise bc and so on for the rest.

Here it should be noted, just as any letter signifies 1 or 0, there emerge another binary expression and another of the number produced. Yet the following can be stated as certain: the value of the penultimate product of a binary square will always be 0, and likewise, whatever is the penultimate of the base, such is the antepenultimate of the square.[13,14]

If the last two [digits] of the base are both 1, the antepenultimate of the square will be 0, otherwise it will be 1. In fact, better, if f is 1, the antepenultimate of the square is always 0. For it is evident if e is 0; if e is 1, then it would be made 2. Therefore 0. If e is 0, the antepenultimate is always 0: if e is 1 and f is 0, then it is always 1. Therefore from these two [digits]—ultimate and penultimate—the last three digits of the square are determined.

---

13. square. ▷ The preantepenultimate is always 0. The antepreantepenultimate of the square is always such as is the antepenultimate of the base. The preantepenultimate in turn is always 0. ◁ *deleted*.

14. The last statement is incorrect. The antepenultimate of the square is 0 if the penultimate of the base is 0, but is the complement of the ultimate of the base if the penultimate of the base is 1, as the text goes on to detail.

If the three last [digits] in the base are 1, there will be in the square three [1s] before the last 0, and generally if there is a base of any number of 1s, its square will be constructed from the terms of which the first and last are 1, and the intermediates 0.[15]

So that we may now investigate the relation between the digits of the base and those of the power, we shall begin with the last digits in order to determine the digits of the power from the digits of the base, that is, according to their origin. But in order to determine the digits of the base from the digits of the power, that is, according to analysis, we shall go in the opposite direction, and begin by considering the first digits. First of all, though, it is certain that the last digit of the base is the same as the last digit of the square.

The penultimate digit of the square is always 0. If the penultimate digit of the base is 0, the antepenultimate digit of the square is 0. For if e is 0, $ef + e$ is also 0. If the last digit of the base is 1, then the antepenultimate digit of the square will be 0. For if f is 1, then e will be either 0 or 1. If with f being 1, e is 0, then the digit with which we are concerned is 0, by the preceding. If with f being 1, e is itself also 1, then $ef + e$ is 2. Therefore, the digit is 0, obviously with a 1 carried into the next column on the left on account of the 2.

If the last digit of the base is 0 and the penultimate digit is 1 (in which case it alone is left over), the antepenultimate digit of the square will be 1, for if f is 0 and e is 1, $ef + e$ will be equal to 1.

If the last digit of the base is 0, then the preantepenultimate digit of the square is also 0. For if f is 0, then df is zeroed out in this preantepenultimate column, nor is anything carried over from the preceding column which, with f being 0, has only e, that is, 1 or 0. If d and f are 1 and e is 0, then only the preantepenultimate [column] of the square is 1.

The preantepenultimate column results in a carry of 1 to the next column only when d, e, f are 1, and then it becomes 0 itself.

Similarly, the antepreantepenultimate causes a carry to the next column only when it is 0, except in the unique case when there are [1s in] d, e, f, and c is 0, that is, when there is [a 1 in] the last of the square and the penultimate and antepenultimate and preantepenultimate are 0.[16] For when it causes a carry to the following [column] it consists only of four or two terms, that is, of 1s or 1. Never of three, which I show like this: it does not receive [any carry] from the preceding column except the 1 it is already assumed not to receive, then in this column there will be 0. For if it does not receive this carry, then by the preceding, either d or e or f is 0. If d [is 0], there will remain only cf, which will not exceed 1. If e [is 0], there will remain $cf + d$, that is, 2, therefore a 1 is carried to the next column and there remains 0. Lastly, if f is 0, there remains $de + d$, that is, either 1 or 2, never 3. If it receives a carry of 1 from the preceding column, then there are [1s in] d, e, f. Therefore this column becomes 4, unless c should be 0, then it becomes 3.

---

15. To be precise, $(2^k - 1)^2 = 2^{2k} - 2^{k+1} + 1$, which in binary is $1^{k-1}0^k1$ (d^n representing n repetitions of digit d).
16. Leibniz has "f, c, e" in place of the correct "d, e, f".

15 March 1679

On the Binary Progression—Part II

On the previous sheet we completed the algorithm, that is, the so-called four kinds of operation were shown in binary progression. From that we proceeded to pure powers[17] and their bases, and we began from the simplest of all, the square. Whereby we have given an outline of such kind:

$$
\begin{array}{c}
a \;+\; b \;+\; c \;+\; d \;+\; e \;+ f \\
\hline
a \;+\; b \;+\; c \;+\; d \;+\; e \;+ f \\
+\, 2af \;+\; 2bf \;+\; 2cf \;+\; 2df \;+\; 2ef \;+ f^2 \\
+\, 2ae \;+\; 2be \;+\; 2ce \;+\; 2de \;+\; e^2 \\
+\, 2ad \;+\; 2bd \;+\; 2cd \;+\; d^2 \\
+\, 2ac \;+\; 2bc \;+\; c^2 \\
+\, a^2 \;+\; 2ab \;+\; b^2
\end{array}
$$

From which, because a binary progression involves a carry to the next column from those which contain 2 or more, and because in place of the powers of digits one simply puts [the digits] themselves, we get:

ab	ac	ad	ae	af	bf	cf	df	ef	0	f
a		bc	bd	be	ce	de		e		
		b		cd		d				
				c						
K	L	M	N	P	Q	R	S	T	V	W X

The algorithm for extracting the square root in any number system breaks the digits into groups of two digits and finds the digits of the root from left to right, one digit of the root for each pair of digits of the base. Leibniz first tries to carry out this algorithm algebraically on a 12-bit number *KLMNPQRSTVWX*, in an effort to infer the 6-bit root *abcdef* by reversing the calculations that produce the square from the root. He abandons this effort when he cannot readily account for the carries. He tries a different method on an 8-bit number *abcdefgh*, noting that the first bit of the root will be 1 (otherwise, the number would have been 6 bits in length rather than 8). He then posits, without explanation, that the second and third bits of the root will be $a + b - 1$ and $\frac{c+d}{a+b}$. (Since the case $a = b = 0$ has been ruled out, $a + b - 1$ is the boolean AND of a and b, 1 if $a = b = 1$, and 0 if $a = 1$ and $b = 0$ or $a = 0$ and $b = 1$.) Of the eight perfect squares expressible in 8 bits but not in 6 or fewer, namely, $8^2, 9^2, \ldots, 15^2$, the formula $a + b - 1$ for the second bit of the root works for all but 12^2 and 13^2. But $12^2 = 144 = 1001\,0000_2$, so $a + b - 1 = 0$ in this case, but the second bit of $12 = 1100_2$ is 1. Similarly, $13^2 = 1101\frac{2}{2} = 1010\,1001_2$. Finding this analysis "difficult," Leibniz moves on to calculating binary fractions.

Analysis or investigation of the base from a given square:

I assume that a, the leftmost digit of the base, is not 0 but 1, otherwise we shall begin not with that but with b. For the same reason, I assume that the first digit of the power is 1, since it cannot be greater, nor can it be smaller, otherwise it would be 0.[18]

---

17. A "power" here is what we would call a term. In an expression such as $x^2 + 6x + 9$, the x^2 term is said to be a *pure* power. The other two terms are called *affected* if the expression is rewritten as $x^2 + 2 \cdot 3 \cdot x + 3^2$ so that the exponents of x and 3 sum to 2 in each addend. The distinction is from the work of François Viète (1540–1603). See Struik (1986, 78).

18. 0 ▷ , because $ab + a$ must become 1 and a is 1. Let b equal 0, therefore the second digit of every quadratic root is 0.

Let the first be marked with a dot according to usual practice, then if K is removed, b is necessarily 0, but let us express the whole thing analytically by calculation: a equals 1, b equals $K + L - 1$, that is, if K is removed then b is 0, if L is removed then b is 1, if KL is removed then b is 10, that is, b is 0 since 1 belongs to a.

But already I see that such calculations are useless, because we do not always know what a column has received from the next one on the right. But if we begin with the right hand columns by comparing f and X, then our calculation does not help except in those cases where we are asking whether the value in question is precisely a square, although by precisely comparing this method with a square, we seem to be able to achieve what is sought by removing as little as possible, so that a given value becomes susceptible to laws. And indeed, this analysis, corresponding to origins, is certain and universal for all pure and affected powers.

Yet there seems to be another method for investigating the root, by general extraction. Therefore let there be

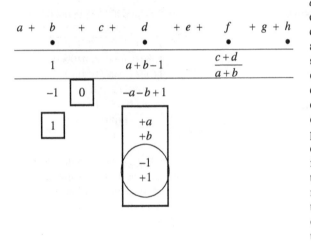

Every analytic quantity can be reduced to a quantity of this form, $a + by + cy^2$ etc. equals $\ell + my + ny^2$ etc. Thus there is no need for the distinction between + and − for there are only affirmatives, even if perhaps some [quantity] vanishes. The period of a sequence may be discovered and explained by a certain kind of character, as in the loci of curves, so one will easily be able to discover the period of a squared cube[19] and what comes about if an infinite periodic formula is multiplied into a finite one: the two sequences must ultimately be made equal on both sides, if indeed that is possible. Let us try the thing out in other problems where we know that there is some equation, which we shall discover in this way.

But I see that this becomes very difficult, whether we wish to pursue the matter generally or with letters. And so, persevering with the first analytical path mentioned above: with any specific number exhibited, it is not difficult to follow the thing.

But leaving powers aside for the time being, let us first of all explain fractional binary quantities in the minutest detail.

On the left of the following text, Leibniz carries out divisions to calculate the binary values of $\frac{1}{2}$, $\frac{1}{3}$, ..., $\frac{1}{9}$, each of which is a possibly empty finite string of 1s and 0s, possibly followed by a repeating pattern of 1s and 0s. Marginalia on the right include the calculation of the first bits of $\sqrt{2}_{10} = 1.011010100\ldots_2$. Leibniz multiplies that number by itself to get the value 01.111111111110100, observing that the infinite numeral .11111...represents 1, so that

---

If the second digit of a power is 0 then the digit of the third base is 0, for if ac is 0 then necessarily a is 1 and c is 0.

If the first and second digits of a power are 1 ◁ *deleted.*

19. "quadrati cubi," that is, sixth power. In Leibniz's day, "period" was sometimes used to refer to a group of three digits of a longer string (Irson 1692, 11), but was also commonly used to refer to the different-sized groups into which digits must be blocked in order to extract roots: 2 for square roots, 3 for cube roots, and 6 for sixth roots—likely the sense in which Leibniz uses the term here. See, for example, Leybourn (1700, 183–184, 193–194).

1.1111... represents 2. The square root algorithm, like that for decimals, breaks the number into groups of two digits and then determines one digit of the result for each pair of digits of the number whose root is being extracted.

$\dfrac{1}{2}$ equals $\dfrac{1}{10}$ equals $\begin{array}{c}10000\\10\end{array}\Big\}$ equals 010000 etc.

$\dfrac{1}{3}$ equals $\dfrac{1}{11}$ equals $\left.\begin{array}{c}1111\\100000\\11101\\111\end{array}\right\}$ 00101010 etc.

$\dfrac{1}{4}$ equals $\dfrac{1}{100}$ equals $\left.\begin{array}{c}1000000\\1000\\10\end{array}\right\}$ 001000000 etc.

$\dfrac{1}{5}$ equals $\left.\begin{array}{c}1111\\1000000\\101111\\1000\\11\end{array}\right\}$ 000110011001100 etc.

$\dfrac{1}{6}$ equals $\dfrac{1}{110}$ equals $\left.\begin{array}{c}11\\100000000000\\1111\\11\end{array}\right\}$ 00010101010 etc.

$\dfrac{1}{7}$ equals $\dfrac{1}{111}$ equals $\left.\begin{array}{c}111\\10000000000000\\111111\\1111\\11\end{array}\right\}$ 0001001001001 etc.

$\dfrac{1}{8}$ equals $\dfrac{1}{1000}$ equals $\begin{array}{c}100000000\\1000\end{array}\Big\}$ 0001000000

$\dfrac{1}{9}$ equals $\dfrac{1}{1001}$ equals $\left.\begin{array}{c}0\\111111\\100000000000\\100111111\\100000000\\100000\\1111\end{array}\right\}$ 000011100011100011100 etc.

$$\begin{array}{c}1\ \ 1\\0\ \ 00\\1111111111111111111\\0100000000000000000000000000000\\\cdot\ \cdot\ \cdot\ \cdot\ \cdot\ \cdot\ \cdot\ \cdot\ \cdot\ \cdot\\\hline 0\ 1\ 0\ 1\ 1\ 0\ 1\ 0\ 1\ 0\ 0\\10001010001000100000\\1100111100110010\\110011110010\\110011110\\11001\\1\end{array}$$

Note: $\frac{01111111}{10000000}$ etc. to infinity is the same as 1. This is true in the decimal progression too. From this it is evident that in $\frac{1}{100-1}f$ [text breaks off]

$$\begin{array}{r}01011010100\\01011010100\\1011010100\\01011010100\\0101101100\\\hline 01111111111110100\end{array}$$

$$\begin{array}{r}110011001100\\101)\ \ \underline{01100110011}\\111111111111\end{array}$$

Now, in order to learn in general the relation between a number and its fraction expressed in binary, general periodic sequences should be taken from the shorter periods moving up to the longer, and a sequence of this sort should be multiplied by a general finite number as much as is sufficient, and [then] it should be determined by what means it can happen that in the product everything is canceled out, that is, becomes 0, except for the first few terms. Perhaps it will be possible to do it better this way: take a general infinite sequence without mentioning periods, then multiply this by a finite number: everything in the product is canceled out aside from the first 1. And in this way one can obtain the method of exhibiting a simple fraction, that is, a fraction whose numerator is 1. Derivatives, whose numerator is a number of these, are obtained by multiplying that number. Although, in this way too one can obtain and generally determine the sum of a given number expressed by recurring periods to infinity. So that, in turn, from a given periodic expression, the sum, or number, may be obtained, just as we said a little earlier that a periodic binary expression is derived from a given number by comparing. Having obtained the period from a given number, one will also obtain the method of expressing by a certain rule the sequence of digits expressing the size of a circle, that is, $\frac{1}{1} - \frac{1}{3} + \frac{1}{5} - \frac{1}{7} + \frac{1}{9}$ etc.[20]

But the main question, and the summit of all analysis, is whether any quantity whatsoever is analytical.[21] Take, for instance, any sequence of digits having a rule according to which it is continued; the question is: if we write in order any number of powers of this sequence, meaning finite and individual powers multiplied by finite numbers (x^0 is also a power, namely 1, whose exponent is 1), and add the products in turn by whatever signs, the question is whether it can ever happen that everything simultaneously cancels each other out. But can we imagine a sequence which can in no way be cancelled out by these methods? And already I see that sequences of this kind, not expressed analytically, can be imagined, for suppose some sequence is constructed like this: take some number constituting the first period of a sequence; the second period is some power and pure or affected root of the former; the third period is again higher than the other, and so on to infinity, with new arbitrary letters taken forever. It does not seem to be always possible hereafter to take such finite affected exaltations and involutions[22] by which that whole mass is cancelled out, for the few arbitrary finite letters which can be taken, it is impossible to satisfy those infinite always-varying letters, which can be taken to be somehow different from one another. Therefore it has been demonstrated that there are transcendental quantities, that is, those that cannot be expressed by an equation of a certain degree. But whether the area of the circle is such is something to be considered in particular.[23] In fact, I already see that it has not even been demonstrated that there are non-analytical quantities, for if a quantity is definite, then at any rate those letters taken for affected powers in place of the following periods cannot be arbitrary, but arise from definite, actually finite, conditions.

There must necessarily be some period after there is no new letter or quantity taken, but since it is possible that in the following period they may be extended in a very long sequence, and in a third [period] again in a longer [sequence], to infinity, I doubt whether finite exaltations can be sufficient to cancel them out.

---

20. Leibniz recognizes that the digits of $\frac{\pi}{4}$ can be calculated by means of his convergent series $\frac{\pi}{4} = \sum_{i=0}^{\infty} (-1)^i \frac{1}{2i+1}$ and that each reciprocal $\frac{1}{n}$ is a repeating binary fraction $B_n \sum_{i=1}^{\infty} 2^{-ik_n}$, where k_n and B_n are determinable integers (see note 11 in the introduction to this book). He apparently reasons that if he sets a sufficiently long finite initial segment of his series equal to an initial segment of $\sum_{i=1}^{\infty} a_i 2^{-i}$, that is, to $\frac{\pi}{4}$ represented as a binary fraction $.a_1 a_2 a_3 \ldots$ with each a_i being 0 or 1, he should be able to derive analytically formulas for the a_i for small i in terms of the k_n and B_n by matching up the 2^{-i} terms on both sides of the equation.

21. For Leibniz, "analytic" meant "calculable" (Knobloch 2015, 92); its precise meaning would evolve gradually.

22. See note 17 of this chapter. In expressing $x^2 + 6x + 9$ as $x^2 + 2 \cdot 3 \cdot x + 3^2$, the last two terms have been "exalted" in order make the expression factorable. An "involution" is the extraction of a root, such as $x + 3$ in this example, as would be done to solve the equation $x^2 + 6x + 9 = 0$. For further details on exaltation and involution, see Jeake (1696, 622) and Barlow (1814, art. "Involution").

23. The transcendence of π was not established until 1882, when it was proved by Carl Louis Ferdinand von Lindemann. The question was significant because the answer settled the ancient question of whether it was possible to square the circle with straightedge and compass.

7 Attempted Expression of the Circle in Binary Progression (c. 1679)

Manuscript:
M: LH 35, 12, 2 Bl. 97. Draft. Latin.

Leibniz's earliest mathematical achievement during his time in Paris was the formula for the arithmetical quadrature (or squaring) of the circle, that is, for determining the area of a circle enclosed by a 1×1 square, a formula he worked out in the first half of 1674. As he wrote in 1674, "With the circumscribed square being 1, the area of the inscribed circle will be $1 - \frac{1}{3} + \frac{1}{5} - \frac{1}{7} + \frac{1}{9} - \frac{1}{11}$ etc." (A VII 6, 111, cf. 89, 174, 600), an infinite series with the value $\frac{\pi}{4}$. In the following text, Leibniz seeks to convert his formula for $\frac{\pi}{4}$ into binary, by comparing the sum of various terms of the binary progression $1 \ \frac{1}{2} \ \frac{1}{4} \ \frac{1}{8} \ \frac{1}{16} \ \frac{1}{32}$ and so on with the sum of the series for $\frac{\pi}{4}$. However, he makes an arithmetic error early on, leading him to conclude incorrectly that $\frac{1}{2} + \frac{1}{4}$ is greater than the area of the circle (which, if true, would make π less than 3). Apparently not spotting the mistake, Leibniz builds on this erroneous result and carries on making further comparisons before giving up. The remainder of the page is filled with tables of binary digits and fractions represented in binary, not always correctly.

Attempted Expression of the Circle in Binary Progression

The binary progression is $1 \ \frac{1}{2} \ \frac{1}{4} \ \frac{1}{8} \ \frac{1}{16} \ \frac{1}{32} \ \frac{1}{64} \ \frac{1}{128} \ \frac{1}{256} \ \frac{1}{512} \ \frac{1}{1024} \ \frac{1}{2048}$

The value of a circle[1] whose diameter is 1 is: $1 - \frac{1}{3} + \frac{1}{5} - \frac{1}{7} + \frac{1}{9} - \frac{1}{11} + \frac{1}{13} - \frac{1}{15} + \frac{1}{17} - \frac{1}{19} + \frac{1}{21} - \frac{1}{23}$

$\frac{1}{2}$ is too small for the value of a circle, because $\frac{1}{2} < 1 - \frac{1}{3}$ (and indeed, $1 - \frac{1}{3}$ is < circle), with a difference of $\frac{1}{6}$

$\frac{1}{2} + \frac{1}{4}$ is not less than $1 - \frac{1}{3} + \frac{1}{5} - \frac{1}{7}$ because by adding $\frac{1}{6}$ to $\frac{1}{5} - \frac{1}{7}$ makes less than $\frac{1}{4}$

$\frac{1}{2} + \frac{1}{4}$ is not greater than $1 - \frac{1}{3} + \frac{1}{5}$

Compare $\frac{1}{2} + \frac{1}{4}$ with $1 - \frac{1}{3} + \frac{1}{5} - \frac{1}{7} + \frac{1}{9}$, that is, $\left(\frac{1}{2}\right) + \frac{1}{3}\left(+\frac{1}{4}\right) + \frac{1}{7}$ with $\boxed{\frac{1}{1} + \frac{1}{5} + \frac{1}{9}}$; $\boxed{\frac{1}{2}}\frac{1}{4}$

the former will be greater than the latter, therefore $\frac{1}{2} + \frac{1}{4}$ is greater than the circle.[2]

Take $\frac{1}{2} + \frac{1}{8}$ and compare it with $\frac{1}{1} - \frac{1}{3} + \frac{1}{5} - \frac{1}{7}$, that is, $\frac{1}{3} + \frac{1}{9}$ with $\frac{3}{8} + \frac{1}{5}$ or $\frac{2}{35}$ with $\frac{1}{24}$; $\frac{1}{2} + \frac{1}{8}$ will not be less than $\frac{1}{1}$ etc. $- \frac{1}{7}$

---

1. That is, the area of a circle of diameter 1 is $\frac{\pi}{4} = \sum_{i=0}^{\infty}(-1)^i \frac{1}{2i+1}$.

2. Leibniz makes an arithmetic error here. In fact, the former is in each case less than the latter, and Leibniz fails to notice that the last statement is equivalent to saying that $\frac{3}{4} > \frac{\pi}{4}$, that is, $\pi < 3$. So everything that follows is based on an incorrect premise.

Take $\frac{1}{2}+\frac{1}{8}$ and compare it with $\frac{1}{1}$ etc. $-\frac{1}{11}$, that is, $\frac{2}{35}$ with $\frac{1}{24}+\frac{1}{9}-\frac{1}{11}$, or $\frac{2}{35}$ with $\frac{1}{24}+\frac{2}{99}$, or $\frac{1}{35}$ with $\frac{1}{48}+\frac{1}{99}$, or $\frac{13}{35\times48}$ with $\frac{1}{99}$, or $\frac{13}{35\times16}$ with $\frac{1}{33}$; the former will be less, [for] if you merely divide 48 by 13, the quotient is greater than 3, which, if you multiply by 35, makes 105, which is greater than 99; the former will be less than the latter, therefore $\frac{1}{2}+\frac{1}{8}$ is less than the circle.

Compare $\frac{1}{2}+\frac{1}{8}+\frac{1}{16}$ with $\frac{1}{1}$ etc. $-\frac{1}{11}$, that is, $\frac{13}{35\times16}$ with $\frac{1}{33}$, and see whether the former is less. That is, $\frac{13}{35}+1$ with $\frac{1}{2+\frac{1}{16}}$, therefore it is not less.

$\frac{1}{2}+\frac{1}{8}+\frac{1}{16}$ with $\frac{1}{1}$ etc. $-\frac{1}{13}$, that is, $\frac{13}{35}+1$ with $\frac{1}{2+\frac{1}{16}}+\frac{1}{13}$ will be greater than the latter, therefore is also greater than the circle. So we may suppose:

$\frac{1}{2}+\frac{1}{8}+\frac{1}{32}$ makes $\frac{1}{35\times16}+\frac{1}{32}$, compared with $\frac{1}{33}+\frac{1}{13}$, the former will not be greater than the latter,

$\frac{1}{2}+\frac{1}{8}+\frac{1}{32}$ with $\frac{1}{1\ \text{etc.}}-\frac{1}{15}$, that is, $\frac{13}{35\times48}+\frac{1}{32}$ with $\frac{1}{99}+\overbrace{\frac{1}{13}-\frac{1}{15}}^{\frac{2}{13\times15}}$. Is the former less?

The relation by arrangement

```
 1  1            1
 2  10           10
 3  11           11
 4  100          100
 5  101          101
 6  110          110
 7  111          111                         11
 8  1000         1000      ½ 0100000        1̷0̷0̷000
 9  1001         1001      ⅓ 0111           1̷1̷1̷1    }01111
10  1010         1010                        1̷1
11  1011         1011
12  1100         1100
13  1101         1101                        01
14  1110         1110                        ──  }0
15  1111         1111                        10
16  10000        10000
```

The calculations that follow appear on the reverse of the manuscript page. Leibniz's results for $\frac{1}{6}$ and $\frac{1}{9}$ are incorrect (his result for $\frac{1}{6}$ is in fact that for $\frac{1}{8}$, while his result for $\frac{1}{9}$ is that for $\frac{5}{32}$).

1) 1.00000

$\frac{1}{2}$ 0100000⌐0.1000 / 1̷0̷0̷ / 1̷

$\frac{1}{3}$ 1̷1 / 01̷0̷0̷000⌐0.01010101 / 1̷1̷1̷1̷ / 1̷1̷

$\frac{1}{4}$ 0.01

$\frac{1}{5}$ 1̷1̷1̷1 / 0̷0̷1̷0̷0̷000⌐0.001100110011 / 1̷0̷1̷1̷1̷1̷1̷1 / 1̷0̷0̷0̷0̷00 / 1̷1̷1̷11

$\frac{1}{6}$ 1̷1 / 0̷0̷1̷0̷0000000⌐0.0010 / 1̷0̷0̷0̷0̷0̷0 / 1̷1̷1̷1̷1 / 1̷1̷1

and thus $\frac{1}{9}$ 000 1̷1̷1̷11111⌐000101 / 1̷0̷1̷1̷1̷1̷1 / 1̷0̷0̷0̷0 / 1̷1̷1

8 Sedecimal Progression (1679)

Manuscript:
M: LH 35, 13, 3 Bl. 23. Draft. Latin and German.

This text presents what is very probably the first account of what is now known as hexadecimal or base-16 notation, an invention for which Leibniz has not yet been recognized, largely because his writings on the subject have remained unpublished until now. In his magisterial *The Art of Computer Programming*, Donald Knuth (1997, II, 201) identifies the inventor of base 16 as the American engineer John W. Nystrom (1825–1885), who, in an 1862 book and a series of articles published in 1863, proposed replacing the familiar decimal system with a base-16 system he called "tonal"—not because of some musical connection, as one might imagine, but because in it the number "10" (that is, decimal 16) is arbitrarily named Ton.[1] Nystrom did not claim base 16 as his own invention, however; he wrote that king Charles XII of Sweden (1682–1718) had considered introducing the base-16 system in Sweden but his objection to the consequent requirement for new symbols led him to prefer the octal (base-8) system instead (Nystrom 1863; 1862, 101). Unfortunately, Nystrom provided no evidence for his assertion, and it is undermined by an eyewitness account by Emanuel Swedenborg (1668–1772) which describes Charles's interest in octal and even base 64, but not base 16 (see Swedenborg 1754). In any case, the following text shows Leibniz constructing a base 16 numeration system in 1679, three years before Charles was born, in what was clearly a fruitful year for Leibniz's forays into number theory.

Base-16 notation is now ubiquitous in computing, because computer words are typically multiples of 4 bits, and the 8-bit byte is standard for information exchange. Knuth (1997, II, 202) notes that the modern term for base 16, "hexadecimal," is a mid-twentieth-century amalgam of Greek and Latin roots and suggests that the "more proper" term "sexadecimal" may have been "too risqué for computer programmers."[2] We adopt Leibniz's equally appropriate "sedecimal," which Knuth also suggests. Sedecimal notation requires six new digits in addition to 0, . . . , 9, which in computing are now universally A, B, C, D, E, and F. In this manuscript Leibniz first employs the Roman letters m, n, p, q, r, and s, but soon switches to the Aretinian syllables ut, re, mi, fa, sol, and la (the Latin precursor of English's sol-fa solmization, and abbreviated by Leibniz to u, r, m, f, s, and ℓ). He then works through the conversion of decimal 1,000, 10,000, and 100,000 into sedecimal.

---

1. In his discussion on the origins of base 16, Knuth overlooks the work of the English schoolmaster and mathematician Thomas Wright Hill (1763–1851), who in 1845, at the age of 83, read a paper proposing a base-16 numeration system at a meeting of the British Association for the Advancement of Science in Cambridge. The paper was posthumously published as "A system of numerical nomenclature and notation, grounded on the principles of abstract utility" (Hill 1860, 63–85). Hill (1860, 69) drew his inspiration from the idiosyncratic use of the term "stone" in the county of Yorkshire, England, to refer to a weight of 16 pounds, rather than 14 as elsewhere in Britain; Hill noted that the Yorkshire practice allowed for more convenient bisections.
2. Technically, "sexdecimal" is the proper term. "Sexadecimal" is a corruption of "sexdecimal," first recorded in 1895. See Whitney (1895, 5535).

						ut	re	mi	fa	sol	la																					
0	1	2	3	4	5	6	7	8	9	10	11	12	13	14	15	16	17	18	19	20	21	22	23	24	25	26	27	28	29	30	31	32
0	1	2	3	4	5	6	7	8	9	m	n	p	q	r	s	10	11	12	13	14	15	16	17	18	19	1m	1n	1p	1q	1r	1s	1t

Value....	Pronunciation ———— Notation
1	eins————— 1
2	zwei————— 2
3	drei————— 3
4	vier————— 4
5	fünf————— 5
6	sechs————— 6
7	sieben————— 7
8	acht————— 8
9	neun————— 9
10	ut————— u
11	re————— r
12	mi————— m
13	fa————— f
14	sol————— s
15	la————— ℓ
16	zehn————10
17	elf————11
18	zwölf————12
19	dreizehn————13
20	vierzehn————14
21	fünfzehn————15
22	sechzehn————16
23	siebenzehn————17
24	achtzehn————18
25	neunzehn————19
26	utzehn————1u
27	rezehn————1r
28	mizehn————1m
29	fazehn————1f
30	solzehn————1s
31	lazehn————1ℓ
32	zwanzig————20
33	einundzwanzig————21
34	zweiundzwanzig————22
35	dreiundzwanzig————23
36	vierundzwanzig————24
37	fünfundzwanzig————25
38	sechsundzwanzig————26
39	siebenundzwanzig————27
40	achtundzwanzig————28
41	neunundzwanzig————29
42	utzwanzig————2u
43	rezwanzig————2r
44	mizwanzig————2m
45	fazwanzig————2f
46	solzwanzig————2s
47	lazwanzig————2ℓ
48	dreizig————30
49	
50	
51	
52	
53	
54	
55	
56	
57	
58	
59	
60	
61	
62	
63	
64	

The last entry above, 1t, should be 20. Leibniz changes digits between the table above and the table to the left, using m, n, p, q, r, and s above and u, r, m, f, s, and ℓ elsewhere. The material below and to the right is scratchwork, not part of the main text.

$$\begin{array}{r} 256 \\ \underline{16} \\ 1536 \\ \underline{256} \\ 4096 \end{array}$$

$$\begin{array}{r} 1 \\ 443 \\ 1679 \\ 256 \\ 165 \end{array} \Big\rangle\ 6$$

Sedecimal Progression 1679

$$\begin{array}{r} 143 \\ 15 \end{array} \Big\rangle\ 8 \qquad 68\ell$$

$$15\,|\,\ell$$

1	1
16	10
256	100
4096	1000

The rule for changing a decimal progression into sedecimal:

Let there be a given number: 1679

The operation happens in decimal numbers to find sedecimals.

The sedecimal numbers of the progression should be laid out in order, namely 1, 16, 256, 4096 etc. From this progression take the closest number that is smaller than the given number, namely 256. The given number is divided by this. The quotient will be 6, which will be the first digit of the sedecimal expression, signifying 600, that is, 6 × 256 because 256 sedecimally is expressed by 100. The remainder, 143, is divided by the next sedecimal, namely 16. The quotient, 8, will be the second term of the sedecimal expression, signifying 80, that is, 8 × 16,[3] because 16 sedecimally is expressed by 10. The remainder will be 15, which cannot be further divided by any sedecimal, and is expressed by ℓ. And therefore the number 1679 is expressed sedecimally by 68ℓ. From this it is apparent that sedecimal expressions are much shorter. The method of operation, abridged and well-ordered in form like division, will be evident from the following examples.

---

3. Leibniz incorrectly wrote "sexies" [6 times] instead of 8.

First manuscript page of "Sedecimal Progression" (LH 35, 13, 3 Bl. 23r).

1	Expressed commonly	Sedecimally[4]
16	10	u
256	100	64
4096	1000	3s8
65536	10000	2710
	100000	186u0

The reverse manuscript page includes a table of powers of 16 in decimal (shown top left), along with another of powers of 10 converted to sedecimal (shown top center). It then includes the workings out of most of these decimal to sedecimal conversions. Some attempts are crossed out in their entirety; the one shown top right here converts 100_{10} to 64_{16}. Individual crossed-out numerals reflect the progress of long divisions. There are also some lines drawn between digits. In what follows, in addition to reproducing Leibniz's conversion calculations, we also reconstruct them in stages. In the figures below, only the rightmost calculation is in the manuscript; the calculations to its left on a gray background are our reconstructions of the stages Leibniz went through to produce the final result. In each case, he is converting a decimal numeral that is a power of 10 to sedecimal by repeatedly dividing by powers of 16. The dividend is to the left of the $\backslash$ sign, the divisors are written below the dividend, and the quotient is to the right. As the dividend is replaced by the remainder for subsequent divisions, the new dividend is written above the old and the digits of the old are crossed out as needed. Subtractions (dividend minus the product of the divisor and one digit of the quotient) are carried out left to right, with borrowing from the partially formed difference, rather than right to left with borrowing from the dividend. This method is popularly known as "left to right subtraction" today.

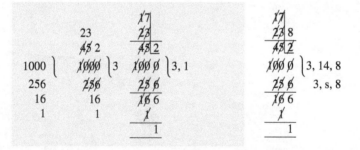

In the example above, Leibniz converts decimal 1000 to sedecimal 3s8. He first writes down dividend 1000 and below it successive powers of 16, which will be the successive divisors. At the first stage, he divides 256 into 1000, yielding 3 with a remainder of 232. The quotient is written to the right of the $\backslash$, and 232 are the digits at the top of the three rightmost columns above the original dividend. The crossed-out 4 and 5 were gotten first, as follows. Leibniz is computing the difference $1000 - 3 \times 256$ from left to right, rather than right to left, as would be conventional today. So he first calculates $10 - 3 \times 2 = 4$ and writes 4 above the first 0 in 1000. He then computes $0 - 3 \times 5 = 5$, which requires borrowing 2 from the column to the left, so the 4 there is crossed out and replaced with 2. He then computes $0 - 3 \times 6 = 2$, which requires borrowing 2 from the 5 in the previous column, which is therefore crossed out and replaced with 3. At the second stage, 232 is divided by 16, yielding a quotient of 14 (s in sedecimal) and a remainder of 8, with the 14 resulting from dividing 16 first into 23 (1, remainder 7) and then finally into 72 (4, remainder 8).

---

4. The third digit of the fourth entry seems to be a "9" corrected to be a "1" (in fact, $10000_{10} = 2710_{16}$).

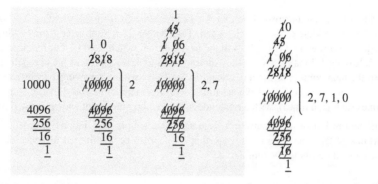

In this example, Leibniz is converting decimal 10000 to sedecimal. He starts by writing the dividend and, below it, the first few powers of 16 in the leftmost stage. In the second stage, 4096 goes into 10000 twice, so 2 is written to the right of the $\lfloor$, and the remainder, $10000 - 2 \times 4096 = 1808$, is written above the dividend and becomes the new dividend. Where 2 and 1 are crossed out in the second row and replaced by 1 and 0 in the row above, borrowing has occurred in the working out of the subtraction. So 1808 is read from left to right, alternating between the first and second rows; these are the top remaining digits. The original dividend and the 4096 are no longer needed, so they are also crossed out. At the third stage, 256 goes into 1808 seven times, leaving a remainder of 16. The difference $1808 - 7 \times 256$ is computed digit by digit from left to right. First, $18 - 7 \times 2 = 4$, so 4 is put in the middle column. Then $0 - 7 \times 5 = 5$, so 5 goes to the right of 4, but that requires borrowing 4 from 4, leaving 0, so the 4 is crossed out. Then $7 \times 6 = 42$ is subtracted from 8, leaving 6, but requiring borrowing 4 from 5, leaving 1. At the last stage, 16 goes into 16 once with a remainder of 0, yielding the result at the right. Another example is worked out just below. (The lines underneath the decimal numerals for the powers of 2 are not underlinings—they are rules separating entries.)

Once again, the dividend and powers of 16 are written first. In the second stage, 65536 goes into 100000 once with a remainder of 34464, working left to right: $10 - 6 = 4$; then $0 - 5 = 5$ (with a borrow from the 4, which is crossed out and replaced with 3); again $0 - 5 = 5$ (with a borrow from the 5 just written down, leaving 4); then $0 - 3 = 7$ (and another borrow replaces the second 5 by 4); then $0 - 6 = 4$ (and a borrow from the 7 leaves 6). In the third stage, 4096 goes into 34464 eight times. Again, the remainder $34464 - 8 \times 4096 = 1696$ is calculated left to right, but now Leibniz uses dots to indicate borrows. First, $8 \times 4 = 32$ is subtracted from 34, yielding 2. Then $8 \times 0 = 0$, so the 4 in the middle of 34464 is let stand. Then $8 \times 9 = 72$ is subtracted from 6, which leaves 4 but requires borrowing 7 from the 4 in the next column, which leaves 7 but requires borrowing 1 from 2, leaving 1. Finally, $8 \times 6 = 48$ is subtracted from 4, which leaves 6 and requires borrowing 5 from the next column to the left, changing the 4 to 9 while borrowing 1 from 7 to yield 6. At the next stage, 256 goes into 1696 six times, with a remainder of 160. Finally, 16 goes into 160 ten times (that is, u times), with a remainder of 0, yielding the result shown at the right.

At the bottom left of the manuscript page are some simple calculations in sedecimal. Leibniz underlines certain numerals to indicate that they are to be interpreted as sedecimal. We have added some underlinings he omitted.

$$\begin{array}{l} \underline{64} \\ \underline{u} \qquad \text{u or 10} \\ \overline{3s8} \end{array}$$

for $\underline{u}$ times $\underline{4}$ equals $\underline{28}$. That is, 40 is $32 + 8$, or $\underline{2}\ \underline{8}$. one writes 8, with 2 remaining. $\underline{u}$ times $\underline{6}$ is $\underline{3}\ m$, that is, 60 is $48 + 12$ or $\underline{3}\ m$.

2 added to $\underline{3}\ \underline{m}$ makes $\underline{3}\ \underline{s}$. For $12 + 2$ is 14, that is $\underline{m} + \underline{2}$ is $\underline{s}$.

9 Binary Progression Is for Theory, Sedecimal for Practice (c. 1679)

Manuscript:

M: LH 35, 3 B 17 Bl. 4. Draft. Latin.

In this short text, written on a small scrap of paper cut from a larger sheet, Leibniz devises two different notations for the sixteen sedecimal digits. For the first, he simply stacks dots and dashes, using a dot for each 0 bit and a dash for each 1 bit, with the most significant bit at the top and the least significant bit at the bottom. For the second, he uses a concave-up curve for 0 (drawn counterclockwise) and a concave-left curve for 1 (drawn clockwise):

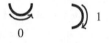

These curves are then chained together so each digit can be written without lifting pen from page. For example, 5 (that is, 101) would be written thus:

Leibniz would later sketch a third alternative notation (chapter 12) and, in a letter to Joachim Bouvet, he develops a fourth (see chapter 22).

When, in the 1960s, the need arose to conveniently represent base-16 (4-bit) numbers in computing, it was natural to use 0–9 and A–F for the sixteen digits, since these symbols were already commonplace.[1] But to avoid confusion in the interpretation of ambiguous numerals, such as 57, where the base could not be inferred from context, some echoed Leibniz in proposing the adoption of an entirely new set of digits, shaped to exhibit their binary equivalents. Martin (1968), for example, suggested using 0 ⅃⊣ ⊒⊣ ⊒⊒⊒⊓, ..., ∃ for the digits corresponding to decimal 0, 1, 2, 3, 4, 5, 6, 7, 8, ..., 15.

The scrap of paper on which Leibniz wrote this text bears a watermark found in paper that he used predominantly in 1679, which suggests that this text most probably dates from that year, although a date of 1680 cannot be ruled out.

---

1. This decision was anticipated by Leibniz in 1703, when he used 0–9 and a–f as his sedecimal character set. See LH 35, 3 B 11 Bl. 11v (reproduced facing the introduction to this book).

Binary Progression Is for Theory, Sedecimal for Practice: Both Are from the Same Source

0	1	2	3	4	5	6	7	8	9	10	11	12	13	14	15	16
0	1	10	11	100	101	110	111	1000	1001	1010	1011	1100	1101	1110	1111	10000

Manuscript of "Binary Progression is for Theory, Sedecimal for Practice" (LH 35, 3 B 17 Bl. 4r).

10 On the Organon or Great Art of Thinking (first half [?] of 1679)

Manuscript:
M: LH 4, 7C Bl. 156–157. Draft. Latin.

Transcriptions:
C: Leibniz 1903, 429–432.
A: A VI 4, 156–160.

During the late 1670s, Leibniz's chief philosophical project was the development of a "general science," at the heart of which was a full-blown logic of discovery, or art of invention, that would enable the quick and accurate discovery of new truths in all areas of knowledge, with the aim of improving the human condition. The following text, which is one of hundreds he wrote to further this project, bears a title that alludes to a number of predecessors who had likewise sought a general method for the discovery of truths, albeit on a less ambitious scale. "Organon," a Greek word meaning "tool" or "instrument," was the collective title given to Aristotle's six logical works,[1] in which, among other things, he laid out his method of syllogisms and the rules for constructing valid arguments. In 1620, Francis Bacon (1561–1626) published his *Novum Organum* [*New Organon*],[2] wherein the term "organon" is taken to refer to a system of or method for philosophical or scientific investigation. Whereas Aristotle's method of discovery was based on deductions from syllogisms and Bacon's on induction from observed data, Leibniz proposed a form of logical calculus governed by a set of rules that would enable one to calculate with ideas or concepts (represented by signs or characters) using just pen and paper, much as mathematicians do with numbers, thereby generating new truths. This approach explains the other allusion to be found in the title of the following text, namely, to Raymond Llull's (1232–1316) *Great Art* [*Ars magna*], a logical system in which discs of paper, inscribed with letters that represent simple notions, are manipulated according to a set of defined, mechanical rules in order to reveal metaphysical and religious truths, specifically those of Christian doctrines, and thereby eliminate the errors of infidels and unify the three Abrahamic faiths. Although Leibniz was critical of some of the details of Llull's *Great Art*,[3] his own project for a general science was cut from the same cloth and shared some of the same aims,[4] though in addition to eliminating religious disagreements, Leibniz hoped his general science would also yield truths throughout a range of disciplines, including physics, jurisprudence, politics, and medicine.[5]

---

1. Namely: *Categories, On Interpretation, Prior Analytics, Posterior Analytics, Topics, Sophistical Refutations*.

2. Bacon (1620). English translation: Bacon (2000).

3. See, for example, Leibniz (2020, 153).

4. As Leibniz himself sometimes acknowledged, even going so far as to state in one text that Llull's *Great Art* was included within his general science; see A VI 4, 527.

5. See A VI 4, 5. English translation in Lewis (2021, 15).

In a suite of writings, Leibniz identified various components of his general science, including a general encyclopedia of facts already established and a formal language of thought—the universal characteristic—the terms of which would, when combined in accordance with a logical calculus, generate and determine new propositional truths. In the following text, however, Leibniz's concern is with a preliminary matter, namely the alphabet of human thought, which consists of an inventory of the most fundamental concepts out of which the rest of our concepts are formed and with which the calculations of the general science would be performed. Having concluded that there must be such fundamental concepts—concepts conceived through themselves rather than through something else—Leibniz first argues that there must be numerous things conceived through themselves, before then contemplating the alternative possibility that there may be only one thing conceived through itself, namely God, aside from whom there is nothing, or privation. Struck by the idea that all our thoughts may derive from God and nothing, Leibniz proceeds to draw a parallel with his binary system, in which all numbers derive from 1 and 0, and then illustrates this with a table of binary equivalents of decimal 1–16. Leibniz would later develop this thought about the metaphysical significance of the binary system, seeing in the expression of all numbers by 1s and 0s an analogy of the Christian doctrine of creation *ex nihilo* by God, an idea that would lead him to inform others of his discovery (see chapters 14–17, 19, and 22).

The dating of the text is based principally on the thematic links between its contents and that of other texts written between the summers of 1678 and 1679.[6] The editors of the ongoing critical edition of Leibniz's writings date it to March–April 1679, partly on account of the aforementioned thematic links to other writings and partly because they take it to have been composed shortly after "On the Binary Progression," which was written 15/25 March 1679 (see chapter 6).[7] However, while Leibniz's claim regarding "the immense advantages of this [binary] progression" in the following text suggests that by the time of writing, he had already developed the fundamentals of the binary system and identified its advantages, this may have occurred earlier than 15/25 March 1679, as is clear from chapters 2–4, which probably predate "On the Binary Progression." Therefore, for the following text, a tentative dating to the first half of 1679 seems appropriate.

On the Organon or Great Art of Thinking[8]

The highest form of man's *happiness* consists in his perfection increased as much as possible.[9]

Vigor, or the state of increased perfection, is as much above *health* as *illness* is below it. Indeed, *perfection* is a more excellent degree of health. Just as *illness* consists of an impaired functioning of the *faculties*, so *perfection* consists in a furtherance of one's power or faculties.

The foremost of human faculties is the *power of thinking*.

The power of thinking can be assisted either by bodily aids or by mental aids.

Bodily aids are bodies which are applied to the corporeal organs themselves, whereby torpor is dispelled, imagination strengthened, and the senses sharpened. But these are not to be discussed here.

The aids prescribed for the mind consist in certain ways of thinking, whereby other thoughts are made easier.

---

6. For example, A VI 4, 79–83 and 83–86 (both likely written summer–fall 1678), and A VI 4, 338–349 (written 15/25 June 1679).

7. See A VI 4, 156. "Critical edition" refers to the ongoing multiseries and multivolume edition of Leibniz's writings being edited by the Berlin-Brandenburgische Akademie der Wissenschaften [Berlin-Brandenburg Academy of Sciences], abbreviated throughout this book as "A."

8. thinking ▷ in Which Are Discussed the True Characteristic, the Kabbalah, Algebra, the Combinatorial Art, Universal Writing, the Language of Nature {1} The organon is the art of assisting reason by the imagination {2} The organon or great art of thinking is the art of assisting thinking ◁ *deleted*.

9. possible. ▷ The foremost of human perfections is the power of thinking. And to increase perfection is nothing other than to further one's power. Just as *illness* consists in impaired function, so vigor, that is, an increase of perfection, which is a certain more excellent degree of health ◁ *deleted*.

The greatest mental aid is if a few thoughts can be found from which infinite other thoughts may arise in order. Just as from the assumption of a few numbers, from 1 up to 10, all the other numbers can be derived in order.

Whatever is thought by us is either conceived through itself, or involves the concept of another.

Whatever is involved in the concept of another is in turn either conceived through itself, or involves the concept of another. And so on.

And so one must either proceed to infinity, or all thoughts are resolved into those which are conceived through themselves.

If nothing is conceived through itself, nothing will be conceived at all. For what is conceived only through others will be conceived in so far as those others are conceived, and so on in turn: and hence we shall be said to conceive something in actuality only when we happen upon those things which are conceived through themselves.

I shall illustrate the matter with an analogy. I give to you a hundred which are to be received from Titius; Titius will send you on to Caius; Caius to Maevius, but if you are continually sent on like this you will never be said to have received anything.[10]

It is necessary that those things which are perceived through themselves are numerous. For let there be a, which is conceived through b, or which involves b. I say that a itself necessarily involves not just b alone, but something else as well; for if it is conceived through b alone, then at any rate nothing else can be conceived in a which cannot be conceived in b, and so there will be no distinction between a and b, but this is contrary to the hypothesis, for we have assumed that a is conceived through another, namely b. It is therefore necessary that a is conceived through two things at least, for example, b and c.

Although the things which are conceived are infinite, it is nevertheless possible that there are only a few things which are conceived through themselves. For infinite things can be constructed through the combination of a few.

Indeed, this is not only possible but also credible or probable, for nature usually accomplishes as much as possible with the smallest possible number of assumptions, i.e., it operates in the simplest way.[11]

It could be the case that there is only one thing which is conceived through itself, evidently God himself,[12] and besides him there is nothing, or privation, which I shall show by a remarkable analogy. We generally count out numbers through a decimal progression, so that when we reach 10, we start again from 1, and I do not now dispute that doing this is convenient. Nevertheless I shall show that instead of this it is possible to use a binary progression, so that as soon as we reach 2 we start again from 1, in this way:

(0)	(1)	(2)	(3)	(4)	(5)	(6)	(7)	(8)
0	1	10	11	100	101	110	111	1000

(9)	(10)	(11)	(12)	(13)	(14)	(15)	(16)
1001	1010	1011	1100	1101	1110	1111	10000

---

10. Titius and Maevius were stock characters used by Justinian (482–565) in his *Digest* when illustrating hypothetical examples of Roman law. Caius (or Gaius) was a Roman jurist, cited often by Justinian. See Justinian (1998).

11. way. ▷ *The alphabet of human thoughts* is a catalog of those things which are conceived through themselves, and whose combination gives rise to the rest of our ideas. ◁ *deleted*.

12. Compare with Leibniz's statement in "On Forms, or, the Attributes of God," from the second half of April 1676: "An attribute of God is any simple form. There necessarily exists simple forms, i.e., perceptions *per se*. For there are things which are perceived through other things; if the latter were again always perceived through other things, then nothing would be perceived" (A VI 3, 514; English translation: DSR 69).

I will not now touch on the[13] immense advantages of this progression: it is sufficient to note how, with this wonderful method, all numbers may be expressed in this way by one and nothing.

Moreover, although there is no hope that in this life men can arrive at this hidden series of things, in which it is apparent how all things proceed from[14] pure being and nothing, it is nevertheless sufficient that the analysis of ideas is advanced as far as the demonstrations of truths require.

Every idea is perfectly analyzed only when it can be demonstrated *a priori* that it is possible. For if we give some definition, and from that it is not apparent that the idea we ascribe to the thing is possible, we cannot trust the demonstrations we have deduced from the definition, because if that idea happens to involve a contradiction it can be the case contradictories are true of it, and at the same time too, and thus our demonstrations will be useless. From this it is clear that definitions are not arbitrary, and this is a secret to which hardly anyone has paid sufficient attention.

Since, however, it is not in our power to perfectly demonstrate the possibility of things *a priori*, i.e., to analyze them down to God and nothing, it will be sufficient for us to reduce the vast majority of them to a certain few, the possibility of which can either be supposed and postulated, or proved by experience. All the lines of motion in the whole of geometry are thus reduced to just two motions, one in a straight line and the other in a circle. For with these two supposed it can be demonstrated that all other lines, for example, the parabola, the hyperbola, the conchoid, and the spiral, are possible. However, Euclid has not taught how a straight line is to be drawn and a circle described, but held that it was sufficient to postulate them. Yet by assuming space, body, a straight line and continuous motion, the possibility of a circle can also be demonstrated. Indeed even a straight line can be demonstrated by assuming space and body and continuous motion. However, what is to be thought of these three continuums seems to depend on a consideration of divine perfection. But geometry does not necessarily have to go this far. For even if there were no straight lines and circles in nature, nor could there be any, it will still be sufficient that there can be shapes which differ so little from straight and circular ones that the error is smaller than any that one cares to mention. This is sufficient for the certainty of a demonstration and equally for its uses. Moreover, it is easily demonstrated that there can be shapes of this kind, merely by granting this one thing: there are some lines.

It is difficult to have from the outset the complete definitions of these ideas (i.e., definitions showing the possibility of the thing *a priori*). In the meantime we shall employ nominal definitions of them; that is, we shall analyze the idea of one thing into other ideas, through which it can be conceived, even though we cannot proceed all the way to the first ideas. And this will be sufficient when it is apparent[15] from experience that the concept is possible. For example, we can define "fire" as a hot and bright vapor, and it is acceptable to define the "rainbow" as a colored bow in the clouds, for it is evident enough from experience that concepts of this kind are possible even if we cannot show from the outset their possibility *a priori*, by explaining their generation or cause.[16]

There are some things for which there are no nominal definitions. Thus, there are no nominal definitions of heat and light themselves, for whoever is ignorant of what is signified by the word "hot" cannot be otherwise helped than by presenting him with the thing being discussed or by referring to equivalent words in a language known to him, or by stirring up his memory of it with another method, if he has experienced heat before. Yet nobody doubts that there is some cause of heat which, if it were known perfectly, would certainly provide a definition of heat.

13. the ▷ wonderful advantages of this expression ◁ *deleted*.
14. from ▷ God ◁ *deleted*.
15. apparent ▷ that a concept of this sort is possible ◁ *deleted*.
16. cause. ▷ Fire is a hot and bright body ◁ *deleted*.

Binary Ancestral Calculations (early 1680s [?])

Manuscript:
M: LH 35, 3 B 5 Bl. 92. Draft. Latin.

It was not uncommon for mathematicians in early modernity to concern themselves with speculative mathematical thought experiments, such as determining how much space would be needed in Noah's Ark for every species of animal (for example, see Wilkins 1668, 162–168) or calculating the degrees of grace of the Virgin Mary (which, according to Schott 1658, 668, was 2^{256}). Leibniz was no exception: in one text, he sought to determine the number of possible words that could be formed from a thirty-two-letter alphabet (see LH 4, 5, 9 Bl. 3) and in another whether all the humans who had ever lived would be able to fit into the valley of Jehoshaphat, traditionally believed to be the site of the Last Judgment (see LGR 318–319).

In the following manuscript, Leibniz takes upon himself to calculate, using a mixture of binary and decimal arithmetic, the number of ancestors of each person who was alive at the time of writing. Along the way, Leibniz makes certain assumptions, draws certain conclusions, and commits several arithmetic errors—so many that we depart from our usual practice of trying to correct them and instead reproduce them, noting where they occur in his manuscript and summarizing a corrected calculation here. Leibniz assumes: (1) 3000 years have passed since the beginning of the tabulation, which (for reasons not stated) is taken to be the siege of Troy; (2) a generation is 30 years, so there have been 100 generations; and (3) 64 million souls have been alive at all times since the beginning of the tabulation, that is, approximately 2^{36}. On the basis of these assumptions, and carrying out the arithmetic correctly: (1) each person living today has 2^{100} ancestors (not all distinct) in generation zero, 3000 years ago, and (2) $\frac{2^{100}}{2^{36}} = 2^{64} \approx 2 \times 10^{19}$, so at least one of these ancestors appears at least 2×10^{19} times in generation zero of each living person's ancestry.

Leibniz's assumption that there were 64 million people alive at the time was far lower than other early modern estimates; for example, Riccioli (1672, 681) estimated the global population of the time at one billion, while Petty (1682, 23) calculated 320 million. (A modern estimate is in the neighborhood of 650 million.) Moreover, Leibniz's assumption that the global population had remained static for thousands of years was out of step with early modern beliefs that the population had increased considerably from thousands of years before. Yet it was one Leibniz was prepared to make elsewhere: in a couple of papers from the early 1680s concerned with life annuities, he made it an explicit assumption "that in a great multitude, the number of people in general, and even that of people of each age in particular, remains approximately the same as it was in the past year" (A IV 3, 464 and LH 34 Bl. 216r; English translation: Leibniz 1683). Since the following manuscript makes the same assumption, one that Leibniz appears to have abandoned in later writings,[1] it may have been written at around the same time. Accordingly, we tentatively date it to the early 1680s. Since the manuscript is undated and the watermark of the paper does not match that of any other in his Nachlaß, a more reliable date cannot be determined.

———————⊸•◖◗•⊸———————

Let us say a generation is 30 years. Let us also suppose that, in the 3000 years since the siege of Troy, the world has been full no less than now. Therefore, since from then to now there have been $3000 \div 30$ generations, that is, 100, and ancestors multiply in a double progression, evidently in a hundred generations 2^{100} ancestors will come forth, or to express it in binary, the number of ancestors will be expressed by a 1 with a hundred zeroes. Let us suppose that there have been 64 million men on the globe both now and 3000 years ago, that is, 2^6 million. Now a million is the sixth power of ten and ten is two times five. Therefore, a million is the sixth power of five multiplied by the sixth power of two.[2] Five expressed in binary is 101. The sixth power of that is 101 1111 0101 1110 0001,[3] a number which multiplied by 100 0000, that is, by 2^6, gives a million, but 64 produce 64 million, if a million is multiplied by 2^6, that is,[4] if 1011 1110 1011 1100 0001 is multiplied by 2^6, that is, by 1 0000 0000 0000,[5] it produces 1011 1110 1011 1100 0001 0000 0000 0000 for 64 million,[6] that is, for the number of men. But let us assume an even higher [number], that is, the next higher [term] of the double progression, that is, 2^{31}, or 100 0000 0000 0000 0000 0000 0000 0000,[7] which is the number of men expressed in binary.[8] The number of genealogical ancestors, which is 2^{100}, when divided by that, will yield 2^{100-31}, that is,[9] 2^{71}, or in binary, 1 with 71 zeroes. Therefore, at least some of the ancestors living 3000 years ago must occur 2^{71} times in the genealogy of a man living today. I say "at least," although all may be supposed to have contributed to him and to have contributed equally, for otherwise there will be those who have contributed much more often.

3000
<u>1700</u>
1300
19
<u>12</u>
31
100
<u>31</u>
79

---

1. For example, in the aforementioned short paper about whether all the humans who had ever lived would be able to fit into the valley of Jehoshaphat, written c. 1715, Leibniz makes very different assumptions, namely, that the global population at the time was 100 million and that it had increased markedly since the earliest humans (see LGR 318–319).

2. That is, $15625 \times 64 = 1,000,000$.

3. This is incorrect; the sixth power of $5 = 15,625$, which in binary is 11 1101 0000 1001. The binary number Leibniz comes up with is equivalent to 390,625 in decimal.

4. Actually, $1,000,000_{10} = 1111\,0100\,0010\,0100\,0000_2$ and $1011\,1110\,1011\,1100\,0001_2 = 781,249_{10}$.

5. This number is of course 2^{12}, not 2^6.

6. In fact, $1011\,1110\,1011\,1100\,0001\,0000\,0000\,0000_2 = 3,199,995,904_{10}$. In the manuscript, Leibniz has four 0s before the final 1, rather than 5 as called for if the multiplication is to be carried out as he states.

7. That is a 1 followed by 30 zeros, not 31, as it should be to represent 2^{31}.

8. $64,000,000,000 \approx 2^{36}$. Leibniz has perhaps lost track of a factor of 2^6.

9. Leibniz's arithmetic here is of course incorrect, as $100 - 31 = 69$, not 71. His arithmetic goes astray in the sum on the right too, as $100 - 31$ is 69, not 79.

12 Sedecimal on an Envelope (c. 1682–1685)

Manuscript:
M: LH 35, 3 B 5 Bl. 77. Draft. Latin.

When exploring ideas, Leibniz wrote on whatever he had to hand, and as our title indicates, the following text was literally written on the back of a used envelope (reproduced as this book's frontispiece). On the left-hand side and at the bottom, Leibniz experiments with a different form of notation for sedecimal digits from that which he had used before (see chapter 9). In this alternative, a concave-up semicircle denotes 0 and a concave-down semicircle denotes 1, with up to 4 bits chained together from left to right to represent 0–15 as sedecimal digits, so that $4 = 100_2 = \smile\frown$, for example.

On the right-hand side of the envelope, Leibniz writes out the sequence of square numbers in binary notation, noting the periodicity of the columns.[1] He had first noted the periodicity of columns of natural numbers expressed in binary in "On the Binary Progression" (chapter 6) and considered it further in December 1679, in a paper entitled "Summum calculi analytici fastigium" [The Highest Peak of Analytical Calculation], near the end of which he sought

> the progression of numbers 1, 3, 5, 7, etc. expressed in binary; this [progression] will be determined as far as possible therefrom because the difference [between the numbers] is 2, that is, 10. The progression of these is evident [from the table on the right]
>
> For while the digits of the first [column] are formed only of 1s, the digits of the second [column] are alternately 0 and 1, of the third alternately 00 and 11, of the fourth are alternately 0000 and 1111, the digits of the fifth are 0000 0000 and 1111 1111, and so on in a geometric progression. (LH 35, 13, 3 Bl. 21v; Zacher 1973, 223)

00001	1	①
00011	10	②
00101	10	②
00111	10	②
01001	10	②
01011	10	②
01101	10	②
01111	10	②
10001	10	②

The periodicity of binary sequences of natural numbers would be of enduring interest to Leibniz, who would return to it often in the following decades in his search for the rule that would enable the periods to be calculated or for tables to be generated simply by applying the rules for the periods. Indeed, a good proportion of his writings on binary are devoted to this endeavor. On this envelope, he indicates that the rule for the periodicity of square numbers might be found simply by extending the sequence of numbers, a strategy he did not pursue.

Although the ideas about sedecimal and the periodicity of binary sequences scrawled on the envelope were ones first treated in 1679 (periodicity in perfect squares and cubes is suggested in chapter 6), this text is of a later date. The watermark indicates that the paper was of a type manufactured in the Harz mountains in 1678, but since Leibniz, along with some of his correspondents who lived in that area, used that paper often between 1682 and 1685, we date the text accordingly.

---

1. Below that table he wrote, and then crossed out, a table of the first few ternary (base-3) numerals.

11
11

———————
11

101 ————— 11
101 ———————
101 1001
0   0
1   1 ———————
2   10 1010
3   11 11001 —————— 111
111
———————
111
0 ⌣   0 ⌣   :111
1 ⌒   1 ⌒   1100   111
2 ⌒⌣ 10 ⌒⌣   11   110001
3 ⌒⌒ 11 ⌒⌒   ———————
4 ⌒⌒⌣ 100 ⌒⌒⌣ 1100   10100
5 ⌒⌒⌒ 101 ⌒⌒⌒ 11   101
6 ⌒⌒⌣ 110 ⌒⌒⌣ ———————   ———————
7 ⌒⌒⌒ 111 ⌒⌒⌒ 100100   10100
8 ⌒⌒⌣ 100 ⌒⌒⌣   1010
9 ⌒⌒⌒ 1001 ⌒⌒⌒ ———————   ———————
10 ⌒⌒⌒ 1010   1001   1100100
11 ⌒⌒⌒ 1011   1001
12 ⌒⌒⌒ 1100   100100
13 ⌒⌒⌒ 1101   1010001
14 ⌒⌒⌒ 1110
15 ⌒⌒⌒ 1111
16   10000

Progression of
squares. They
[i.e., 0 and 1]
always alternate
in the end
[column]. The
penultimate
[column] is
always 0. In the
third [column],
1 and three 0s
always alternate.
In the fourth
column perhaps
only 0s follow
after a few
others, as
happens in the
periods of
decimals.

000	
001	1
100	4
1001	9
10000	16
11001	25
100100	36
110001	49
1000000	64
1010001	81
1100100	100

Proof by casting
[out] is useless
here because when
a 1 is cast out, no
remainder is ever
left over in that
which has been
cast out.

Hence at
the be- ⎰ 100
ginning, ⎱ 1001
one pair

Then   ⎰ ⎰ 10000
two   ⎱ 11001
pairs ⎰ 100100
       ⎱ 110001

1   2   4
8 16 32
   64 128
       8
       4
       1

3141592

Perhaps now the threes

The sequence of squares should be continued to reveal the rule

0 0
1 1
2 2
3 10
4 11
5 12
6 20
7 21
8 22
9 100

	0	u	simple
⌒	1 eins	u	inverted
⌒⌣	2 zwei	u	antecaudate
⌒⌒	3 drei	n	antecaudate
⌒⌒⌣	4 vier	w	antecaudate
⌒⌒⌒	5 funf	u	bicaudate
⌒⌒⌣	6 sechs	n	bicaudate
⌒⌒⌒	7 sieben	m	antecaudate
⌒⌒⌣	8 acht	iw	antecaudate
⌒⌒⌒	9 neun	w	bicaudate
⌒⌒⌒	10 ut	uu	antecaudate
⌒⌒⌒	11 re	un	antecaudate
⌒⌒⌣	12 mi	nu	antecaudate
⌒⌒⌒	13 fa	ni	[bicaudate]
⌒⌒⌒	14 sol	m	bicaudate
⌒⌒⌒	15 la	mi	antecaudate

In the two right-hand columns, Leibniz presents each of his sedecimal digits using one or two letters, generally extended by "tails"; some have a tail on the left-hand side alone (antecaudate) while others have tails on both sides (bicaudate). In some cases it takes some imagination, and shrinking the tails, to see the indicated shapes. For example, the symbol for 14, which Leibniz characterizes as "m bicaudate," might be pictured thus: ⌒⌒⌒, with the arcs of the "m" in the middle and diminutive arcs serving as tails fore and aft. It is likely that the discrepancy between Leibniz's drawings of the characters and some of his descriptions thereof is due to his drawing the characters first and only afterwards attempting to work out descriptions for them.

13 Remarks on Weigel (1694–mid-March 1695)

Manuscript:
M: LH 4, 1, 6 Bl. 12–15. Draft. Latin.

Transcription:
FDC: Leibniz 1857, 146–168.

On 20 June 1663, just days after completing his bachelor's degree, Leibniz enrolled for the summer semester at the University of Jena, where he attended the mathematical lectures of Erhard Weigel (1625–1699). He would maintain an intermittent correspondence with Weigel thereafter, as well as an enduring—if often critical—interest in his works. In 1673, Weigel published a short book, *Tetractys*, in which he outlined the quaternary (base-4) system and advocated for its adoption (see Weigel 1673). This work is sometimes thought to have inspired Leibniz to develop the binary system,[1] though there is no evidence of Leibniz having read it before the first half of 1683, when he sketched out brief notes on it.[2] The notes are mostly descriptive, with Leibniz content to summarize some of Weigel's claims, but at the end, he offered his own assessment: "I think the binary is best absolutely, the ternary in planes, and the octal in solids" (A VI 4, 1163).

Leibniz returns to the comparative merits of binary over the quaternary system in the following set of critical remarks on one of Weigel's later books (namely, Weigel 1687e). Much of the text concerns Weigel's proof for the existence of God, which Leibniz had also critiqued elsewhere (see, for example, Leibniz 1857, 168–170), but near the end, Leibniz turns to a discussion of binary, sparked by Weigel's advocacy of the quaternary system. Here Leibniz insists that if common practice were to be reformed, as Weigel had urged, then either the base-12 or his own sedecimal system would be preferable, since they allow for more convenient calculations than the quaternary. Leibniz allows that if theoretical investigation is the aim, then the quaternary should be considered preferable to the decimal system but also insists that binary is better still, since it goes back to first principles (i.e., 0 and 1).

The text is notable for containing the first mention of the analogy between the representation of all numbers by 1 and 0 and the theologically orthodox idea of creation *ex nihilo*, that is, the creation of all things out of nothing by (one) God. In this analogy, God is clearly intended as the analogue of 1 and nothingness the analogue of 0, and the fact that the binary system shows that all numbers depend upon and originate from 1 and 0 is taken to reflect the doctrine that all created things depend upon and originate from God and nothingness. To understand Leibniz's thinking here,

---

1. See, for example, Couturat (1901, 473), who suggests that Leibniz's invention of binary "had probably been suggested to him by his old master Weigel's *Tetractys*, published in 1673."
2. See A VI 4, 1162–1163. Leibniz also made notes on *Tetractyn*, the companion work by the Pythagorean Society (1672); see A VI 4, 1163–1164. Both sets of notes have been dated to the first half of 1683 by their watermarks.

note that in his metaphysics, all created beings contain the same perfections as God (namely, power, wisdom, goodness), albeit not to the same degree, because, qua created beings, they have limitations or privations. In this sense, all created beings originate both from God, the source or principle of their perfections, and from nothingness, the source or principle of their privations or imperfections. The analogy between binary and creation, which clearly builds on ideas found in an earlier text, "On the Organon or Great Art of Thinking" (see chapter 10), was one that Leibniz would employ often in subsequent writings when outlining the binary system (see, for example, chapters 19 and 22).

It is often claimed that the binary-creation analogy was the brainchild of Rudolph August, Duke of Brunswick and Lüneburg (1627–1704), with whom Leibniz discussed it in May 1696 (see chapter 14).[3] The claim is incorrect, since the analogy appears in the following text, which was written at least a year before the face-to-face discussion between Leibniz and the duke. Indeed, the manuscript of this text contains two watermarks that indicate a date range between 1692 and 1695. Internal evidence suggests a date near the end of this range: Leibniz's claim that his infinitesimal calculus is "now used by the most eminent men everywhere" would fit with a dating of 1694 or 1695, by which time he had become aware of its use by such luminaries as Christiaan Huygens (1629–1695) and Guillaume-François-Antoine de L'Hospital (1662–1704). However, the text cannot have been written any later than 28 February–19 March 1695, which is when one of Leibniz's correspondents, Justus Christoph Böhmer (1670–1732), wrote to Leibniz thanking him for letting him borrow it (see A I 11, 335).

I very much approve the admirable plans of our Weigel to instill useful notions[4] in the tender minds of children so that the perpetual practice of virtues may be added at the same time. And he is to be praised for his virtue and constancy, indeed for his zeal and charity, which, by making him scorn inadequate opinions,[5] breaks the ice, and what others desire by pointless vows he really strives after for God's glory and the public good. Certainly, if I can do something not only by expressing my approval in the public sphere but also by private exhortations to my friends and patrons so that they are inflamed by these excellent works, I shall never stop. Sometimes he uses analogies between mathematical and moral things; these are most elegant and apt to fix in minds the truths of each, so that when the opportunity is given they burst into action.

Yet I would wish that while he appropriately transfers the truths he has already discovered into practice by his teaching, at the same time he also not stop giving to the public his new discoveries, with which he increases our treasure, in order to save them from extinction. For both of these tasks should be considered as important by those to whom God has given the power to accomplish them. I also strive to contribute something in my own limited way, and I work continually on the art of inventing itself (which mathematicians call analysis) in order to extend it beyond its previous limits. For the science of quantity in general, or of estimation, as our most distinguished Weigel calls it, seems to me to be treated only halfway. For only the part that deals with finite quantities is known, but that leaves the more sublime part of general mathematics, namely the science of infinity, often necessary for the investigation of finite quantities themselves, and which I was perhaps the first to equip with analytical precepts, not least by proposing a new kind of calculus which is now used

---

3. The claim may have originated with Zacher (1973, 36–37), who does not mention "Remarks on Weigel" at all. Zacher appears to have credited the analogy to Duke Rudolph in part because of the duke's noted piety and in part because of a (noneyewitness) account that emerged in 1718, two years after Leibniz's death and twenty-two years after the discussion between Leibniz and the duke: "Leibniz shared [his] admirable discovery [i.e., the binary system] with the most gifted, most serene Prince Rudolph August, Duke of Brunswick and Lüneburg, of glorious memory; the most devout prince revered it as the most elegant symbol of creation, in which one divine essence composes all things from nothing" (Wideburg 1718, 21). Zacher's erroneous claim that the analogy was devised by Duke Rudolph has since been repeated often, for example, by Aiton (1985, 206) and Antognazza (2009, 357).
4. Reading "notitiis" (M) in place of "notionibus" (FDC).
5. Reading "judiciis" (M) in place of "judicibus" (FDC).

by the most eminent men everywhere. Huygens, who is an excellent judge of these things, himself admits that through this he obtained access to things which otherwise we would scarcely have been able to penetrate.[6] But this is said in passing, and I have actually noted it because I understood that our Weigel, who may not have sufficiently grasped my intention, feared some kind of excess in this investigation, as if we could be excessive in perfecting the art of inventing and in equipping ourselves more and more to obtain great truths.

In metaphysics too, I have put more than a little effort into discovering truths that are undoubtedly the most important of all and most valid for the true moral science, and all the more do I esteem the devotion of those who diligently consider these things. But I think that the art of demonstrating in metaphysics requires unusual[7] caution and care, greater even than in received mathematics. This is because in numbers and figures and the notions which depend on them, our mind is guided by a sort of Ariadnean thread of the imagination and of examples, and it has at hand comprobations[8] (which arithmeticians call proofs), which easily disprove paralogisms. But in metaphysics[9] (as has been bequeathed to us thus far) we lack these aids and are forced to supply, by the very rigor of reasoning, that which is lacking in comprobations or examinations. And so, although many eminent men have promised us demonstrations in metaphysics, I think for the most part they have been too readily led astray, and that we have very few and very rare demonstrations of this kind which are really[10] worthy of the title.

Weigel's opinion that divine conservation is the continuous creation of other things I consider as most true and not inconsistent with received doctrine, and I think it proceeds from the notion of depending, since one depends upon God in the first moment of one's existence no less than in all the others. Therefore creation and conservation differ only extrinsically in the connotation of a similar operation, either pre-existing or not, such that creation is conservation started, just as conservation is creation continued. Yet I admit I am somewhat at a loss with Weigel's method of proof and still desire something to complete and perfect the demonstration of divine existence deduced from it, which proceeds thus in *The Mirror of Virtues*, which he himself edited in Vienna.[11]

There is a God, that is, a creator of heaven and earth.[12] Demonstration: Because the existence of the essence of this world is reborn at each moment (by observation 1), and yet this cannot happen by means of its previous existence (axiom 2), which of course has already been extinguished, nor by means of the nothingness into which it has already fallen (axiom 3), it follows that besides the things of this world, which are of course transitory, there is something enduring, which at every instant produces the existences of the things of this world from nothing, that is, that there is a creator of heaven and earth.

This demonstration seems to me a little too short and too liberal in its assumptions to be able to satisfy the mind. It makes use of one observation and two axioms that I will lay out so that the force of the whole thing may be discerned. Observation 1 was: when one finds in anything an *actual essence* that is not changed but such as it was before, then the thing which is produced afterwards is the same as that which existed previously, whether the existence is the same or different.[13] Axiom 2 was: what is nothing consequently does nothing.[14] Axiom 3: the existence of this essence cannot be born spontaneously of itself, from nothing.[15] And I think the axioms are obvious, but the observation

---

6. See Huygens (1693) and Huygens's letters to Leibniz of 1 September 1691 and 17 September 1693; A III 5, 161 and 634–635.

7. unusual ▷ attention and care ◁ *deleted*. Note that none of Leibniz's deletions are recorded in transcription FDC.

8. Reading "comprobationibus" (M) in place of "comprobatione" (FDC).

9. metaphysics ▷ for the most part ◁ *deleted*.

10. Reading "omnino" (M) in place of "animo" (FDC).

11. Namely, Weigel (1687e).

12. The proof Leibniz goes on to outline here comes from Weigel (1687b); the conclusion is quoted from page 16.

13. Weigel (1687b, 8–9).

14. Weigel (1687b, 14).

15. Weigel (1687b, 15).

seems partly obscure and partly insufficient for proving what is deduced from it: obscure, because the distinction between *actual essence* and existence is not adequately explained, and insufficient for proving because from the fact that, with the essence remaining unchanged it is judged that the same thing remains, there is no way to infer that existence itself is reborn at each moment.

The meaning of the first observation cited is this: if existence changes but essence does not, the thing remains the same. But from this hyperbolic assertion how can one infer unconditionally that existence in fact always changes? Perhaps the printer, or some other person, committed a citation error when the *first observation* was cited, since the *second* [observation], being more to the point, should have been cited: it asserts that the existence of the world is continuously produced, moment by moment.[16] And certainly the first observation is scarcely necessary for the question, for whether or not things remain the same when existence has changed and the essence remains, it would be sufficient that the necessity of the creator can be inferred from the changed existence: it would be sufficient, I say, if it were adequately proved.

But it is apparent, if we like appropriate rigor in demonstrating, that a difficulty remains even after this correction. For adversaries could still doubt whether it is true that the existence of worldly things is produced anew at each moment, in the sense that the things themselves are annihilated and created at each moment. And I am astonished that among the principles of demonstrating we have erected one which itself needs demonstration, as in this lies the crux of the difficulty. It is indeed true that modes of existence are continuously renewed, by reason of time, place,[17] qualities, and circumstances. It is one thing for us to have been or to have existed today, another for us to have been yesterday, for it is one thing[18] for us to have been or to have existed in a garden and another to be in a house, one thing for us to have been healthy and another to be sick, and it can even be said[19] that our existence yesterday is one thing and today's is another, likewise our being in a garden or being in a house, being healthy or sick. But the change of these respective existences, or of these modes of being, does not prove the change of absolute existence, such that it follows that this[20] very thing is annihilated. Undoubtedly there are many respective existences, and even simultaneous, according to different respects; so while we were in the garden last summer, our mode of existing in summer can be conceived as different from our mode of existing in the garden, that is, temporal existence differs from local, even if by accident such a time coincides with such a place. And indeed, temporal existence flows perpetually by the force of its nature, while local, qualitative, or circumstantial existence sometimes changes and sometimes doesn't. But absolute existence is only one and the same, and not manifold like respective existence. Consequently, it was to be shown that the latter is changed by the lapse of time alone and therefore that the thing is annihilated and then created anew.

I see in some places in the scholia and discussions that the illustrious man puts forward some things which are not to be despised to prove this renewal of absolute existence, but since the crux of the difficulty consists in that, one would wish that this very thing had been put in the form of a demonstration, such as when he says (in the scholia of the demonstration, p19)[21] that time is nothing other than the very existence of things, that is, their actuality, so that even existence dies and is renewed by the lapse of time. But this assertion in turn lacks proof. Nor is it sufficient to say that opponents do not understand the nature of time, for even if[22] they do not understand it, it is up to the demonstrator to assist understanding by clear proofs. Nor does someone with doubts about promised arguments always have need for reasons for doubting, since when it is a matter of a complete demonstration it is sufficient for doubt that there is something not yet proved. And out of

---

16. Weigel (1687b, 10–11).
17. Reading "loci" (M) in place of "locis" (FDC).
18. Reading "aliud enim est" (M) in place of "aliud est" (FDC).
19. Reading "dici" (M) in place of "diu" (FDC).
20. Reading "hanc" (M) in place of "inde" (FDC).
21. Weigel (1687b, 19).
22. Reading "etsi" in place of "esto".

an abundance [of caution], some reason for doubt can profitably be raised, so that the demonstrator is given the opportunity to complete the demonstration more thoroughly. So,[23] in this case, if time was the same as the existence of things, it seems to follow that there will be as many times as there are things, or the existences of things,[24] and therefore those things which exist simultaneously do not exist at the same time, from which it seems possible to infer that there is the greatest distinction between time and the existence of things, and therefore that it is not yet established that absolute existence perishes by the lapse of time, such that a thing is annihilated.

Another doubt of no less importance still remains. For if we grant that the existence of the mundane things we see is perpetually perishing, and that therefore transitory things are continually created by an enduring thing, it does not follow that this creator is the creator of heaven and earth, and it follows even less that it is God. For an opponent will say that there can be different, indeed innumerable creators, that is, that there are different enduring beings for the different transient things being created. And a new demonstration will be needed to prove that there is only one common enduring being for all the transitory beings that are continuously produced. Indeed, there will be those who say that essences themselves always produce new existences, for essences themselves are not annihilated but endure, the contrary of which must be demonstrated.

I have mentioned these things not because I deny that the aforementioned things are most true, nor because I disapprove of the praiseworthy thoughts of an eminent man, but because I wish that what is missing is added to them so that they deserve to be called a mathematical demonstration. He, along with any fair critics, will recognize, I hope, that I have conducted myself openly, and that I am not seeking to cause annoyance or looking for difficulties where none exist. The Cartesians strive, after their master, to prove the continuous production of things, on the grounds that our future existence does not follow from our present existence. But this argument does not[25] resolve the matter either, for an opponent will say that the future generally follows from the present unless some impediment intervenes. Finally, some more recent Cartesians remove from things the power to act, as if God alone does everything and things are only occasional causes,[26] a view that seems to please our Weigel somewhat, though I don't support it at all. For although I admit that in metaphysical rigor there is no influence of one created substance on another, I nevertheless think that creatures act and have the power to act, things which have been explained more abundantly elsewhere.[27]

Otherwise, I find many things clearly and elegantly said in the *Mirror of Vienna*, for example that the supreme excess of charity is called apotheosis.[28] Indeed, it is very true that men sometimes love creatures to the point of making them their gods, just as Holy Scripture speaks of those who made their god of their belly.[29] Similarly, servility is rightly considered as an excess of humanity, and unanimity as an excess of concord,[30] and he gives a good many of these most excellent notions.

He quite rightly insists that all strict reasoning is a kind of computation, on which I shall give new thoughts (God willing) that will be useful for the cultivation of this computation, beyond those things which may be easily suspected. In the meantime, we shall please ourselves with the elegant analogies of our Weigel; for example, he agreeably observes that, when acting according to self-sufficiency, just as it does not matter whether the same fraction or proportion is expressed by large or small numbers, so it does not matter to he who enjoys a contented mind whether he has reached the summit of this earthly bliss by great wealth and armies of slaves or by resources that are accessible and ready to hand.

---

23. Reading "Sic" (M) in place of "Nec" (FDC).

24. Reading "res seu rerum existentias" (M) in place of "res rerum existentias" (FDC).

25. Reading "nec" (M) in place of "vel" (FDC).

26. Leibniz is referring here to the philosopher and theologian Nicolas Malebranche (1638–1715), who argued in numerous writings that God alone was a true cause.

27. Leibniz may be referring here to his essay "Whether the Essence of Body Consists in Extension," published in June 1691 (Dutens II, 1, 234–236; English translation: SLT 123–125).

28. Weigel (1687e, 5).

29. See Philippians 3:19.

30. Weigel (1687e, 6).

Penultimate manuscript page of "Remarks on Weigel" (LH 4, 1, 6 Bl. 15v).

As for the rest, I (although not an eyewitness) add my calculus to those things the eyewitness J. M. P. P., whoever he was, has called Aretologistica in accordance with Weigel's method of teaching.[31] And indeed, it was good to note this briefly on the *Mirror of Vienna*, especially since I had not read this book before, nor the other one, *Aretologistica*,[32] which extends even further into metaphysical matters, although on the basis of the abstract he seems to sketch out notions more than explain them; but it is above all in practice itself that he elegantly shows us the use of the arithmetic[33] of virtues.

As far as quaternary arithmetic is concerned,[34] I think that if anything were to be changed in practice, it would be to use the duodecimal or sedecimal instead of the decimal, for the larger the numbers used by a progression, the more convenient the calculation (provided that the fundamental Pythagorean tables are committed to memory). But since these things accepted by all may be changed only with difficulty, we can be satisfied with the Cato we have in the case of the usual calculations of decimal arithmetic.[35] As for theory and the discovery of extraordinary truths in arithmetic, which may likewise be very useful for practice, I think that not only should the quaternary be preferred to the decimal, but also that preferable to the *quaternary* in turn is the *binary*, which is the most perfect of all, nor does it suppose anything but instead completely resolves numbers.

According to the binary, all numbers are expressed by the digits 0 and 1 alone, by unity and nothing, a remarkable analogy of the origin of created things[36] from God and nothing, with creatures having their own perfections from pure, positive actuality, or God, and their imperfections or limits from the negative, or nothing. We shall give a specimen of this expression:

Decimal	Binary
1	0
1	1
2	10
2	11
4	100
5	101
6	110
7	111
8	1000
9	1001
10	1010
11	1011
12	1100
13	1101
14	1110
15	1111
16 etc.	10000 etc.

Example of
Binary Calculation

Addition

$$
\begin{array}{rr}
+5 & +101 \\
+7 & 111 \\
\hline
12 & 1100 \\
\end{array}
$$

Multiplication

$$
\begin{array}{rr}
5 & 101 \\
3 & 11 \\
\hline
 & 101 \\
 & 101 \\
\hline
15 & 1111 \\
\end{array}
$$

From this it deserves to be noted that the binary expressions of numbers proceed by a certain law to infinity, which cannot happen in[37] decimal and other systems, since only in the binary do the digits express their dependence on or origin from unity and nothing, that is, from first principles, and thus express the deepest nature of number. From this it also follows that all numerical theorems must

31. Leibniz is referring here to Weigel (1687c).

32. Leibniz is referring here to Weigel (1687a).

33. Reading "arithmeticae" (M) in place of "arithmetica" (FDC).

34. Weigel outlines his system of quaternary arithmetic (base 4) in Weigel (1687d, 350–452).

35. Leibniz here adapts a favorite proverb of Caesar Augustus, *contenti simus hoc Catone* [let's be satisfied with the Cato we have], used to encourage others to put up with the present situation. See Suetonius (1914, 254).

36. things ▷ , from the divine unity and nothing, that is, negative or limit ◁ *deleted*.

37. in ▷ binary ◁ *deleted*.

be apparent in the sequences of numbers because of these very digits, and innumerable secrets and even truths of the utmost utility for practical calculation can be elicited from this. It is also worthy of note that this binary expression presents to us in numerals what weighers of money and similar craftsman have for a long time shown us in weights: they showed how, by using[38] double geometric progression,[39] many other numbers are formed with the fewest weights. Thus, with five weights of values 1, 2, 4, 8, 16, combined in various ways in one pan of a pair of scales, they can form all the weights from 1 to 31. And with six weights of 1, 2, 4, 8, 16, 32, they can form all weights from 1 to 63. The demonstration of this is evident at first sight from the binary representation. Therefore, they reckon for many reasons that this [i.e., the binary system] is preferable to the quaternary for meditating upon the secrets of numbers and increasing knowledge.

---

38. using ▷ arithmetic progression ◁ *deleted*.
39. Reading "progressionis" (M) in place of "progressivis" (FDC).

14 Leibniz to Duke Rudolph August (7/17–8/18 May 1696)

Manuscripts:
M1: LBr F 17 Bl. 17. Discarded fair copy. German.
M2: Gotha A 449 Bl. 87. Draft. German.

Transcriptions:
LV: Loosen and Vonessen 1968, 138 (following M1).
Z: Zacher 1973, 235 (following M2).
A1: A I 12, 65 (following M1).
A2: A I 12, 65 (following M2).

In the years immediately following his development of binary arithmetic, Leibniz made little attempt to publicize it, aside from a few very brief remarks to a handful of his mathematical acquaintances in 1680 and 1682.[1] On the basis of his surviving papers, it appears he then kept his invention to himself for more than a decade. However, as the following letter attests, this changed in May 1696. Near the end of a sojourn in Wolfenbüttel that lasted from 23 April to 21 May 1696, Leibniz had occasion to discuss the binary system in person with Rudolph August, Duke of Brunswick and Lüneburg, for whom he had worked as director of Wolfenbüttel library since 14 January 1691. Following the discussion, and apparently at the duke's request, Leibniz wrote the following letter as well as a longer paper with which it was enclosed (see chapters 15 and 16). Two drafts of Leibniz's letter are extant, the first written on 7 May 1696, the second a day later. The dispatched version of the letter has now been lost, so it is unclear whether it differed much if at all from the second draft. The two drafts are very similar in terms of length and content, and it is clear that Leibniz intended the letter as little more than a covering note for the longer paper with which it was enclosed.

The draft letters are notable for their focus on the binary-creation analogy first outlined in Leibniz's "Remarks on Weigel" (see chapter 13) and foreshadowed in his "On the Organon or Great Art of Thinking" (see chapter 10). The analogy clearly won the duke's admiration, and Leibniz would return to it when writing to the duke at the start of the following year (see chapter 17). Curiously, Leibniz does not explicitly claim the analogy as his own in any of his letters to or writings for the duke, or indeed in most of the subsequent writings for others in which the analogy was employed. However, in a letter to the mathematician Johann Christian Schulenburg (1668–1732) of 29 March 1698, Leibniz does indicate that the idea was his:

> And this is the origin of things from God and nothing, positive and privative, perfection and imperfection, value and limits, active and passive, form (i.e., entelechy,[2] endeavour, energy) and matter or mass which is in itself inactive, except insofar as it has resistance. I have

---

1. See the introduction, this volume, for the specifics.
2. Introduced by Aristotle, the term "entelechy" has been used in different ways by different philosophers. Leibniz often used the term to refer to the principle of action or power of actualization possessed by all individual substances, such as human beings or animals. For Leibniz, then, the entelechy was the source of a substance's own actions, external or internal, i.e., its active force.

made those things clear to some extent by the origin of numbers from 0 and 1, which I have observed is the most beautiful symbol of the continuous creation of things from nothing, and of their dependence on God. (A II 3, 427; English translation: SLT 39)[3]

Although the circumstances of the discussion between Leibniz and the duke are unknown, a remark Leibniz later made to another correspondent, the Jesuit missionary Claudio Filippo Grimaldi (1638–1712), suggests that he had brought his binary system to the duke's attention because it provided suitable material for pious thoughts (presumably on account of the analogy between binary and creation *ex nihilo*):

For a while now I had this thing [i.e., the binary system], like so many others, buried among my papers, when by chance I happened to mention it to the most serene Prince Rudolph August of Brunswick-Lüneburg, Duke of the Wolfenbüttel line, as something appealing and useful for pious meditations, in which the exceptional prince was accustomed to take pleasure. (pp. 109–110 below)

[M1]

Most Noble Duke, Most Gracious Prince and Lord

Since Your Serene Highness graciously wished to be reminded of the meditation I mentioned concerning numbers, which represent creation, as it were in a mirror, or the origin of things from God and nothing else, and which therefore serve to elucidate to some extent such mysteries unknown to pagans (who set matter and God side by side), through the light of reason, I wanted all the more keenly to sketch this out briefly in the attached paper, since I am aware that Your Serene Highness judges everything most commendably for the glory of God and knows that the greatest use of human sciences is when one notices the radiance of divine wonders therein. I remain in constant devotion,

Your Serene Highness   most humble servant
Wolfenbüttel, 7 May 1696   Gottfried Wilhelm Leibniz

[M2]

Most Serene Duke, Most Gracious Prince and Lord

Since Your Serene Highness graciously wished to be reminded of the meditation I mentioned concerning numbers, I should not fail to sketch it out in the attached paper, all the more since I am aware that Your Serene Highness judges everything most commendably for the glory of God and knows that the greatest use of human sciences is when one notices the radiance of divine wonders therein. Such as how numbers represent creation, as it were in a mirror, or the origin of things from God and nothing else, and thus serve to elucidate to some extent such mysteries unknown to pagans, through the light of reason, and to indicate how wrong pagan philosophers were to set matter side by side with God as co-fountainhead, so to speak. I remain in constant devotion,

Your Serene Highness   most humbly, G. W. L.
Wolffenbüttel, 8 May 1696

---

3. Here Leibniz puts a twist on the binary-creation analogy by suggesting an analogy between binary and the medieval theological idea of continued creation, which holds that as the universe of created beings has no power to maintain itself in existence, it is therefore conserved in existence by God's continually creating it anew at each moment.

15 A Wonderful Expression of All Numbers by 1 and 0 Representing the Origin of Things from God and Nothing, or the Mystery of Creation (7/17 May 1696)

Manuscripts:
M1: Gotha A 449 Bl. 86–87. Draft. Latin.
M2: LH 35, 3 B 5 Bl. 93. Partial fair copy of M1, in the hand of Leibniz's amanuensis. Latin.

Transcription:
Z: Zacher 1973, 225–228 (following M1).

There exist two distinct versions of Leibniz's paper intended as a supplement to his letter to Duke Rudolph (see chapter 14), one in Latin (translated here), the other in German (for which see chapter 16). The Latin version is likely the earlier of the two and was perhaps drafted on 7/17 May 1696, the date Leibniz wrote the first of his two draft letters to the duke. Two manuscripts of the Latin version of the paper survive: Leibniz's own draft and a partial fair copy in the hand of his amanuensis, left incomplete possibly because the amanuensis had faithfully reproduced the errors Leibniz had made in his draft (and which were subsequently corrected). The Latin version was never sent to the duke; Leibniz subsequently decided to revise and redraft the paper in German (see chapter 16).

The title of the paper highlights the analogy between binary numeration and the creation of all things from nothing by God, which Leibniz and the duke had discussed during their face-to-face conversation earlier that month. Despite the prominence given to the binary-creation analogy in the title, it plays a relatively small part in the paper itself, which is otherwise devoted to outlining the representation of numbers in the binary system, the repetitive bit patterns in the sequence of consecutive numbers, and attempts at the basic operation tables for the four arithmetic operations acting on the operands 0 and 1.

A Wonderful Expression of All Numbers by 1 and 0 Representing the Origin of Things from God and Nothing, or the Mystery of Creation

[M1]

Just as in the common method for writing numbers we use only ten digits: 0, 1, 2, 3, 4, 5, 6, 7, 8, 9, and when we get to *ten* we begin again from 1, discarding only 0,

And just as some people begin again when *four* is reached and use only the digits 0, 1, 2, 3,

So I have concluded that the simplest way, and the one most in keeping with the nature and origin [of numbers], is rather to begin again when we reach *two*; thus there is need for only two digits, 0 and 1.

The foundation is this: just as usually

$\qquad$ 10 signifies[1] ten
$\qquad$ 100 signifies ten times ten
$\qquad$ 1000 signifies ten times ten times ten
$\qquad$ 10000 ten times ten times ten times ten
$\qquad$ 100000 ten times ten times ten times ten times ten

So in the new notation
$\qquad$ 10 signifies two
$\qquad$ 100 signifies four, or two times two
$\qquad$ 1000 signifies eight, or two times two times two
$\qquad$ 10000 signifies sixteen, or two times two times two times two
100000 signifies[2] thirty two, or two times two times two times two times two.
And so on.

Hence all other numbers can now be constructed from these alone. For example:

3 is constructed from two	10	2
one	1	1
Therefore 3 is	11	3

6 is constructed from four	100	4
two	10	2
Therefore 6 is	110	6

7 is constructed from four	100	4
two	10	2
one	1	1
Therefore 7 is	111	7

1	1
10	2
11	3
100	4
101	5
110	6
111	7
1000	8
1001	9
1010	10
1011	11
1100	12
1101	13
1110	14
1111	15
10000	16
10001	17
10010	18
10011	19
10100	20
10101	21
10110	22
10111	23
11000	24
11001	25
11010	26
11011	27
11100	28
11101	29
11110	30
11111	31
100000	32
and so on	

This most perfect arrangement of numbers proceeds by the law of geometric progression. For in the first column is contained a simple alternation: 0.1.0.1.0.1.0.1. etc. And in the second column through twos:
0011 0011 0011 etc.
And in the third column, through two by two, i.e. fours:
0000 1111 0000 1111 etc.
In the fourth column, by eights:
0000 0000 1111 1111 0000 0000 1111 1111 etc.

---

1. signifies ▷ ten, and 100 signifies a hundred times a hundred, and 1000 [signifies] a hundred times a hundred times a hundred, and 10000 [signifies] a hundred times a hundred times a hundred times a hundred, etc. ◁ *deleted*. The errors here were repeated by Leibniz's amanuensis in M2.

2. M2 ends here.

Mira Numerorum omnium expressio per 1. Et o. repræsentans rerum originem ex deo Et Nihilo, seu Mysterium creationis

Quemadmodum in communi Notandi Numeros ratione, non nisi decem adhibemus characteres : 0, 1, 2, 3, 4, 5, 6, 7, 8, 9, Et ubi ad decem progrediendum est rursum incipimus ab 1, adjiciendo tantum.

Et quemadmodum aliqui rursus incipiunt ubi ad quatuor perventum est et adhibent tantum Characteres 0, 1, 2, 3,

Ita simplicissimum judicavi rursus incipere ubi pervenitur ad duo; ita duobus tantum opus est characteribus o Et 1.

Fundamentum hoc est quemadmodum vulgo 1.0. significat decem, Et 100 significat centies centum, et 1000 centies centies centum Et 10000 centies, centies, centum Etc.

Ita in nova notatione

10	Significat duo
100	Significat quatuor Seubis duo
1000	Significat octo Seubis bis duo
10000	Significat Sedecim Seubis bis duo
100000	Significat

First manuscript page of the partial fair copy of "A Wonderful Expression of All Numbers by 1 and 0" (LH 35, 3 B 5 Bl. 93).

But in the usual ordering of numbers, no fixed progression of digits is apparent, except after very long intervals.

From which it follows that, according to the new notation, all things proceed in an orderly way and by the most beautiful harmony.

This is also useful for the discovery of extraordinary and useful truths in numbers, which till now have been unknown.

From this we also learn that this notation is the true way of explaining things when they are brought all the way back to their origin, which in numbers is 1 and 0.

No prior knowledge is required to complete the four kinds [of arithmetic operation], unlike in the common arithmetic.

For one needs to know only these following things, which are manifest in themselves:

For addition, 0 plus 0 is 0. And 0 plus 1 is 1, and 1 plus 1 is 10.

For subtraction, 0 minus 0 is 0, 1 minus 0 is 1, and 1 minus 1 is 0.

For multiplication, only this: 0 times 0 is 0, 0 times 1 is 0, and 1 times 0 is 0, and 1 times 1 is 1.

For division, I have 1 in 1 once, and I have 0 in 0 once.

Examples
of addition

8	...	1000
24	...	.11000
32	...	100000

3	...	11
7	...	111
15	...	.11!1
25	...	11001

Examples
of subtraction

+27	...	11011
-19	...	10011
8	...	1000

+32	...	100000
-19	...	10011
13	...	1101

Examples
of multiplication

3	...	11
3	...	11
		11
		11
9	...	1001

9	...	1001
3	...	11
		1001
		1001
27	...	11011

11	...	1011
3	...	11
		1011
		1011
33	...	100001

Examples of division

33...100001
3...11
11...1011

32...100000
7...111
4...100

And just as in multiplication there is no need for the Pythagorean table, which is commonly called *Einmaleins* [once one], so in division there is no need to try to find the quotient, since there cannot be any except 1 and 0.

Meanwhile, this method of calculating was not invented so that it may be transferred to common use, but so that it may serve for meditation. For it discloses the wonderful secrets and shortcuts of numbers, which are also useful afterwards in the common calculus.

But since the principal use is for human knowledge, such that this knowledge leads us to God by the hand, as it were, and reveals the wonderful vestiges of the divine author, I venture to say that scarcely anything discovered in the mathematical sciences better raises the soul to God.

For since among the principal wonders of God is the mystery of creation, unknown to pagans (who put matter to God as if they were side by side), and the essences of things are like numbers, as philosophers have wisely said, therefore nature itself here shows us, behind the wonderful image in numbers, that all things come forth not from God and matter, but from God and nothing.

16 Wonderful Origin of All Numbers from 1 and 0, Which Serves as a Beautiful Representation of the Mystery of Creation, since Everything Arises from God and Nothing Else (8/18 May 1696)

Manuscript:
M: Gotha A 448–449, Bl. 84–85. Draft. German.

Transcriptions:
Z: Zacher 1973, 229–234.
A: A I 12, 66–72.

The following text is likely the second draft of Leibniz's paper intended as a supplement to his letter to Duke Rudolph of 8/18 May 1696 (see chapter 14). Unlike the first draft of the paper (for which see chapter 15), which was written in Latin, the second draft was written in German, the language Leibniz and the duke used in their correspondence. Only Leibniz's own draft of the German version is still extant; it is likely a fair copy was made and dispatched to the duke, but no such copy has survived. The paper is significant for it being the first time Leibniz formally presented a detailed account of his binary system to someone else.

There are various differences between the Latin and German versions of the paper. The German version is slightly revised and expanded, uses different examples to illustrate the arithmetic operations in binary, and opens with Leibniz's assurance that he did not intend binary to replace the decimal system in everyday use, a point repeated later on in the paper, perhaps to assuage a concern raised by the duke in their face-to-face discussion. The two versions of the paper also bear different but related titles. The German version of the paper also features four Latin epigrams written at the top of the first page, apparently as an afterthought. The epigrams are as follows:

"Essentiae rerum sunt sicut numeri" [The essences of things are like numbers], a phrase Leibniz sometimes uses to mean that just as no two numbers are the same, so no two essences are identical either,[1] though in the context of the binary-creation analogy, his point is more likely to be that just as all numbers derive from the first principles of 0 and 1, so also all essences derive from the first principles of God and nothingness.[2]

"Cuncta Deus numeris" [God [does] all things by numbers], which is likely derived from Wisdom 11.20: "you have set all things in right order by proportion: by measure, by number, and by weight."

---

1. "The essences of things are like numbers. Just as two numbers are not equal to each other, so two essences are not equally perfect." A VI 4, 1352.

2. This appears to be Leibniz's meaning when he writes elsewhere, "All creatures are from God and nothingness. Their self-being is from God, their non-being from nothingness. This is also borne out in numbers in a wonderful way, and the essences of things are like numbers. No creature can be without non-being, or else it would be God. Even angels and saints must have it." LGR 81, cf. 277–278.

"Unum necessarium" [One thing is necessary], which is a quotation from the Vulgate version of Luke 10.42.[3]

"Numero Deus impare gaudet" [God delights in an odd number], a quotation from Virgil's *Eclogues* 8.75.

All four epigrams, which are used elsewhere in Leibniz's work,[4] are here clearly intended to provide scriptural and literary support for the study of the binary system.

Despite the duke's enthusiasm for the binary system, neither this paper nor the covering letter elicited a response from him.

——————◦○◦◦————

Wonderful Origin of All Numbers from 1 and 0, Which Serves as a Beautiful Representation of the Mystery of Creation, since Everything Arises from God and Nothing Else

The essences of things are like numbers
God [does] all things by numbers
One thing is necessary, 1, 0, nothing but one
God delights in an odd number (in accordance with my cyclometric arithmetic, by 1, 3, 5, 7, 9, 11, etc.)[5]

First and foremost, it should be remembered that this way of calculating is not meant to be introduced into common usage, but rather serves for the consideration of both the nature of numbers itself and many fruitful and useful properties so concealed therein, as well as the wonderful example of creation which can be found within.

Just as one does not need more than ten basic digits in the common method of counting, namely 0, 1, 2, 3, 4, 5, 6, 7, 8, 9, and so as soon as one comes to ten when counting on paper one starts again from 1 and adds a 0, and thus ten is denoted by 10;[6]

And just as some learned people, instead of the progression with ten, use the progression with four, in such a way that they need only four basic digits, namely 0, 1, 2, 3, and thus start again at four, and therefore in place of the digit 4 in such a calculation they put 10;[7]

So I considered the most natural, most original, and simplest thing to do was to use the very first progression, namely with 2, that is, after doubling, one starts again from the beginning, because the form (which seems strange) necessarily implies that one may write all the numbers without needing any other basic digits than 0 and 1, and instead of 2 one writes 10.

---

3. "Et respondens dixit illi Dominus: Martha, Martha, sollicita es, et turbaris erga plurima, porro unum est necessarium" (Luke 10.41–42). The English Standard Version renders this as: "But the Lord answered her, 'Martha, Martha, you are anxious and troubled about many things, but one thing is necessary.'"

4. For example, "The essences of things are like numbers" is a phrase Leibniz first uses in his bachelor's thesis of 1663, *Disputatio metaphysica de principio individui* [*Metaphysical Disputation on the Principle of Individuation*] (A VI 1, 19), and can be found in various other writings, such as two short texts on metaphysics from the late 1670s (A VI 4, 1352 and 1389) and a text on mystical theology from the mid-1690s (LGR 81). The expression has Scholastic roots; see Pasini (2016, 326–327). A variation of the passage from Wisdom 11.20 can be found in a work of metaphysics from 1714, "The Principles of Nature and Grace" (Strickland 2014, 276). "Unum necessarium" is used in a dialogue on the problem of evil from 1695 (LGR 277). "God delights in an odd number" is used in a table in a published article announcing his formula for the quadrature of the circle from 1682 (Leibniz 1682, 43).

5. Cyclometry is the measurement of circles. Leibniz is referring here to his formula for the circle of diameter 1,
$\frac{\pi}{4} = \frac{1}{1} - \frac{1}{3} + \frac{1}{5} \dots$.

6. 10; ▷ ¶ And just as some learned people, instead of the progression with ten (which probably stemmed from our having 10 fingers), suggested other progressions, sometimes greater, as with 12, sometimes smaller, as with 4. And the latter ◁ *deleted*. With regard to base 4, Leibniz is likely referring to Weigel (1673), and with regard to base 12, Leibniz is probably referring to Blaise Pascal, who had briefly entertained the idea in Pascal (1665, 47–48).

7. Leibniz is likely referring here to Weigel (1673).

Progressions have the *base digits*	so 10 means
Decimal …0, 1, 2, 3, 4, 5, 6, 7, 8, 9	…ten
Quaternary …0, 1, 2, 3	…four
Binary …0, 1	…two

It also follows from this that if 10 is the same as 2, then 10 times 10, or 100, in this new notation must mean the same as two times two, or 4, and 1000 the same as two times 4, or 8, and so on.

According to the common method:

10	is ten
100	ten times ten, or a hundred
1000	ten times a hundred, or a thousand
10000	ten times a thousand, or ten thousand
100000	ten times ten thousand, or a hundred thousand
etc.	etc. etc.

According to the new notation:

10	is two
100	twice two, or four
1000	twice four, or eight
10000	twice eight, or sixteen
100000	twice sixteen, or thirty-two
etc.	etc. etc.

And this is the whole basis for the following notation, as one can justify or try out such a thing through addition or combining, because once one accepts that when arriving at two one starts again at the beginning, and thus designates zero with a 0, one with a 1, and two with 10, one can lay out the following example of addition thus:

Namely, 2 is denoted by	1 0	‖	2
add 1	1	‖	1
0 and 1 is 1, yields A	1 1	‖	3
1 is 1, yields B	BA	‖	

So 3 must be designated by	1 1	‖	3
add 1	1	‖	1
	1 0 0	‖	4
	CBA	‖	

Such addition is carried out thus:

1 and 1 is 10, write 0 underneath A and keep 1 in mind, or denote it with a dot under the following column, so that it is the same as if 1 was written there. Then, in the following or next column there appears 1 again along with a dot, that is, 1, which makes 10, so write 0 for B and keep 1 in mind, or because of this, mark a dot below the following column; and if there is nothing other than this dot or 1, one writes 1 under C, moreover

Table of numbers continuable ad libitum[8]

A one is necessary

1	1
10	2
11	3
100	4
101	5
110	6
111	7
1000	8
1001	9
1010	10
1011	11
1100	12
1101	13
1110	14
1111	15
10000	16
10001	17
10010	18
10011	19
10100	20
10101	21
10110	22
10111	23
11000	24
11001	25
11010	26
11011	27
11100	28
11101	29
11110	30
11111	31
100000	32

---

8. That is, "as you desire." To the right of the table of the numbers, Leibniz wrote and then deleted the following: "Examples of the four kinds [of arithmetic operation]

	1000	8	11 3
Addition:	11000	24	111 7
	100000	32	1111 15
			11001 25

Subtraction:"

$$
\begin{array}{c|c}
100 & 4 \\
\hline
1 & 1 \\
\hline
101 & 5 \\
\hline
.\,1 & 1 \\
\hline
110 & 6 \\
\hline
1 & 1 \\
\hline
111 & 7 \\
\hline
\ldots 1 & 1 \\
\hline
1000 & 8
\end{array}
$$

Whereas no order is apparent in the digits according to the common notation, one sees here instead a beautiful and completely perfect order and harmony in the progression of all numbers, namely in the first column one finds alternately 010101 etc., in the second column one finds 0011 0011 0011 etc., in the third column 0000 1111 0000 1111 etc., in the fourth 0000 0000 1111 1111 0000 0000 1111 1111 etc., and so the alternation is at first between one 0 and one 1, then between two 0s and two 1s, then between four 0s and four 1s, and then between eight 0s and eight 1s etc., so always in a twofold or doubling progression.

It now follows from this that, when following this new notation, there must come forth, in an orderly way, many beautiful properties in these numbers which are not easy to find when following the common method, because where the basic digits are orderly, there must also be order in all that flows therefrom.

With this example one also sees that there is a beautiful order in all things in the whole world, as soon as one arrives at their proper origin, namely 0 and 1. One and nothing else.

To carry out the four kinds [of arithmetic operation] one does not need anything in advance, as one does in the common method, since one must know certain tables by heart or have them before oneself, such as the "Eins und Eins" [one-and-one] for adding, and the "Einmaleins" [one-times-one] for multiplying;[9] for example, one must know that 7 and 8 make 15, and that 7 times 8 make 56, and the like. But here I don't need to know anything other than 1 and 1 makes 10, and 1 times 1 makes 1. So the whole "one-times-one" here consists purely of the first words, one times one is one, and nothing else is needed. Hence what is elsewhere presupposed, or laid down as a foundation, is here found by itself.

Otherwise, one uses throughout the method and advantage of the common calculus, as in adding a 0 at the end when multiplying and dividing. For example, when multiplying 101 by 10, I can just write 1010.

And because the punctuation under the columns serves here for addition, one can represent 2 with a dot under the first column along, and 4 with a dot under the second, 8 with a dot under the third, and so on.

Examples of the four kinds [of arithmetic operation]

Addition				Subtraction			
1001	9	11	3	+11011	+27	+100000	+32
11000	24	111	7	−10011	−19	−10011	−19
..		1111	15	1000	8	1101	13
100001	33	::.					
		11001	25				

---

9. In Germany, the "Eins und Eins" and "Einmaleins" are the names of addition and multiplication tables that schoolchildren have to learn by heart.

Multiplication   Division

```
  11│3        1001│ 9            ̸1̸1̸1                      1
  11│3          11│ 3       33│100001 ⟌1011       32│1̸0̸0̸0̸00 ⟌100 100⁄111
  ───          ────          3│ 1̸1̸1̸1̸1              7│ 1̸1̸1̸1̸1
  11           1001                 ̸1̸1̸1                    ̸1̸1̸1
  11           1001          11│1011                       1
   ..          ─────
 ─────        11011│27
 1001│9
```

$$4\tfrac{4}{7}\,\Big|\,100\tfrac{100}{111}$$

One can see from this that it is almost entirely a matter of addition, because in multiplying one only writes down and in dividing one does not have to try to find the quotient, which instead gives itself because it can be nothing but 0 or 1.

But as noted at the beginning, there is, however, no intention to introduce this type of calculus into common use, and instead one knows very well how to make do, because one already knows the "One-times-one" and like tables by heart, so that one should proceed more quickly than here, where everything from the very start falls into place, just as if one did not know anything about calculating at all. In the meantime, however, this new method serves well for knowledge and contemplation, and also shows beautiful things in numbers, so that afterwards it can also be useful in common reckoning. More about this elsewhere.

But the greatest utility I was looking for in it goes much higher. For, since the chief aim of human science is supposed to be that it guides us to God, and lifts our soul up with the help of stepping stones, so to speak, thus showing the divine footprints of the Creator in creatures by means of the wonderful order and harmony of things, so I dare to say that in all of mathematics it is not easy to find anything more beautiful and more suitable to that end.

Since it is known that the mystery of creation, although surely grounded in reason, was nevertheless unknown to pagan philosophers, who put matter as it were side by side with God because they could not understand how it was possible that everything arises from one alone, yet at the same time [reason] itself may recognize that the essences or natures of things are comparable to numbers, and with this calculus is revealed quite wonderfully how all numbers are written with 1 and 0, so this beautiful representation provides an agreeable and lofty consideration of the *unum necessarium*, namely how from God alone, as the most perfect and simplest one, and nothing else, all other things arise.

God [does] all things by numbers.

(For things are considered by the calculation of measures, and a certain weight is a measure of power.)

Final manuscript page of the draft of Leibniz's letter to Duke Rudolph August, 2/12 January 1697 (LBr F 15 Bl. 19v).

17 Leibniz to Duke Rudolph August (2/12 January 1697)

Manuscripts:
M1: LBr F 15 Bl. 18–19. Draft. German.
M2: LBr F 15 Bl. 16 and 24–25. Draft medal designs.
M3: Gotha A 448–449 Bl. 92–94. Copy of no longer extant dispatched letter. German.

Transcription:
A: A I 13, 116–121 (following M1).

Having returned to Wolfenbüttel again at the end of December 1696, Leibniz decided to pen a further letter to Duke Rudolph August about the binary system. Written on the occasion of the New Year, the letter was motivated by Leibniz's desire to have the binary-creation analogy commemorated in the form of a medal to be struck at the duke's command.[1] Much of the letter is devoted to Leibniz's lengthy description of what he thought fitting to feature on the medal. He suggested that one side feature the Latin phrase "Imago creationis" [image of creation] and a table of binary-decimal equivalents flanked by examples of binary addition and multiplication. A background of light and darkness (representing the spirit of God hovering over the waters in the Genesis account of creation) would suggest binary's 1 and 0 as analogs to God and nothingness, while the motto "OMNIBUS. EX NIHILO. DUCENDIS. SUFFICIT UNUM" [To draw all things out of nothing, one thing is sufficient] would be inscribed across the top. For the reverse, Leibniz suggested a portrait or signature of Rudolph, his initials either side of the Greek letter Φ (which resembles a 0 with a 1 through its middle), and the phrase "Unum autem necessarium" [And one thing is necessary].

Leibniz would have enclosed a sketch of his medal design with the New Year's letter, but this is no longer extant. However, the manuscripts of three other sketches, different from the above description and therefore likely earlier draft designs, have survived and are reproduced at the end of this chapter: one is drawn by Leibniz, another possibly by Balthasar Ernst Reimer (1669–1697), who was in Leibniz's employ at the time, and a third is by an unknown hand. Two further medal designs were published in 1718 and 1734. The first, published as the frontispiece of a small book on binary arithmetic by Johann Bernhard Wideburg (1687–1766), is by Wideburg's own admission based on a specification given to him by Rudolph Christian Wagner (1671–1741), who was Leibniz's secretary from 1697 to 1699 (see Wideburg 1718, frontispiece and 21). Meanwhile, the second design, published as the frontispiece of a small book on the binary-creation analogy by Rudolph August Nolte (1703–1752), very closely resembles the description of the medal Leibniz gives in his New Year's letter to the duke (only the phrase "unum autem necessarium" is missing) and so is likely either a

---

1. Possibly around the same time, Leibniz also sketched out a medal design to commemorate his formula for the quadrature of the circle, discovered more than twenty years earlier. On this design, Leibniz added a table of binary equivalents of the numbers 1–16 (which took four attempts to get right; the first three attempts were aborted and then deleted due to errors). See LBr F 15 Bl. 22r; English translation: Leibniz (1696). The medal design, without the binary tables, is reproduced in A IV 6, 626–627.

copy of the design that would have been enclosed with Leibniz's letter or a reconstruction based on Leibniz's description therein (see Nolte 1734, frontispiece). The proliferation of designs gives some indication of how seriously Leibniz took the idea of commemorating the binary-creation analogy on a medal. In fact, Leibniz may have had the idea for a medal more than a week before writing to the duke. On a spare page of a copy of a letter he had written to Caspar Büssing on 24 December 1696/3 January 1697 (see LBr 137 Bl. 6v), Leibniz wrote several decimal numbers and their binary equivalents and sketched a number of binary calculations, including the example of multiplication (shown immediately below) later found on one of the draft medal designs for the duke,[2] specifically the one possibly in Reimer's hand.

$$\begin{array}{r} 101 \\ 11 \\ \hline 101 \\ 101 \\ \hline 1111 \end{array}$$

Despite the design efforts of Leibniz and those in his employ, no medal was ever struck. Suggestions circulating shortly after Leibniz's death indicated that this was on account of Rudolph's "premature death" (Wideburg 1718, 21), but this can be ruled out by the fact that the duke did not die until 26 January 1704, seven years after Leibniz sent his New Year's letter with a medal design. It appears more likely that Leibniz abandoned his idea for a commemorative medal when the duke found a different way of commemorating the binary-creation analogy (see chapter 18).

In addition to outlining the desired features of a commemorative medal, Leibniz's New Year's letter sees him explore the binary-creation analogy in some detail and even identify its potential to convince non-Christians of the doctrine of creation *ex nihilo*. To that end, Leibniz proposes communicating these ideas to the Jesuit Grimaldi, in the hope it might support his missionary efforts in China. His letter to Grimaldi shortly thereafter (see chapter 19) was the first in a series of such communications over the years to other Jesuit missionaries. Leibniz also sent one of the medal designs (the one originally drawn by an unknown hand) to one of these missionaries, Joachim Bouvet (1656–1730), most likely sometime after he first informed Bouvet of the binary system in 1701 (see chapter 22).[3]

Leibniz's New Year's letter to the duke circulated in his lifetime,[4] probably from copies made from the no-longer extant dispatched letter. Unsurprisingly, this led to it appearing in print soon after Leibniz's death, including no fewer than six different books published between 1720 and 1768,[5] making it one of the few of Leibniz's writings on binary available in the eighteenth century.

[M1]

To His Highness Duke Rudolph August of Brunswick and Lüneburg
Most Noble Duke, Most Gracious Prince and Lord

I hope Your Serene Highness will graciously note that I, on the occasion of the New Year and according to both custom and my very own nature, offer you my best wishes from a faithful heart for continued good health and the fulfilment of all your royal desires fruitful for common and personal interests, for this and many subsequent years.

---

2. On the spare page, Leibniz also wrote out three sums (namely, $100 + 1 = 101$; $1111 + 1 = 10000$; and $1000 + 100 + 10 = 1110$) and multiplication (namely, $10 \times 100 = 1000$), which are not found on any of the extant medal designs.

3. Although there is no mention of the medal design in any of the extant correspondence between Leibniz and Bouvet, the medal design is mentioned and also faithfully reproduced by Bouvet in a letter of 12 December 1707 to some fellow Jesuits in China, with Bouvet remarking that the design was something "Mr. Leibniz sent to Sien Seng [i.e., Xiansheng, Master = Bouvet]." The letter is held at the Bibliotheca Apostolica Vaticana, in the collection Borgia Latino 515 (p. 205). We thank Claudia von Collani for bringing this to our attention.

4. For example, in a letter to Leibniz dated 12 January 1707, Caspar Neumann (1648–1715) explained that he had had in his possession "for a long time" a copy of Leibniz's New Year's letter to Duke Rudolph (LBr 683 Bl. 6).

5. Namely, Leibniz (1720, 103–112), Nolte (1734, 1–16), Ludovici (1737, I: 132–138), Leibniz (1739, I: 234–239), Leibniz (1740, 92–100), and Leibniz (1768, III: 346–348).

And so that I do not come empty-handed this time, I enclose a design of what I had the good fortune of discussing with you recently. It is in the form of a memorial coin or medal, and although my sketch is small and needs to be improved by your judgment, the matter is itself such that it would well be worth displaying in silver for posterity, if such a thing were struck by your Highness's orders. For one of the main points of the Christian faith, and one of those that have least purchase with the philosophers and are not easy to teach to pagans, is the creation of all things out of nothing through God's omnipotence. Now it can be said that nothing in the world better introduces and simultaneously demonstrates this than the origin of numbers as represented here, by expressing them only and solely with one and zero, or nothing, and a better representation of this mystery will hardly be found in nature or philosophy. Therefore I have put on the proposed medal: IMAGO CREATONIS.[6]

But it is still no less worth considering how beautiful it appears not only that God made everything out of nothing, but also that God made everything well and that everything he created was good, as we can see here, with our own eyes, in this illustration of creation. Because, instead of there being no perceivable order nor clear sequence to the general presentation of numbers in terms of their digits or designations, here on the other hand is shown, as one is confronted with their innermost reason and origin, a wonderfully beautiful order and harmony that cannot be improved. Because there is a consistent rule of alternating progression, one can continue writing as long as one wishes, even without calculation or the aid of memory. If one keeps on putting one beneath the other alternately 010101 etc. in the first column on the right hand side, or in the last position, then in the next column from the right comes one beneath the other 0011 0011, etc.; in the third we put 0000 1111 0000 1111 etc; in the fourth 0000 0000 1111 1111 0000 0000 1111 1111 etc. and so on that the period or alternating cycle always doubles each time. And such harmonious order and beauty can also be seen in the small table on the medal up to 16 or 17, since there is not sufficient room for a larger table up to 32. And one can deduce from this that the disorder one imagines in the works of God is only apparent, that if one looks at things from the right point of view, as in proper perspective, their symmetry is revealed. This inspires us to praise and to love more and more the wisdom, goodness, and beauty of the highest good, from which all goodness and beauty have flowed. Therefore I am now writing to Father Grimaldi of the Jesuit Order in China, who is currently president of the Mathematical Tribunal there, with whom I became acquainted in Rome, and who wrote to me on his return trip to China from Goa, as I thought it good to share with him this idea of numbers in the hope that perhaps this illustration of the mystery of creation could serve to show the monarch of this mighty empire more and more the excellence of the Christian faith, since Grimaldi told me himself that he [the Chinese emperor] was a very great lover of arithmetic and had learned the European way of calculating from Father Verbiest, Grimaldi's predecessor.[7]

But so that I may explain the rest of the medal, I have marked the main places, namely 10 or 2, 100 or 4, 1000 or 8, 10000 or 16, with * or an asterisk, because if one just takes note of these, one will see the grounds for the other numbers. As an example, why 1101 stands for 13 is demonstrated thus:

$$
\begin{array}{r|l}
1 & 1 \\
00 & 0 \\
100 & 4 \\
1000 & 8 \\
\hline
1101 & 13
\end{array}
$$

and the same with all other [numbers]

I have also given an example of addition and one of multiplication on the medal on the sides of the table so that one can also observe the basis of the operations and how the normal rules or types of arithmetic also apply here. Whatever one's opinion, this type of calculation is under no circumstances to be used in everyday life, apart from observing and discovering the secrets of numbers.

But so that creation would be better depicted and the medal itself would not just contain numbers, and also so that it has something else pleasing to the eyes, light and darkness are also sketched on it,

---

6. "Image of creation."

7. For Leibniz's letter to Grimaldi, see chapter 19 of this volume.

or according to human depiction the spirit of God over the water. For "darkness was over the surface of the deep, and the Spirit of God was hovering over the waters. And God said, 'Let there be light,' and there was light."[8] And such [depictions] are all the more appropriate since the empty deep and the desolate darkness belongs to zero and nothing, but the spirit of God with its light belongs to the almighty one.

I have thought for a while about the words of the symbol or the *Motto dell'impresa*[9] and have finally found it good to write this inscription:

2, 3, 4, 5 &c.   0
OMNIBUS.   EX NIHILO. DUCENDIS. SUFFICIT. UNUM.[10]

since this clearly indicates what is meant by the whole symbol and why it is an image of creation. This inscription can also be justifiably divided into two parts, brought out both by means of the differences between the letters as well as the small spaces in between, so that the last part of it, SUFFICIT UNUM, being recognized as the main phrase on the right, which is as it is required in such things, contains some subtlety and profundity. For this "sufficit unum," whether it is actually already being said here of the numbers and the creation they imply, goes even further, namely to our doctrine, and holds in itself the chief rule of our life and of Christianity—that the unique good is sufficient for us, if we only hold on to it correctly. Above "omnibus" are placed the digits 2, 3, 4, 5 etc., and above "nihilo" is placed 0, so that everyone can more easily relate the meaning of the inscription to the number table.

1	1
10	2
100	4
1.000	8
10.000	16
100.000	32
1.000.000	64
10.000.000	128
100.000.000	256
1.000.000.000	512
10.000.000.000	1024
100.000.000.000	2048
1.000.000.000.000	4096
10.000.000.000.000	8192
100.000.000.000.000	16384

What could go on the still-empty side of the medal depends, like everything, on Your Serene Highness. It could be a portrait, a signature, or something else pleasing, at your discretion. The Φ between R and A as R Φ A with a crown over the Φ, the 0 with 1 struck through, would perhaps not be unfitting, because the Φ implies the Greek phi or ph, as found in Your Serene Highness's name, namely at the end of the first word of the two comprising the name.[11] The UNUM.AUTEM.NECESSARIUM, as Christ himself told us,[12] would be a symbol perhaps not bad for this, or whatever else may be suitable.

Lastly, because many secrets of numbers are found in this idea, I would like them to be written in such a way up to 16000, or rather to 16384, that is, in this way to

100.000.000.000.000   16384

which would not fill more than a [page of] writing paper or book paper. It would be easy to write out, since one may write 0 and 1 in a certain order off the top of one's head just as quickly and even more quickly than when copying something. A number written like this will not be more than 3 or 4 times longer than when written in the usual way.[13] But contained within are, I think, so many wonderful and useful observations for the increase of knowledge that the Hamburg Mathematics Society, whose industry and determination is commendable, would find, if some members wanted to devote themselves to thinking about it with enthusiasm, such things, as I can assure, that would bestow not inconsiderable renown upon it and to the German nation, because the matter was first brought to light in Germany. Because from this way

---

8. Genesis 1.2–3.

9. "imprinted/impressed motto."

10. "To draw all things out of nothing, one is sufficient."

11. That is, the final two letters of "Rudolph."

12. Luke 10.42.

13. The binary representation of a positive integer requires about $\log_2 10 \approx 3.32$ times as many digits (rounded up) as the binary representation.

of writing numbers I see wondrous advantages arising, which will afterwards also come into use in common arithmetic, from which time onwards more might be mentioned.

As far as the medal is concerned, it would be so much easier to fit into the coin stamp of the metal worker since it mostly consists of letters and numbers, and the workers have their numbers and alphabets so that they can hammer them into the still unhardened iron. But I leave everything to the most gracious discretion of Your Highness, and remain forever,

Your Serene Highness

Your most submissive, loyal, and obedient servant, Gottfried Wilhelm Leibniz

Wolfenbüttel, 2 January 1697

To the right (LBr. F 15 Bl. 24r), the design on the left is in Leibniz's hand and the design on the right in Reimer's. The inscription on the left design reads *OMNIA.vt.ex.NIHILO fiant. VNVm esse necesse est. SVFFICIT IPSE SIBI. SVFFICIT ILLE MIHI.* = "In order that all things are made from nothing, it is necessary that there is one thing. It is sufficient unto itself. It is sufficient unto me." The inscription on the right design reads *VNVM PRO CVNCTIS PRAETERAQUE NIHIL* = "for all things, one and nothing else [is sufficient]." Underneath these designs, in Leibniz's hand, is written *Corona cum equo saliente* = "A crown with a horse rearing up." *IMAGO CREATIONIS. INVEN. G.G.L.* = "Image of creation, devised by G. W. L." *Zahlen bis 10000* = "Numbers up to 10000." *VNVM pro CVNCTIS praetereaq NIHIL* = "for all things, one and nothing else [is sufficient]."

Below, the design on the left (LBr. F 15 Bl. 16v) is in Reimer's hand. Its inscription reads *VNVM pro CVNCTIS praetereaq NIHIL* = "for all things, one and nothing else [is sufficient]." *IMAGO CREATIONIS INVEN. G.G.L.* = "Image of creation, devised by G. W. L." The design on the right (LBr. F 15 Bl. 25r) is in an unknown hand. Its inscription reads *NIL PRAETER VNVM IN OMNIBUS. VNVM AVTEM NECESSARIVM.* = "Nothing except one in all things. And one thing is necessary."

smi ducis Rudolphi Augusti
emblemata occasione Arithmeticae
meae Binariae, quae non alias habet
notas quam o et 1

Signum Solis seu circulus in
cujus medio punctum videtur mihi
optime exprimere # 1 per punctum
in centro, et O per vacuum
interceptum inter centrum et
circumferentiam

Circellus intimus in
quo inscriptum unum
posset esse luminosus
seu radians, inter-
ceptum inter hunc
circellum et circum-
ferentiam posset esse
tenebrosum, nisi ubi
literae inscribuntur. Circumferen-
tia quaeby circulus inclusa erit
candida id est nec lucida nec
tenebrosa

VRVM
GENIKIKO
OMNIA BENE

21

Leibniz's comments on a letter from Duke Rudolph August to Johann Urban Müller,
5/15 January 1697 (LBr F 15 Bl. 21r).

Manuscript:
M: LBr F 15 Bl. 20–21. Fair copy. German and Latin.

Transcriptions:
Z: Zacher 1973, 236.
A: A I 13, 127–129.

Duke Rudolph August did not respond directly to Leibniz's suggestion of commemorating the binary system on a medal (see chapter 17). However, he did send the following letter to his Chancellery Secretary, Johann Urban Müller (?–1708), asking Müller to give Leibniz three engraved wax seal impressions that commemorated the binary-creation analogy. Moreover, the duke reports that he had had the seals made immediately after his discussion with Leibniz about binary, a discussion that had occurred in May 1696, but had forgotten to let Leibniz know until Leibniz's letter of 2/12 January 1697 (see chapter 17) jogged his memory. Upon receipt of the duke's letter, Leibniz added a series of remarks in Latin under the seal impressions, describing his own preferred design for a wax seal, which he then proceeded to sketch. The fact that the duke had already seen fit to commemorate the analogy between binary and creation in wax seals appears to have satisfied Leibniz, who made no further attempts to convince the duke to have a medal struck.

Further, Leibniz was sufficiently pleased by the duke's wax seals that he occasionally mentioned them when outlining his binary system to his correspondents, for example Grimaldi, to whom Leibniz wrote shortly after receiving the duke's seals (see chapter 19). Leibniz even mentioned them four years later, when outlining his binary system to the mathematician Johann Bernoulli (1667–1748) in a letter written on 19 April 1701:

> When I showed it [i.e., the binary system] to the most serene Duke Rudolph August several years ago, it pleased him to such an extent that he saw in it an image of creation, that is, of the origin of all things from nothing by God in accordance with [the origin] of numbers from 1 and 0, so that he even arranged for 0 and 1 to be engraved on a seal, with which he usually sends me sealed letters. (A III 8, 639)[1]

---

1. While Leibniz could be read here as claiming that the duke devised the binary-creation analogy, this should not be taken at face value, since the binary-creation analogy can be found in Leibniz's work before his discussion with Duke Rudolph in May 1696, namely, in "Remarks on Weigel" (see chapter 13). Accordingly, it is likely Leibniz is doing little more here than using the duke's name to show that the binary system had received princely approval.

Dear Urban, yesterday I forgot to give this to Mr. Leibniz. Greet him on my behalf, thank him for his New Year's wish, and the beautiful present attached, and wish him in return enduring and eternal welfare. And give him these three imprints, with the message that I had them made one after the other immediately after we spoke to each other about 1 and 0; should he, in connection with this, have anything of which to remind me, he might want to reveal his opinion to me. Farewell.

R.A. 5/15 January 1697, 8.30am

[Leibniz's comments]

The embossed seals of the Most Serene Duke Rudolph August on the occasion of my binary arithmetic, which uses no other digits than 0 and 1.

The image of the sun, or the circle with a dot in the middle, seems to me the best way to express a 1 by a dot in the center and a 0 by an interstitial space between the center and the circumference.

The innermost ring in which is inscribed "unum" could be bright or radiating. The interstitial space between this ring and the circumference could be dark, except where the letters are inscribed. The circumference enclosed by the two circles will be clear, that is, neither bright nor dark.

VNVM

EX NIHILO

OMNIA BENE

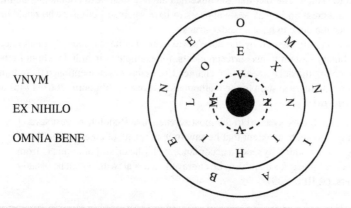

The inscription on the first of the duke's seals reads *unus ex nihilo omnia fecit* = one has made all things from nothing.

The inscription on the second and third of the duke's seals reads *unus ex nihilo omnia bene fecit* = one has made all things well from nothing.

The inscription on Leibniz's sketch reads *vnvm ex nihilo omnia bene* = one [made] all things well from nothing.

19 Leibniz to Claudio Filippo Grimaldi (mid-January–early February 1697)

Manuscripts:
M1: LBr 330 Bl. 15–18. Draft. Latin.
M2: LBr 330 Bl. 20–26. Copy of dispatched letter. Latin.
M3: LBr 330 Bl. 28–34. Copy of M2. Latin.

Transcriptions:
A: A I 13, 516–528 (following M2).
W: W 74–102 (following M2).

In his letter to Duke Rudolph of 2/12 January 1697 (see chapter 17), Leibniz mentioned that he was in the process of writing to the Jesuit missionary Claudio Filippo Grimaldi to share the analogy between binary and the creation of all things from nothing, in the hope that it would support the missionary work of the Jesuits in China, where Grimaldi was based at the time. A meeting with Grimaldi during a visit to Rome in the summer of 1689 had intensified Leibniz's interest in China[1] and through Grimaldi he quickly built a network of correspondents among the Jesuit missionaries there. Leibniz kept the missionaries informed of recent political and scientific developments in Europe, requesting in return information on a host of Chinese technical and scientific matters and on the rules governing the construction of Chinese characters.[2] These exchanges mirrored what Leibniz hoped would be a broader exchange of knowledge between Europe and China, a trading of intellectual wealth he termed a "commerce of light" (A I 14, 838).[3]

Given the distances involved, these correspondences were often intermittent, with years sometimes passing between letters. This was the case with Grimaldi, who wrote to Leibniz only once, on 6 December 1693, while in Goa en route to China to take up his role as head of the Imperial Bureau of Mathematics (see W 48–50). Leibniz received Grimaldi's letter in November 1695 (see A I 12, 154), but it apparently took the stimulus of the binary-creation analogy to finally prompt him to reply, almost a year later, with the following letter. Wide-ranging in scope, the letter sees Leibniz detailing various scientific developments that had occurred in Europe since Grimaldi's departure for China in 1693 as well as a number of his own advances in mathematics—chief among them the binary system, which he introduces through the binary-creation analogy and its potential use by missionaries. He then explains the binary representation of numbers before offering numerous examples of addition and multiplication, albeit with little detail of the method involved. He also notes the periods found in the columns of natural numbers expressed in binary.

Although the initial draft of Leibniz's letter bears the date 20 December 1696 (i.e., 30 December in the Gregorian calendar), he did not dispatch it straightaway, instead instructing his amanuensis to make two fair copies, both of which bear the same date as the initial draft. Leibniz then continued

---

1. For Leibniz's notes on their meeting, see W 2–4.
2. For an exhaustive collection of Leibniz's correspondences with the Jesuits in China, see W.
3. Or, as he put it elsewhere, "a commerce of doctrine and mutual light" (A I 5, 484).

to work on the fair copies, making some corrections and additions (see A I 13, 111–112). It must have been completed after 2/12 January 1697, because it makes reference to Leibniz's letter to Duke Rudolph August of that date (see chapter 17), but before 25 January/4 February 1697, when Leibniz informed another correspondent, Christophe Brosseau, that it had been dispatched (see A I 13, 532), hence our dating of mid-January–early February 1697. Leibniz received no response to his letter, and it is uncertain whether it ever found its way to Grimaldi. The dispatched copy is no longer extant.

Despite the value Leibniz placed on the binary-creation analogy as a possible tool for convincing non-Christians of the dogma of creation *ex nihilo*, it would be another four years before he mentioned the binary system in any way to his other Jesuit missionary correspondents; its next appearance was in a letter to Joachim Bouvet of 15 February 1701 (see chapter 22). Leibniz's letter to Grimaldi was published shortly after his death (see Leibniz 1718, 18–24), albeit only in part, with all of the material on binary omitted. This material was first published only in 1987, when the letter was finally published in full (see A I 13, 516–528).

[M2]

To the Most Reverend Father, Claudio Filippo Grimaldi, SJ, President of the Mathematical Tribunal in the Chinese Empire, from Gottfried Wilhelm Leibniz

I was extremely heartened by your letters from Goa dated 6 December 1693,[4] which reported that you had landed there safely, and heartened also by the expectation that you would remember me and my questions. At that time it was also reported that you had arrived in the Chinese capital but with weakened health and a sickness due to the stress of the journey, which troubled me greatly. For in my judgment, what was endangered in you was not the commerce in wealth between the two most cultivated parts of the world but the commerce of minds. But since nothing travels faster than a rumor of adverse events, I have great hope for your restored health, by the very fact that we have heard nothing further. May God grant this be so.

But I would like you to think, distinguished Sir, and I do not doubt that you do, that God has granted to you, but to very few mortals, the ability to accomplish the greatest things which exalt his glory and increase the goods of humankind.[5] But since wisdom and virtue are the true goods of man, and since the more someone has a proper understanding of God's wonderful work, the more roused he will be to cultivate virtue and to love God himself, it follows that he who can shine a bright light on knowledge can also contribute greatly to piety. And your China is like another world from which ours could be enriched with the most excellent knowledge, while in turn, knowledge of heavenly and divine things, in which we excel by God's favor, is transplanted there. Moreover, just as Emperor Caligula, the worst of men, wished that the whole Roman people had one neck so that he could cut through it in a single blow,[6] so it is a great and desirable thing that there be a single wise monarch of vast empires who will do good to all if the ardor for great things can be aroused in him, because the goods of a prince are spread upon all. I have learned from our conversations in Rome[7] and from the testimony of others that the emperor of China has great virtue and a great love of wisdom,[8] and those who have come from these lands afterwards add that his son,[9] who is destined to inherit such a great empire, follows his father in every respect. So you, who enjoy the highest favor with the father, and have (as is credible) no less with the son, have been presented with a wonderful opportunity from

---

4. See Grimaldi to Leibniz, 6 December 1693 (W 48–50).

5. humankind. ▷ Now it is agreed that whatever improves men also increases the glory of God, who has decreed that we should express his own image. ◁ [M1] *deleted.*

6. See Suetonius (1914, 464).

7. Leibniz met Grimaldi in Rome in July 1689; for his record of their meeting, see W 2–4.

8. Namely, Kangxi (1654–1722).

9. That is, Yunreng (1674–1725), born Yinreng, who was Crown Prince at the time of writing but was deposed in 1712.

heaven to bless both parts of the world. I know that one has to proceed step-by-step in winning over the minds of men, and that with those things that God has revealed to humankind through Christ it is not easy to persuade through proofs those people to whom our sacred and profane history is insufficiently known and demonstrated, and this lack of aids can be quickly compensated for only by divine inspiration. But in the meantime, God's benevolence is so great that he can help those who are not yet sufficiently familiar with his revelation by another kind of his never-ending grace, but only if they are not lacking a good will. For moved by the contemplation of nature and aided internally from on high, they can love above all things that which they understand to be superior to all things in beauty and perfection, until the light of faith shines in their prepared minds. Consequently, for as long as it does not seem good to God to give the full weight to the outward preaching of the Gospel that we wish, we must endeavor to arouse the love of God, upon which Christ especially insisted and reason itself commends. But it is certain that no one can love someone whose beauty is unknown, and the divine wisdom and power (in which the beauty of the supreme intelligence consists, as far as we can grasp) cannot be known better than by knowledge of the things he has made. From which we deduce that three things have to be done to augment the natural light of God within us: first, evidently, gather knowledge of all the extraordinary things already discovered; second, look for that which is still to be discovered; lastly, recount all things, either invented or still to be invented, in hymns to be sung to the supreme author of things and so secure an increase in divine love, to which charity towards men is connected. But if mortals were sufficiently fortunate to ever have a great king (especially if he had neither a love nor need for war, as we know yours has), he would adopt these three things, as it were making them his own, and we would accomplish more in one decade for divine glory and the benefit of humankind than we would otherwise in many centuries. At the present time, I do not see in the world a prince who, without danger, could succeed in this better than the Chinese one, and there is no-one around him more suitable or more effective at persuading him than you alone. If only I, whose ardor for procuring the public good is second to no-one's, were not separated from you by vast spaces, it would have been possible to support your pledges and labors in this matter! Now at least I am trying to contribute something to the advancement of the sciences, so that we may gradually make progress in knowing the secrets of things.

The third version of my arithmetic machine is already being built. I would like it if the inspection of it could satisfy you as much as the description of it pleased you. For I have no doubt that you would get great pleasure from seeing the longest multiplications and divisions carried out at wonderful speed in almost the same time as shorter ones, so that they yield the correct products, quotients and remainders and numbers even up to twelve digits and even further if desired. Nor is there need to be assisted by any additions when multiplying. Accordingly, humankind is now almost freed from this burden of calculating, and now the most exacting calculations can be done by a small boy, as if he were playing a game with very little attention and the least concern. No one discerns more correctly than you how important are things of this kind. Nor do the calculations [of my machine] have anything in common with logarithms or Napier's bones.[10]

And there is another discovery of mine in numbers that seems to have the power to arouse a sense of piety as much as anything else in the whole of mathematics or even philosophy; it is also in itself most worthy of admiration. And I don't think it's easy to encounter anything in nature that offers a more beautiful and vivid image of the creation of all things that God has drawn out of nothing. And since some think this dogma is not the least problematic of the beliefs our faith proposes to us, it will be so much more useful if the most indisputable of the sciences also provides new and brilliant evidence for Christian dogmas by this discovery, and it will perhaps have some efficacy among your gentlemen in augmenting the light of truth. Indeed, for a while now I had this thing, like so many others, buried among my papers, when by chance I happened to mention it to the most serene Prince Rudolph August of Brunswick-Lüneburg, Duke of the Wolfenbüttel line, as something appealing and

---

10. Invented by John Napier (1550–1617), Napier's bones were a set of inscribed rods manipulated in such a way that multiplication and division were reduced to a series of guided additions and subtractions; see Napier (1617, 1990). They are unrelated to another of Napier's inventions, logarithms.

useful for pious meditations, in which the exceptional prince was accustomed to take pleasure. At first he was amazed, but as soon as he understood it accurately, he took so much delight in it that he ordered seals engraved with the mystical Φ as a symbol of all things created from nothing by the unity.[11] But I think it worthwhile to explain the thing itself.

numbers[12]

0	0
1	1
10	2
11	3
100	4
101	5
110	6
111	7
1000	8
1001	9
1010	10
1011	11
1100	12
1101	13
1110	14
1111	15
10000	16
10001	17
10010	18
10011	19
10100	20
10101	21
10110	22
10111	23
11000	24
11001	25
11010	26
11011	27
11100	28
11101	29
11110	30
11111	31
100000	32

addition

1	1
10	2
11	3
10	2
101	5
111	7
110	6
1001	9
1111	15
1001	9
1101	13
.	
10110	22
100	4
101	5
110	6
1111	15
111	7
1001	9
1011	11
11011	27

multiplication

11	3
11	3
11	
11	
1001	9
101	5
11	3
101	.
101	.
1111	15
110	6
101	5
110	
1100	
11110	30
111	7
111	7
111	
111	
111	
110001	49
1001	9
111	7
1001	
1001	
1001	
111111	63

Everyone knows it is up to men to decide which progression of numbers they wish to use.[13] The decimal, with no other privilege than that it corresponds to [the number of] fingers, seems to be accepted, and indeed for that reason it is rightly retained. Nevertheless, it is apparent that some considered the duodecimal to be more useful while others took pleasure in the Pythagorean tetractys.[14] At some point it occurred to me to consider what would ultimately be revealed if we used the simplest of all [progressions], namely the dyadic or binary. Then with my mind turned to this, I immediately saw it was necessary that all numbers could be written using only these two very simple symbols of unity and non-being, 1 and 0. Because just as in the decimal progression all [numbers] can be written by values below the number ten, namely by 0, 1, 2, 3, 4, 5 6, 7, 8, 9, and in the quaternary by values below the number four, namely 0, 1, 2, 3, so it was necessary in the binary for everything to be able to be written using values below the number two, which are none other than 0 and 1. So when it pleased me to really try this out, the result was such a table of numbers and other products as you see here up to 32, together with the operations of addition and multiplication illustrated by examples, not for the sake of practical use but for contemplation.[15]

---

11. See Leibniz's letter to Duke Rudolph August of 2/12 January 1697 in chapter 17 of this volume.

12. In the manuscript, the number at the bottom left is incorrectly "10000," and the number just above 27 near the bottom of the middle column is incorrectly "1."

13. This echoes Pascal's claim that the decimal system "has been established not as a result of natural necessity, as the common man thinks, but as a result of human custom, and quite foolishly, to be sure" (Pascal 1665, 42).

14. With regard to the tetractys (i.e., base 4), Leibniz is likely referring to Weigel (1673), and with regard to the duodecimal, Leibniz is probably referring to Pascal (1665, 47–48).

15. contemplation. ▷ And the table itself emerges by the continuous addition of 1s ◁ [M1] *deleted*.

Here, a 1 in the first position from the right signifies 1, in the second position it signifies 2, or the base, in the third 4, or the square of the base, in the fourth 8, or the cube, in the fifth 16, or the biquadrate, in the sixth 32, or the surdesolid and so on. From this, the following things are evident:

First, all numbers can be written by 1 and 0.

Second, just as the essences of things are like numbers, so also the creation of things from nothing by God is represented through the expression of numbers with the help of mere unity joining with non-existence.

Third, this expression is the simplest and most natural of all. For the reduplicative progression is the first and easiest of them all, as also is the division into two equal parts.

Fourth, in this progression nothing is assumed, but everything is demonstrated. For example, during calculating everyone assumes that 1 and 2 is 3, likewise that three threes are nine. Yet here it is demonstrated, as is obvious in the attached examples.

Fifth, in this progression there is no need for the Pythagorean table or any other assistance. For multiplying or dividing, one needs to know only that multiplying 0 by 0, or 0 by 1, or 1 by 0, will be 0, and multiplying 1 by 1 will be 1.

Sixth, when dividing according to this progression, there is no need to try to find the quotient.

Seventh, in this progression everything is resolved down to the first principles, 0 and 1. From which it happens that demonstrations are more readily apparent there, and from this it also follows that,

Eighth, this progression is of the utmost benefit in discovering theorems and their rules. In addition,

Ninth, everything in this progression proceeds in an orderly manner. For while no connection is apparent in the natural series of numbers expressed in the usual way, 0, 1, 2, 3, 4, etc., the greatest regularity is immediately apparent here. For in the natural series of numbers expressed in this new way, the last numerals are 0101, etc., the penultimate ones 0011 0011, etc., the third last are 0000 1111 0000 1111, etc., the fourth last 0000 0000 1111 1111 0000 0000 1111 1111, etc., where the length of the period is always doubled.

Tenth, at a stroke we can see from this the basis of that remarkable property that has long been observed in the double geometric progression, namely that any number from the numbers (posited only once) of a double geometric progression can be constructed by simple addition, if needs be with a unity added. For example, 6 is 2 + 4, and 7 is 1 + 2 + 4, and 10 is 2 + 8, and 20 is 4 + 16, and 29 is 1 + 4 + 8 + 16. And this is immediately clear from our expression. For because 20 is 10100, and a 1 in the third position from the right is 4 and in the fifth position is 16, it follows that 20 is 4 + 16; and because 29 is 11101, and a 1 in the first position from the right is 1, in the third 4, in the fourth 8, in the fifth 16, it follows that 29 is 1 + 4 + 8 + 16. And so on for the other numbers. Examiners of metals, having observed the usefulness of this property, subdivide their little weights in doubles, since in this way they can weigh out the greatest number of masses to be weighed with the fewest weights. It would therefore be in the public interest to consider coinage in a similar descending progression.

However, I repeat what I have already said, that this calculus should not be used in everyday practice, in which it is preferable to assume many things that have already been grasped by one's memory than to assume nothing and to repeat everything from the beginning. From which, those who have a larger Pythagorean table in their memory can calculate faster than those who have only the ordinary or smaller one.[16] Therefore, the real use of this calculus lies in it revealing the mysteries of numbers, perfecting analysis, and revealing the sources. In this matter I have many unexpected [ideas] that I shall reveal at some point.

But leaving aside the analysis of numbers, I say that since you departed from Europe, Most Reverend Father, there have been wonderful advances in *analysis speciosa*, that is, in the mathematical

---

16. At the time, the ordinary Pythagorean table contained the products of the single digits 1–9 arranged in a square.

art of inventing. This has happened especially through my work and that of some friends,[17] who have put into practice a certain new method I proposed of calculating with letters.[18] In doing this, they have provided what Descartes himself admitted was missing in his analysis,[19] and he tried in vain to present it as mechanical since he could not capture it with his calculus.

I have invented a new form of expressing in a calculus the infinitely small increases of motions and the very rudiments of the increases which are infinitely, infinitely small.[20] For since motion, which of course takes place in time, is like a regular line, it must be that impetus, as the momentary beginning of motion, is like an infinitely small or infinitesimal line. Whereas conatus (for example, gravity or the force receding from the center), since it ultimately constitutes impetus through infinite repetitions, will be an infinitely, infinitely small quantity. When I transferred these notions of the infinite into a geometrical calculus, solutions were found to physico-mathematical problems which previously did not appear to be in our power, but which now appear easier with infinitesimal calculus, because nature everywhere involves something of the infinite, as it displays vestiges of the immeasurable author. Galileo inquired in vain into the catenary line, which of course is formed when the most flexible chain consisting of the smallest links is suspended from two given points.[21] At the request of friends I tackled this [problem], was the first to solve it and—what is surprising—to discover that it depends upon logarithms.[22] Such that, if there was no table of logarithms or divisions of apparatus drawn from them, the effects of logarithms could still be ingeniously elaborated with the help of such a chain, presented with the same accuracy as the length of the chain can afford. Others later confirmed the truth of this too. On this model, the most ingenious mathematicians used my calculus to determine the shapes of sails and [solve] various optical and mechanical problems. So I also came up with a new kind of lines, which I call kissing[23] as they are tangents of a tighter nature, and I show their notable use in practical mechanics.[24] Having observed these things, the recently-deceased Huygens, second-to-none of the mathematicians of our century, publicly and privately acknowledged that with the help of the calculus I put forward we could now achieve results to which access was otherwise scarcely possible.[25] And Mr. Marquis de L'Hospital, who recently published in France an extraordinary book on this new infinitesimal analysis,[26] recognizes that I started where everyone else had left off, and that this kind of calculus has in a way broken through the wall that separates physics from mathematics. And the art of invention has now advanced to the point that it is clear that nothing will be insuperable for us when the givens are sufficient, as long as men do not betray themselves. I have also begun to put together a new science of dynamics,[27] and having discovered Descartes's errors I brought forth the true laws of motion demonstrated by considering what leads to absurdity, because if these laws are not accepted then the equality between entire cause and full effect would be violated to the extent that a perpetual motion could be caused, which I assume to be absurd.[28] Here, too, the harmonies show themselves to be wonderful, from which it is readily apparent that nature is the work of a supreme wisdom. The greatest fruit of our meditations, indeed of our entire life, lies in our recognizing and honoring this one thing, for we were made to love and to honor divine wisdom.

---

17. Leibniz means Jacob Bernoulli (1655–1705), Johann Bernoulli (1667–1748), and Guillaume-François-Antoine de L'Hospital (1662–1704).

18. See Leibniz (1684).

19. See Descartes (1898, 511).

20. Leibniz is referring to his invention of the infinitesimal calculus.

21. See Galilei (1638, 146). English translation: Galilei (1914, 149).

22. See Leibniz (1692a).

23. Leibniz is referring here to osculating curves.

24. See Leibniz (1692b).

25. See Huygens (1693) and Huygens's letters to Leibniz of 1 September 1691 and 17 September 1693; A III 5, 161 and 634–635.

26. L'Hospital (1696).

27. See Leibniz (1695b). English translation: PPL 435–450.

28. Leibniz (1686a, 1691a).

More and more advances have been made in astronomy, too. Learned men are putting great effort into elucidating lunar movements. And certainly some deviations seem to have their origins in the mutual interaction of planetary bodies; it is probable these deviations occur to all planets, but they are easier to observe in the moon, closest to us. However, if these deviations are carefully observed, they can be used to predict more accurately than has happened before. Huygens left a posthumous work entitled *Cosmotheoros*, in which he explained many things that are generally little known about celestial matters.[29] He also left a planetarium different from the one that Rømer built,[30] a specimen of which I think the French Fathers brought to China.[31] I myself have discovered that the elliptic motion of the planets can be explained mechanically if we imagine that the bodies of the planets orbit the sun harmoniously, that is, that their orbiting speeds are in a harmonic progression, since the different distances of one and the same planet from the sun are in arithmetic progression.[32]

As for optics, the art has now been invented of making mirrors from glass that are considerably larger than the ones before, for we already have some that are 80 inches (according to the Paris foot), that is, $6\frac{2}{3}$ feet in diameter. A few years ago, even the Venetians in Murano would have declared such a thing impossible. With regard to the method of production, so far I have not been told anything sufficiently reliable and definite, but I think the art will be widely known soon.

In Kassel, in our neighborhood, an experiment with an underwater boat has been made, observed by the prince Landgrave of Hesse;[33] it had some success, such that there can be no doubt that one day it will be put to some use, as soon as workers become more familiar with it. The inventor had set up the experiment so that, enclosed in the boat, he could estimate the depth of the dive with the help of mercury, as in a barometer.

Recently, the bones of a very large animal from the elephant genus have been excavated in Gotha, Thuringia.[34] Curious folk have been wondering where it came from in these regions, but similar things have also been discovered in the region of Brunswick, Brabant, and elsewhere in different places. I have no doubt that things not seen on the earth today are sometimes excavated by the Chinese, news of which would be useful for Europeans to set up a comparison.

Little has happened in science ever since the war, which involves almost the whole of Europe, has been going on,[35] with men turning to the arts of harming. Only now are they beginning to think seriously about peace. For with the consent of both sides, the King of Sweden has been named mediator and has already negotiated with regard to the place of the assembly,[36] but since preparations take a lot of time, I'm afraid that nothing will be concluded by this winter, so at the moment a bitter war awaits us next summer. I hope this does not undermine the hopes for peace with new calamities. Truly, the war is conducted with so much military might on both sides it is a wonder that the nations can afford the immense costs of it. So it happens that even the greatest princes look like the destitute. But everything may be put back into order once peace is made (if our sins do not already prohibit us hoping for this).

---

29. Huygens (1698).

30. The Danish astronomer Ole Rømer (1644–1710) designed a planetarium to show the orbit of planets around the sun. It was built in 1678–1679 by Isaac Thuret. See Anon (1682).

31. See Tachard (1686, 10).

32. See Leibniz (1689, 1993).

33. See Papin (1695, 126–137).

34. In 1695, some bones were found in Tonna, near Gotha in Thuringia, which numismatist and historiographer Wilhelm Ernst Tentzel (1659–1707) claimed were from an elephant. See Tentzel (1696). During 1696 and 1697, it became a topic in his correspondence with Leibniz. Tentzel (2006) includes English translations of the relevant parts of the correspondence. For discussion and analysis, see Strickland (2005).

35. Leibniz is referring here to the Nine Years' War, which ran from 24 September 1688 to 20 September 1697.

36. The belligerents in the war had accepted Charles XI of Sweden (1655–1697) as mediator in 1693, with negotiations beginning in Rijswick in 1695.

You know that the Germans first established a society for those curious as to nature; the great emperor of the Christians has now taken over its patronage.[37] Later, the English founded a Royal Society, in which there are and were many distinguished men, but the society has enjoyed only limited favor of its royal founder.[38] For Charles [II] was a prince of great talent, but had little concern for serious matters. So I believe that William [III], who reigns now, will be more of a patron, even if he takes less pleasure in these things, since he knows how to place value on things. Further, after the Germans and the English, the French, under the auspices and also in the name of their great king, set up an academy of sciences,[39] which has no what no other society has, namely the king covering the expenses of experiments and members.

It is not surprising that this has been thrown into disorder by the war, but I think it has not been entirely disrupted and it will soon be restored with greater effort. There is also hope that the emperor himself and, standing out by his power among the great German princes, the Elector of Branden-burg,[40] as well as others will ensure that affairs are settled so that Germany is not inferior to France in these matters either. At that time, a copious supply of new things will be available to us, which I shall share with you too, if God prolongs my life. And I shall be able to do that more easily because I have been a member of the English society for more than twenty years and the French Academy has pretty much considered me as one of their own.[41] For even if people are on bad terms in many other things, common to them all is the pursuit of war against a nature that opposes us and steals us away. We hunt it in its own hiding places with various arts, and accomplish something every day, only slowly at the moment, but faster in the future if the great princes show that the matter is more dear to them. But if your people were to persuade the Chinese monarch of this, humankind could be promised much greater things, whether he were to order the boundless crop of knowledge of his empire be gathered together or prefer it to be worked on in accordance with foreign investigators. For there is nothing within human power that could not be achieved by the ardent desire of such a prince who, ruling unconditionally over the most flourishing peoples across immeasurable areas in the deepest peace, should fear nothing besides God. Therefore, if (to return to where this letter began) you aroused him to such a love of divine glory (for God cannot be known by the natural light more truthfully and effectively than through the marks of wisdom and power imprinted on things), there would be nothing greater you could do, except the propagation of the Christian faith.

For the rest, may you live as long as possible, and may happy successes accompany your years for a long time yet, and may you love me. Sent from Hanover, December 20, at the end of 1696.

P. S. With your permission, I take the opportunity to attach other letters from a man interested in your work. I hope Reverend Father Laureati and many other travel companions have entered China happily. I give hope that these men be received as complements to you and that they follow the advice and example of your men. But if you do not have the time to write to us in more detail on worthy matters of general interest, perhaps there will be one among them to whom you could delegate this task. Again, farewell. Thus I pray, as above.

Gottfried Wilhelm Leibniz.

P. S. I am writing these letters in the hope that they will reach you and find you flourishing and thriving, but if I had heard something from you since you arrived in Peking, I would add many things [and write] with greater confidence. In the meantime, it occurs to me that, should God grant us the opportunity of further communication, in some [letters] we could use a sort of coded script, such as I am adding here. As soon as I have heard that you have received it, nothing will be able to

---

37. The Academia Naturae Curiosorum was founded in Schweinfurt in 1652; it became the Sacri Romani Imperii Academia Caesareo-Leopoldina Naturae Curiosorum in 1687, after Emperor Leopold I declared it an Imperial Academy.

38. The Royal Society was founded in London in 1660. Leibniz was elected as a member in 1673.

39. The Académie Royale des Sciences was founded in Paris in 1666 by Louis XIV, following a plan of Jean-Baptiste Colbert.

40. That is, Frederick III.

41. Leibniz was elected as a foreign member of the Académie Royale des Sciences on 13 March 1700; see LH 41, 8.

prevent me from writing more. I have found out the many things that otherwise happen to letters, and what may surprise you, to those that Reverend Father Gerbillon sent to French friends on 22 August and 2 September 1689 from the camp at Nischou or Nipchou,[42] a region of Eastern Tartary beneath the meridian of Peking but 300 French leagues from there. He had gone there in the retinue of the Chinese envoys to negotiate a treaty with the Russians. These letters, intercepted by the Russians, were passed on to their addressees through me, although late.[43] Again, farewell.

```
L a b y r i n t h u s c d e f g k m o p q w x z
a b c d e f g h i k l m n o p q r s t u w x y z
```

Accordingly, "uorfg" would be written in place of the word "Pekin," for example. But it is sufficient to do this just for some words and, to make the key much less easy to crack, one can omit most of the vowels, which a reader in possession of the key easily supplies. It is also sufficient for repeated vowels to be put down once, and it is not necessary to separate the beginnings of words, since it can be written as if it were a single word.

---

42. That is, Nerchinsk.

43. The 22 August letter was for François de la Chaise, royal confessor of Louis XIV, while the 2 September letter was for Antoine Verjus, procurator of French missions to the Far East. Leibniz got hold of the letters in 1695 and forwarded them to their intended recipients. For the story of how the letters came to be in his possession, see Carhart (2016).

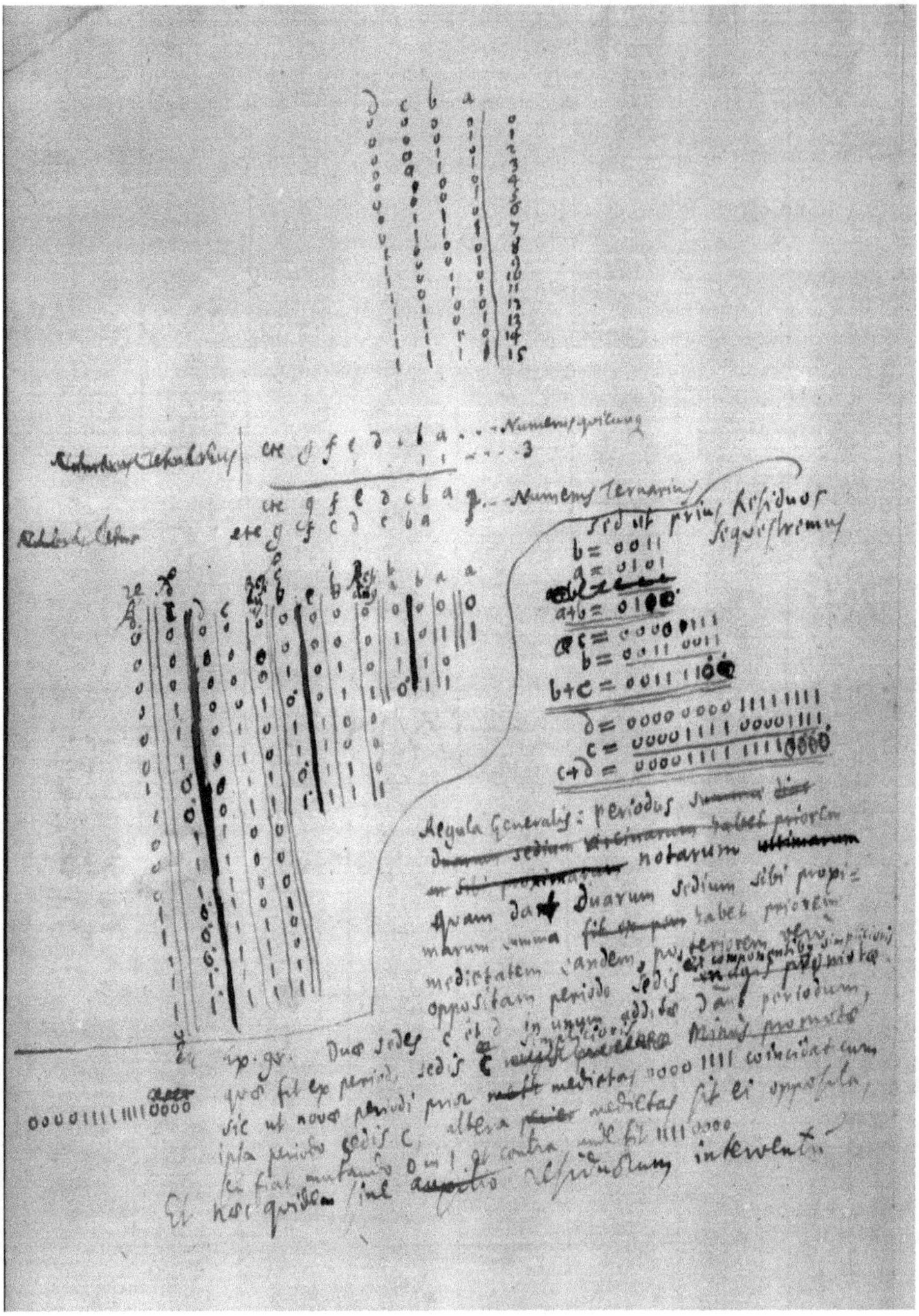

First manuscript page of “Periods” (LH 35, 3 B 3 Bl. 5v).

Manuscript:
M: LH 35, 3 B 3 Bl. 5–6. Draft. Latin.

Although Leibniz had begun to communicate details of his binary system to others in the second half of the 1690s, his accounts tended to consist mostly of ideas he had worked out in the late 1670s and early 1680s, as he had done relatively little work on binary in the interim. However, his renewed interest in binary in the late 1690s did lead to a few advances, such as the realization that when using binary notation, the periods of any arithmetic progression are divided into complementary half-periods, so that (for example) if the period of the first half is 0011 then the period of the second half will be 1100. Leibniz first noted this in a letter of 17/27 May 1698 to the mathematician Johann Christian Schulenburg, stating that, when looking at various arithmetic progressions expressed in binary, it was clear that "half of any period always has digits that are the complements of the corresponding digits of the other half of the same period" (A II 3, 451).[1] The following text develops this observation into a general rule and applies it to the particular case of the multiples of 3, which are calculated in binary by adding a shifted copy of a number to itself. Since it builds upon what Leibniz had observed when writing to Schulenburg in May 1698, it may date from around that time. However, it is unlikely to have been written after the first half of January 1701, by which time Leibniz was certainly treating the observation about periods as a general rule; for example, describing it as a "beautiful theorem" in a letter to Philippe Naudé of 15 January 1701 (see chapter 21). Leibniz would mention the rule in a number of other writings thereafter, such as his letter to Joachim Bouvet of 15 February 1701 (see chapter 22) and the "Essay on a New Science of Numbers" (see chapter 23). In a further manuscript, likely dating from the first half of 1701, Leibniz referred to this rule as both "the opposition of bisection" and "the law of the bisection of the column" (LH 35, 3 B 5 Bl. 57r), but acknowledged that it would not help with finding a general rule by which the column periodicity of various arithmetic progressions could be worked out.

---

1. The same observation is to be found in a short paper enclosed with the letter to Schulenburg, wherein Leibniz writes: "It is also clear that this observation always remains true, that the first half of the digits of a period is the complement of the second half." LH 35, 3 B 5 Bl. 10.

d	c	b	a	
0	0	0	0	0
0	0	0	1	1
0	0	1	0	2
0	0	1	1	3
0	1	0	0	4
0	1	0	1	5
0	1	1	0	6
0	1	1	1	7
1	0	0	0	8
1	0	0	1	9
1	0	1	0	10
1	0	1	1	11
1	1	0	0	12
1	1	0	1	13
1	1	1	0	14
1	1	1	1	15

$$\begin{array}{c} \text{etc. } g\ f\ e\ d\ c\ b\ a \ \dots \text{any number whatsoever} \\ \hline 1\ 1\ \dots 3 \end{array}$$

$$\left.\begin{array}{c} \text{etc. } g\ f\ e\ d\ c\ b\ a \\ \text{etc. } g\ f\ e\ d\ c\ b\ a \end{array}\right\} \dots \text{ternary number}$$

carry	$\frac{d}{c}$	d	c	carry	$\frac{c}{b}$	c	b	carry	$\frac{b}{a}$	b	a	a
0	0	0	0	0	0	0	0	0	0	0	0	0
0	0	0	0	0	0	0	0	1	0	1	1	1
0	0	0	0	1	0	1	0	1	1	1	0	
0	1	0	0	1	0̇	0	1	1	0̇	1	1	
0	1	0	1	0	1	1	0	0				
0	1	0	1	0	1	1	0	0				
1	0̇	0	1	1	0̇	1	1	0				
1	0̇	0	1	1	1̇	1	1	1				
0	1	1	0	0								
0	1	1	0	0								
0	1	1	0	0								
1	0̇	1	0	1								
1	0̇	1	1	0								
1	0̇	1	1	0								
1	1̇	1	1	1								
1	1̇	1	1	1								

Leibniz is working out algebraically the pattern of bits in the successive columns of the numbers $3n$ for $n = 0, 1, \dots$. Under $a, b, c, \dots,$ are the values of the rightmost or least significant bit of successive values of n, the next to rightmost, and so on. The fourth from the right or $\frac{b}{a}$ column is the sum of the a and b columns, with 0̇ indicating that $1 + 1 = 0$ with a carry of 1. So the rightmost "carry" column shows a pattern of carry bits for $n = 0, \dots, 7$. The $\frac{c}{b}$ column shows the sum of $b, c,$ that "carry" column, and so on.

But in order for us to separate the carries first[2]

---

2. In the left margin, Leibniz wrote the number 0000 1111 1111 0000.

$$
\begin{array}{rl}
b = & 0011 \\
a = & 0101 \\
\hline
a+b = & 011\overset{.}{0} \\
\hline\hline
c = & 0000 \ 1111 \\
b = & 0011 \ 0011 \\
\hline
b+c = & 0011 \ 11\overset{.}{0}\overset{.}{0} \\
\hline\hline
d = & 0000 \ 0000 \ 1111 \ 1111 \\
c = & 0000 \ 1111 \ 0000 \ 1111 \\
\hline
c+d = & 0000 \ 1111 \ 1111 \ \overset{.}{0}\overset{.}{0}\overset{.}{0}\overset{.}{0}
\end{array}
$$

General rule: The period of the string [of digits] given by the sum of the two adjacent columns has the same first half while the second half is the complement of the period of the simpler of the components. For example, the period of two positions c and d now added into one: this is formed from the period of the simpler position c in such a way that the first half of the new period, 0000 1111, coincides with the very period of position c [while] the second half is the complement of it, that is, it comes about by changing 0s into 1s and vice versa, from which we get 1111 0000. And these indeed without the occurrence of remainders.

From this it is now evident that the remaining 1 for the carries is

first: in any fourth position
second: in any 8th and 7th position
third: in any 16th, 15th, 14th, 13th
fourth: in any 32nd, 31st, 30th, 29th, 28th, 27th, 26th, 25th

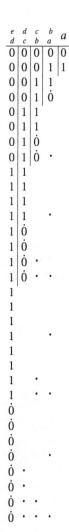

e/d	d/c	c/b	b/a	a
0	0	0	0	0
0	0	0	1	1
0	0	1	1	
0	0	1	0̇	
0	1	1		
0	1	1		
0	1	0̇		
0	1	0̇	·	
1	1			
1	1			
1	1			
1	1		·	
1	0̇			
1	0̇			
1	0̇	·		
1	0̇	·	·	
1				
1				
1				
1			·	
1				
1				
1		·		
1		·	·	
0̇				
0̇				
0̇				
0̇			·	
0̇	·			
0̇	·			
0̇	·	·		
0̇	·	·	·	

This table recapitulates the results above (extended to include $d + e$), but in vertical columns and using the headers from the table on p. 118 above. Each column's period is given only once, and thereafter simply a dot is recorded in the each position where 0̇ would have appeared had the periods been written out in full. (1̇ cannot occur because, as on p. 118, each column is the sum of just two bits since possible carries-in from the right are being ignored.)

Manuscript:
M: LBr 679 Bl.1–2. Draft. French.

Transcriptions:
Z: Zacher 1973, 239–242.
A: A III 8, 507–510.

During his stay in Berlin from 10 May to 22 August 1700, Leibniz was granted permission by Frederick III, Elector of Brandenburg (1657–1713), to establish the Kurfürstlich Brandenburgische Societät der Wissenschaften [Electoral Brandenburg Society of Sciences], the society being founded on 11 July 1700 and Leibniz appointed its president for life a day later. Although an institution existing only on paper in its early years, this society provided Leibniz with the perfect framework to seek assistance to develop some of his own projects that he'd had insufficient time to pursue as far as he would have liked. One such was the binary system, which at some point while in Berlin he had outlined to the mathematician Pierre Dangicourt (1664–1727), in the hope that he would agree to develop it further. Leibniz's efforts were initially unfruitful; in mid-September–October 1700, another of Leibniz's acquaintances, Philippe Naudé (1654–1729), at the time a professor at the Academie der Mahler-, Bildhauer- und Architectur-Kunst [Academy of Painting, Sculpture, and Architecture] in Berlin, wrote to him to say that "having promised you to prompt Mr. Dangicourt to work on the [binary] tables about which you had spoken to him, and seeing that for want of having understood you correctly, or an inability to bring himself to work on something for which his mind is not engaged, so to speak," Naudé felt compelled to work on it a little himself (A III 8, 480). To that end, he enclosed extensive tables of the binary representations of arithmetic progressions, powers, and figurate numbers.[1] In the following letter, which is his reply to Naudé, Leibniz makes it clear that his interest lies not in producing tables of numbers but in discerning the rules for the periods of numbers representing different arithmetic progressions, and outlines his procedure for working out the patterns in each bit position for the sequence of multiples of 3 represented in binary, which he had been exploring (see chapter 20). Leibniz concludes by expressing his wish for a general rule that would enable the periods of any column of any arithmetic progression to be worked out without the need for induction.

In his reply, written sometime between the start of February and mid-March 1701, Naudé made no comment on the binary system other than to state that, after showing Leibniz's letter to Dangicourt, the latter had composed a lengthy paper "On the Periodic Intervals of Binaries in the Columns of

---

1. The tables provided by Naudé consist of columns with thousands of entries, capturing the binary representations of even numbers 2–80, odd numbers 1–71, multiples of 3 from 0–105, "simple or first numbers" 0–149, square numbers 1–1,521, cubic numbers 1–59,319, biquadrate numbers 1–2,313,441, and natural numbers 0–1,023. See Gotha A 448–449 Bl. 118–120.

Natural Numbers,"[2] which Naudé enclosed with his letter (see A III 8, 546). Leibniz scribbled a number of remarks on Dangicourt's paper and then drafted two separate sets of comments on it, one in French (see Göttingen G 2 Autographen Leibniz Nr. 7 Bl. 9), the other in Latin (see Göttingen G 2 Autographen Leibniz Nr. 8 Bl. 10–12). Dangicourt later revised his paper, and it was eventually published under the title "On the Periods of Columns in a Binary-Expressed Sequence of Numbers of an Arithmetic Progression" in 1710, in the first volume of *Miscellanea Berolinensia*, the journal of the Electoral Brandenburg Society of Sciences (see Dangicourt 1710).

Hanover, 15 January 1701

To Mr. Naudé
Mathematician at Berlin

Sir,

You will have heard that my trip to the Teplice baths and beyond, and also a bout of illness, have resulted in my absence, and that your very nice letter couldn't be delivered to me until my return, along with a lot of others, in accordance with what I told Counselor Chuno in my previous one.[3] I now reply to you, Sir, without further delay, and am quite obliged for your favorable remembrance, and for the effort you have taken on my behalf, hoping to be able to find the occasion to offer you the same in some matter of your choosing.

I am thinking of putting together my meager thoughts, now detached and dispersed in many journals and other books, since you think that might be useful. As to the specious division,[4] I proceed by analogy with numbers, which has elsewhere been very useful, for understanding the one by the other. If one had to divide 144 by 12, it is as though one were dividing $1xx + 4x + 4$ by $1x + 2$, taking x to be 10. Here it is the same if x had any other value, and as one would be needlessly burdened in the numbers if one were to change the order of the powers, it would be the same and for a stronger reason in the specious, and as you request, Sir, a rule for how to do it, I believe that what I just said may serve, and that that rule will never fail if the division is possible. For in following the order of degrees, one inevitably diminishes them, and making the large ones vanish, one descends always to the remainder, which one cannot make disappear, since exact division is not feasible.

For the dyadics or binaries, I should be quite upset, Sir, were one to believe that I am advising that anyone take upon himself difficulties no soul need assume, doing calculations that one could have done by a little boy. My thought is to research with my friends the series of periods, since the natural numbers have their periods:

	The Naturals	The Ternaries[5]
in the last position	01	01
in the penultimate	0011	0110
in the antepenultimate	0000 1111	0010 1101
in the preantepenultimate	0000 0000 1111 1111	0001 1100 1110 0011
	and so on	etc.

---

2. This paper can be found in A III 8, 534–545.

3. Leibniz is referring here to his letter to Johann Jacob Julius Chuno of 31 December 1700; see A I 19, 308–311.

4. "Specious" means algebraic. Leibniz is replying to a question from Naudé. When dividing one polynomial by another, for example, $10ab - 2a^2 - 3ab - 15 \div a + 5b$, which terms should be divided first? In this case, the result should be $2a - 3b$, but starting with $10ab \div a$ won't make progress. Leibniz responds with an example of dividing one monomial by another, which seems to avoid Naudé's question.

5. For the last two elements of this column, Leibniz incorrectly wrote 0101 1010 and 0111 0011 1000 1100, which are cyclic rotations of the correct periods.

It is a question of finding by reasoning how the ternaries, quinaries, and others[6] should continue their periods, as well as the squares, cubes, etc. That is what I hope will be discovered to some extent without too much difficulty, and I beg you and Mr. Dangicourt to think about it at your leisure.

When the opportunity presents itself, be kind enough to pay my compliments to Mr. Ancillon, Mr. Teissier, Mr. Chauvin, Mr. Dangicourt etc. but especially to Mr. de Margas and Mr. de la Croze, and above all to your excellent ministers, Mr. Lenfant and Mr. Beausobre, to whom I add that learned man who is with Mr. von Schwerin, with whom I dined at Mr. Lenfant's, and whose name I cannot remember at present.[7]

Have you not continued to think about the principles of justice and morality (of which I gave you my opinion)[8] and to talk about these things with friends?

To return to binary, it's a matter of considerable importance that all series not only have their periods, that is their first digits repeat at certain intervals; but also that the intervals are always the same in all these series, namely after 2 in the last position, 4 in the penultimate, 8 in the antepenultimate, 16 in the preantepenultimate, and so on. Now the rules of these very digits should be discovered.

Here is how I go about it without doing induction tests. Take whatever number you like, etc. $fedcba$.[9] Multiplying by 11 (or 3) gives every ternary number. Now the last digit a is 01 (that is, 0 or 1), b is 0011, c is 0000 1111, d is 0000 0000 1111 1111 and so on.

$$
\begin{array}{r}
f\ e\ d\ c\ b\ a \\
1\ 1 \\
\hline
\text{etc.}\quad f\ e\ d\ c\ b\ a \\
\text{etc.}\ f\ e\ d\ c\ b\ a
\end{array}
$$

Having noted the repetitive patterns in the various bit positions of the sequence of binary numbers—01 01 01... in the last position, 0011 0011 0011... in the next-to-last position, and so on—Leibniz works out the patterns in each bit position for the sequence of multiples of 3, that is, the binary representations of 0, 3, 6, 9, ... If the number being tripled has binary representation $fedcba$, where $a, b, c, d, e, f \in \{0, 1\}$, then the last bit of the tripled number is a and therefore follows the pattern 01 01 01...; the next-to-last position is $a + b$ mod 2 and therefore follows the pattern 0110 0110 0110.... The antepenultimate position is

$$b + c + \text{(a possible carry-in from the penultimate column) mod 2.}$$

That carry-in is a 1 only if $a = b = 1$, and as a result, the position follows the pattern 00101101 00101101 00101101..., again with potential carryouts to the fourth-from-last bit position.

---

6. That is, the multiples of 3, 5, and so on.

7. Leibniz here gives a roll call of prominent Huguenots living and working in Berlin, all of whom were known to him from his travels there: Charles Ancillon (1659–1715), at the time historiographer to Frederick III, Elector of Brandenburg; he also assisted Leibniz in founding the Berlin Academy in 1700; Antoine Tessier (1632–1715), a jurist who was best known at the time for his translations of various works by Samuel Pufendorf; Etienne Chauvin (1640–1725), editor of *Nouveau Journal des Savants* from 1694–1698 but at the time a pastor and professor of philosophy of the Französisches Gymnasium [French School] in Berlin; Mathurin Veyssière de la Croze (1661–1739), a Benedictine historian and orientalist, at the time a royal librarian in Berlin; Jacques Lenfant (1661–1728), a Protestant divine and prolific author of works on church history; and Isaac de Beausobre (1659–1738), at the time court preacher in Berlin. Little is known about de Margas aside from that he was an instrument maker and Naudé's neighbor (see A III 8, 509). The only non-Huguenot mentioned is Otto von Schwerin (1645–1705), who was at the time minister of state for Brandenburg-Prussia.

8. This would have been in person, during Leibniz's stay in Berlin over the summer of 1700.

9. The "etc." denotes missing higher-order digits.

$$
\begin{array}{r|l}
0101 & a \\
0011 & b \\
\hline
011\dot{0} & \\
\hline
\end{array}
$$

$011\dot{0} = \frac{0110}{0001}$, where 0001
is saved for the next column.

$$
\begin{array}{r|l}
0011\,0011 & b \\
0000\,1111 & c \\
0001\,0001 & \text{from that} \\
 & \text{which was saved} \\
\hline
0010\,110\dot{1} & \\
\hline
\end{array}
$$

$011\dot{0}\,110\dot{1} = \frac{0110\,1101}{0001\,0011}$, where $0001\,0011$
is saved for the next column.

$$
\begin{array}{r|l}
0000\,1111\,0000\,1111 & c \\
0000\,0000\,1111\,1111 & d \\
0001\,0011\,0001\,0011 & \text{from that} \\
 & \text{which was saved} \\
\hline
0001\,11\dot{0}\dot{0}\,1110\,\dot{0}\dot{0}1\dot{1} & \\
\end{array}
$$

So the last digit of the ternary number is a, that is to say the same as the last digit of the natural number, namely 01, or indeed 010101 etc.

The penultimate digit has to be made from a and b, or $a + b$, that is from $0101 + 0011 = 011\dot{0}$, where the dot above the last 0 in the calculation in the margin[10] signifies that a 1 should be saved for the next or antepenultimate column $c + b$ in the fourth place of its period. Consequently there will be in this column $0011\,0011 + 0000\,1111 + 0001\,0001$, which gives $0010\,1101$, the dots above the digits in the margin marking again that which needs to be saved. In the same way it will be found following the calculation in the margin that the preantepenultimate column, where there is $d + c$, will give as period $0001\,1100\,1110\,0011$.

I wish I could already give the general rule to infinity of these periods for each column, and I have no doubt it will be found, but these things don't happen all of a sudden. In any case, it seems to me that I have already mentioned a beautiful theorem, that one half of the period is always the complement of the other, where when I say complement, I mean that where one has 0 the other has 1 or vice versa. For example, 0110 has two halves $\left\{ {01 \atop 10} \right.$, and 0010 1101 has two halves $\left\{ {0010 \atop 1101} \right.$, and finally 0001 1100 1110 0011 also has two halves $\left\{ {0001\,1100 \atop 1110\,0011} \right.$.

A thousand pardons, Sir, for a letter written in such confusion. Wishing you a good century and good year with all kinds of satisfaction, I am sincerely, Sir,

Your very humble and very obedient servant,

Leibniz.[11]

---

10. The material to the right of the table above is in the margin of the original manuscript. Here Leibniz uses a dot to indicate a carryout of a digit position, whereas at other times he uses a dot to indicate a carry-in.

11. Underneath his signature, Leibniz wrote (in Latin) "please turn over."

22 Leibniz to Joachim Bouvet (15 February 1701)

Manuscript:
M: LBr 728 Bl. 94–97. Draft. French.

Transcriptions:
A: A I 19, 401–415.
W: W 300–324.

On 18 October 1697, during a stay in Europe after almost a decade in China, the Jesuit missionary Joachim Bouvet wrote to Leibniz to congratulate him on the publication of *Novissima Sinica* [*The Latest News from China*], a collection of writings by missionaries in China that Leibniz (1697, partial English translation in WOC 45–59) had compiled, edited, and translated (see A I 14, 614–615). This letter initiated an occasional correspondence between the two that would span ten years, at least from Leibniz's side.[1] The early exchanges largely consisted in Bouvet providing information about and from China, with Leibniz posing questions and giving a general overview of his own studies and achievements in philosophy and the sciences. Leibniz's first two letters, sent on 2/12 December 1697 and 30 January 1698 (see A I 14, 826–835 and A I 15, 247–248), contained no mention of the binary system, probably because he was not working on it around those times so it was not foremost in his mind. However, the binary system is a central feature of Leibniz's third letter, of 15 February 1701, which is reproduced below. This letter serves as a reply to Bouvet's letter of 19 September 1699 (see A I 17, 490–496), which was written shortly after Bouvet had landed back in China. Owing to the long delay in transporting letters across such great distances, Bouvet's letter would have arrived in Hanover in the second half of 1700, but as Leibniz was away from Hanover from May to 30 December of that year, he would not have seen it until his return. By the time Leibniz wrote the following reply, the binary system was again at the forefront of his thoughts, as in the early weeks of 1701 he was working on what would become the "Essay on a New Science of Numbers" (see chapter 23) for the Académie Royale des Sciences [Royal Academy of Sciences] in Paris, which likely sparked him to write a relatively detailed account of binary in his letter to Bouvet (much as Leibniz's lengthy letter on binary of 2/12 January 1697 to Duke Rudolph August resulted in Leibniz explaining his binary system in detail to the missionary Claudio Filippo Grimaldi, to whom he was writing at around the same time; see chapter 19).

Leibniz's discussion of binary starts and ends with the binary-creation analogy and its perceived value for missionary purposes, with Leibniz urging Bouvet to present the binary system and the binary-creation analogy in person to the Chinese emperor, Kangzi (1654–1722), known to be fond of mathematics. Of the binary system itself, Leibniz offers an outline of binary notation before turning to the periodicity of columns of multiples of 3, of perfect squares, and of perfect cubes, for which

---

1. The last extant letter from Bouvet was written 8 November 1702, after which Leibniz wrote a further five letters, the last on 13 December 1707.

various tables are provided to show the periods. Leibniz also mentions the sedecimal system as offering a greater harmony between theory and practice; although no details are provided, he does sketch out notation for sedecimal digits different from that devised in the 1670s and 1680s (see chapters 9 and 12). The remainder of the letter consists of Leibniz reporting recent scientific discoveries from Europe, outlining his own discoveries in the sciences along with his plans for further work, and posing a litany of questions about China.

—◦◦◖◗◦◦—

My very Reverend Father   Brunswick, 15 February 1701
     I am extremely grateful for the kindness you have had to maintain all the way in another world the remembrance of a man who infinitely honors you but who is of little use to you.
     Your long and pleasant letter got to me late, because of a long journey from which I only recently returned. The Elector of Brandenburg (now crowned King of Prussia the 18th of January of this year)[2] had me take a trip to Berlin to help found a new society of Sciences,[3] of which His Majesty wants me to take charge even from afar, since I cannot always be present.
     An observatory will be built there,[4] and other measures will be taken to research both nature and art. This concern occupied all of last summer and a part of the fall; after that I went to the baths of Teplice in Bohemia because of my health,[5] where I thought I would find Reverend Father Kochański,[6] but he had died several months earlier.[7] Nonetheless I have since sent the letter Your Reverence had written for him to the Fathers of his Province.[8] This is a loss, for he assuredly was one of the ablest men of your company. From Teplice I also made an excursion in upper Germany,[9] and during that time, as I had no fixed residence, I gave an instruction for my letters to be kept in Hanover.
     So, as no one in Paris got a response from me straightaway, I have still not heard anything about what Reverend Father Fontaney brought back.[10] I shall write to this Father myself,[11] as well as to Reverend Fathers Verjus and Le Gobien about that,[12] and I hope I will be apprised of it as per the letter of Your Reverence. But if Your Reverence wants to send me something in the future intended for me personally, it should be included in the letter.
     Europe has recently produced several admirable discoveries. My new analysis of infinitesimals has been pushed forward considerably. I myself discovered a remarkable advancement of this science; and by this means problems which formerly stumped algebra and analysis are [now] within our power; for example, we have determined the shape of vessels which have the least resistance in a fluid, the path along which a heavy body would go from one point to another in the shortest possible time, and a thousand like problems; and as nature everywhere preserves the character of the infinity of its author, it is only this science of the infinite that really provides the passage from mathematics

---

2. Frederick III of Brandenburg (1657–1713) was crowned King Frederick I on 18 January 1701. Note Leibniz's incorrect styling of Frederick William, whose official title was King in Prussia rather than King of Prussia.

3. Namely, the Kurfürstlich Brandenburgische Societät der Wissenschaften [Electoral Brandenburg Society of Sciences]. Leibniz was appointed the president of the Society on 12 July 1700.

4. Construction of the observatory began in the early fall of 1700 and was completed in 1709.

5. Leibniz stayed in Teplice in the second half of September 1700.

6. Adam Adamandy Kochański (1631–1700) was a mathematician and a clock-maker. He and Leibniz exchanged almost fifty letters from 1680 to 1699. For details of the correspondence, see Heinekamp (1978).

7. Kochański died in Teplice on 17 May 1700.

8. That is, Bohemia.

9. Leibniz went from Teplice via Prague to Vienna at the end of September 1700.

10. Jean de Fontaney (1643–1710), a teacher of mathematics, led a Jesuit mission to China in 1687. In an earlier letter to Leibniz, of 19 September 1699, Bouvet had informed Leibniz that Fontaney was returning to Europe, though ultimately, Fontaney did not land there until 1702. See A I 17, 490.

11. Leibniz wrote to Fontaney in mid-February 1701. See A I 19, 417–419. English translation: Leibniz (1701).

12. That is, Antoine Verjus (1632–1706), a Jesuit teacher and missionary, and Charles Le Gobien (1653–1708), a Jesuit teacher in Tours and Alençon. Both were correspondents of Leibniz's.

to physics. I also found that the catenary—that is, the curve that a chain would assume if it had very small links—yields logarithms without any calculation and without tables.[13]

In chemistry, an able German friend of mine has found a wonderful liquid; it is called, with good reason, *spiritum igneum [fiery spirit]*, because this spirit, when mixed with some oils, bursts into flame and vanishes into carbon. But the inventor has not yet published the secret.[14]

I have often advised testing at sea the extent to which a barometer could serve to predict storms,[15] since Mr. Guericke observed that great storms make it fall.[16] Finally it was tried during a long voyage,[17] and I learned that a future storm can be anticipated about 12 hours in advance, which is very often sufficient to protect oneself, either by taking shelter, or at least by getting away from the banks and the rocks and making all preparations in the vessel so as not to be taken by surprise, so that in the future this will save a countless number of ships.

I am quite upset about the war stirred up between the Russian Czar and the King of Sweden, because I fear it will harm my plan to take advantage of the caravans that go from Moscow to China with the help of the Court of Brandenburg, which is on good terms with the Czar. For Mr. Golovin,[18] who was at the time Russian Envoy for [the negotiation of] the border treaty with China, when passing through this region with the Czar, for whom he is now one of the principal ministers, made us hope that he [i.e., the Czar] would favor these plans; but as the Czar's army was utterly defeated at Narva because of the Russians who threw down their arms without obeying the foreign officers,[19] this prince is said to be angry towards foreigners, as if they had served him badly. But it is thought he will change his mind. If peace is made, trade between Europe and China may become more frequent, from this side to there.

I don't know if I have already mentioned on other occasions when writing to your Reverence the new numerical calculus I've invented not for common practice but for the theory of the science, because it opens a vast field for new theorems; and more than anything the calculus provides a wonderful representation of Creation. Following this method all numbers are written as mixtures of 1s and 0s almost as all creatures come exclusively from God and from nothing.[20] Nothing in mathematics seems to me more beautiful for[21] usage by religion,[22] and to confirm one of the most notable [Christian] doctrines which non-Christian philosophers have been accustomed to reject unanimously. Also it is not said in vain that essences are like numbers, and all the imperfections of things consist only in negations, which is why St. Augustine said very well that evil comes from nothing.[23] Here is what this calculus involves. As we are usually accustomed to make use of progression by ten, and as some have used other progressions, I wanted to consider what

ℵ	
1	1
1	
10	2
1	
11	3
·· 1	
100	4
1	
101	5
· 1	
110	6
1	
111	7
··· 1	
1000	8

---

13. See Leibniz (1692a).

14. Leibniz is referring here to Friedrich Hoffmann (1660–1742), who was at the time professor of medicine at the University of Halle. Hoffmann published a brief description of his fiery spirit in Hoffmann (1700). It was also a recurring theme of Leibniz's correspondence with Hoffmann throughout 1701 and 1702; see Dutens II, 2, 97–101.

15. In 1697, Leibniz had commissioned Rudolph Christian Wagner to construct an aneroid barometer (that is, a barometer without mercury) of his own design; see A III 7, 683–684.

16. Leibniz is referring to Otto von Guericke (1601–1674) and his barometrical observations. See Guericke (1672, 98–100).

17. voyage, ▷ made by an Englishman and the secretary of the Royal Society of England ◁ *deleted*.

18. That is, Fjodor Aleksejevič Golovin (1650–1706).

19. The Battle of Narva took place on 30 November 1700. It saw the Swedish forces defeat a much larger contingent of Russian siege forces.

20. nothing ▷ without any originating matter being needed ◁ *deleted*.

21. for ▷ illuminating revealed religion ◁ *deleted*.

22. religion, ▷ especially since most philosophers have rejected Creation ◁ *deleted*.

23. See Augustine (1952, 201: XI.9): "evil has no positive nature; what we call evil is merely the lack of something that is good."

would be the simplest progression possible, which is the binary progression or the double geometric progression, and I saw first of all that while the decimal system uses just ten digits, namely, 0 through 9, the binary would use just two, namely, 0 and 1. And just as in the decimal system 10 signifies ten, and 100 signifies a hundred, 1000 a thousand, 10,000 ten thousand, 100,000 a hundred thousand, 1,000,000 a million, and so on, I concluded that in the binary system 10 would be two, and 100 would be four, and 1000 would be 8, and 10000 16, and 100000 32, and 1000000 64; etc.[24] This is what the calculus itself also shows when just two digits are used, 0 and 1, since $1 + 1$ is 10, and 3 is $10 + 1 = 11$ and 4 is $11 + 1 = 100$.[25] Because $1 + 1$ is 10, that is, 0 underneath the column leaves 1 for the following column, marked with a dot. And in the next one along, $1 + 1$ is 0, leaving 1 for the next column after that, which gives 100. As shown in the margin under the sign ℵ, where I have marked these reserved digits by dots.[26]

In the same way, 5 will be $100 + 1 = 101$ and 6 will be $101 + 1 = 110$ and 7 will be $110 + 1 = 111$ and 8 will be $111 + 1$, that is, 1000, and so on. But to represent the series of these numbers up to 32, see the table under ⌐, where the empty positions above the columns are filled with small 0s to better show the periods, of which I shall soon speak.[27]

⌐

₀₀₀₀0	0	₀1010	10	10100	10	11110	30
₀₀₀₀1	1	₀1011	11	10101	21	11111	31
₀₀₀10	2	₀1100	12	10110	22		
₀₀₀11	3	₀1101	13	10111	23		
₀₀100	4	₀1110	14	11000	24		
₀₀101	5	₀1111	15	11001	25		
₀₀110	6	10000	16	11010	26		
₀₀111	7	10001	17	11011	27		
₀1000	8	10010	18	11100	28		
₀1001	9	10011	19	11101	29		

In this table is found at first glance a wonderful harmony, which is that there are regular periods in each column, in the first 01, for there is 010101 and always 0 and 1 alternately. In the second column there is 0011 which repeats always, and in the third 0000 1111, and so on with the others, so that the table can be continued without any calculation merely by continuing to write. This order in the original construction of numbers according to this expression opens up a new science,[28] of which till now no one has even dreamed, since as the analysis of all numbers into 0 and 1 is the most perfect and most continued to its limit, don't be surprised that we can go farther by this means. Now I find that all numbers that are ternary, quinary, septenary, etc., that is, [numbers] divisible by 3, by 5, by 7, etc., have similar periods which recur always to infinity, for example, for the ternaries in the first column

---

24. etc. ▷ Now, all numbers are compounded solely by the addition of these numbers of the double geometric progression, without ever needing to repeat the same one in the summation, which is an advantage of this progression, as has already been demonstrated by many: for example, 23 is $16 + 4 + 2$ and 113 is $64 + 32 + 16 + 1$ etc. ◁ deleted. Leibniz's summation of 23 is incorrect, as he missed out the final $+1$.

25. $100 ▷ \frac{..1}{100}\ \frac{...1}{1000} ◁$ deleted. Note that this deletion is not recorded in transcriptions A or W.

26. dots ▷ under the following column where they should be carried ◁ deleted. Note that this deletion is not recorded in transcriptions A or W.

27. Leibniz originally added another row "100000 32" to the table immediately below but then deleted it.

28. science ▷ of numbers ◁ deleted. This recalls the title of the "Essay on a New Science of Numbers" (see chapter 23), which Leibniz was working on at the same time as this letter.

the period is 01, in the second 0110, in the third is 0010 1101, in the fourth is 0001 1100 1110 0011 and so on in the following columns, as I am able to determine it; all this is clear in table ⅃.[29]

⅃

[Ternaries]		Squares[30]		Cubes	
00 0000	0	0 0000	0	00	0
00 0011	3	0 0001	1	01	1
00 0110	6	0 0100	4	1000	8
00 1001	9	0 1001	9	1 1011	27
00 1100	12	1 0000	16	100 0000	64
00 1111	15	1 1001	25	111 1101	125
01 0010	18	10 0100	36	1101 1000	216
01 0 101	21	11 0001	49	1 0101 0111	343
01 1000	24	100 0000	64	10 0000 0000	512
01 1011	27	101 0001	81	10 1101 1001	729
01 1110	30	110 0100	100	11 1110 1000	1000
10 0001	33	111 1001	121	101 0011 0011	1331
10 0100	36	1001 0000	144		
10 0111	39	1010 1001	169		
10 1010	42	1100 0100	196		
10 1101	45	1110 0001	225		
11 0000	48	1 0000 0000	256		

To see what is going on, let's start with the simplest example, the multiples of 3. Any multiple of 3 can be written as $m = 3n = 3(n' + 2b + a) = 3n' + 6b + 3a$ for some nonnegative integer n, where $a, b \in \{0, 1\}$ and $n' = n - (n \bmod 4)$ is divisible by 4. The low-order bit of m is $3a \bmod 2 = a$, which explains the alternating pattern $010101\ldots$ in the rightmost position of the Ternaries column. Since $3n'$ is a multiple of 4, its value cannot affect the two low-order bits of m. So (taking into account a possible carryout from the low-order bit position) the next to low-order bit of m is $3b + a \bmod 2 = a + b \bmod 2$, resulting in the pattern $0110\ 0110\ldots$ in the next-to-last bit position.

Now more generally consider the multiples of any odd k, and for $n = 0, 1, \ldots$, look at the bit in the jth bit position from the right, starting from 0, in the binary expansion of $m = kn$. (The previous paragraph explored the $k = 3$ case, with $j = 0$ and 1.) We can express n as $n' + \sum_{i=0}^{j} a_i 2^i$, where n' is a multiple of 2^{j+1}. Then $m = k \cdot (n' + \sum_{i=0}^{j} a_i 2^i)$. Because $n' = 0 \bmod 2^{j+1}$, the coefficient of 2^j in the binary expansion of m depends only on k and $a_j, \ldots, a_0$. So the sequence of jth bits of $k \cdot 0$, $k \cdot 1, \ldots$ repeats itself with period 2^{j+1}. Moreover, consider a block of length 2^{j+1} from $n = p \cdot 2^{j+1}$ to $n = (p + 1) \cdot 2^{j+1} - 1$, and let the values of the jth bit of $m = k \cdot n$ in this sequence be $b_0, \ldots, b_{2^{j+1}-1}$. For the first 2^j of those values of m the coefficient $a_j = 0$, and for the second 2^j values $a_j = 1$, while the values of $a_{j-1}, \ldots, a_0$ run from $000\ldots0$ to $111\ldots1$ and then again $000\ldots0$ to $111\ldots1$. Because the low-order bit of k is 1, $k \cdot a_j 2^j$ contributes 0 to b_j for the first 2^j members of the sequence and 1 for the second 2^j, so $b_{2^j} \ldots b_{2^{j+1}-1}$ is the complement of $b_0 \ldots b_{2^j-1}$.

---

29. In the Ternaries column, Leibniz added underlines to mark some (but not all) of the points at which the periods repeat.

30. To the right of the squares column, Leibniz wrote and then deleted another column of decimal numbers representing the differences between the squares, (3, 5, 7, 9, 11, 13, 15, 17, etc.), and to the right of the cubes column, he wrote and then deleted another column of numbers representing the differences between the cubes (namely, 1, 7, 19, 37, 61, 91, 127, etc.). To the right of the latter column, he also added—and then deleted—another column recording the differences between the differences (i.e., 6, 12, 18, 24, 30, 36, etc.).

The same basic principles apply to the patterns in the tables of squares and cubes: only the last few bits of the number being squared or cubed can affect the last few bits of the square or cube, so the bits in position j (starting with 0 as the rightmost) are cyclical with a period of 2^{j+2} at most. For example, consider the square of an integer $m = n' + 4c + 2b + a$, where $a, b, c \in \{0, 1\}$ and $n' = 0 \bmod 8$. Then $m^2 = (4c + 2b + a)^2 \bmod 8$, and $(4c + 2b + a)^2 = 16(c^2 + ab) + 8ac + 4(ab + b^2) + a^2$. So the table of squares of 0, 1, 2, 3, ... has a^2 or 010101... in the low order column, 000... in the next, and in the next $ab + b^2 = ab + b \bmod 2$, that is, 1 if $ba = 10$ and 0 otherwise, so the repetitive pattern is 0001 0001.... For cubes, if $m = 4n + 2b + a$, then the last two bit positions of m^3 depend only on a and b, where $(2b + a)^3 = 8b^3 + 12ba^2 + 6ba^2 + a^3$. So the low order bit of m^3 is $a^3 = a$ and therefore alternates 010101..., while the next bit is $\frac{12b^2a + 6ba^2 \bmod 4}{2}$, which is 1 if $a = b = 1$ and 0 otherwise, so has a period of length 4 with the pattern 0001 0001....

0	I
1	I
2	U
3	U
4	W
5	W
6	W
7	W
8	UU
9	UU
10	UU
11	UU
12	UU
13	UU
14	UU
15	UU

It works the same with quinaries and septeneries, novenaries or with multiples in general, where one always finds that one half of the period is the complement of the other, that is 0 opposite 1 and *vice versa*, etc. But what is more, these periods are also found in squares, cubes, and other powers. For example, for the squares the period of the first column is 01,[31] the second has only 0, the third has as its period 0010, the 4th has 0001 0100, the 5th 0000 1101 0101 1000. For cubes, the period of the first column is 01, of the second 0001, of the third 0000 0101; and so on with the others in such a way that by the extraordinary means of periods, the tables of powers of higher degrees can be written down almost without calculation.[32] Aside from contemplation, these will have great uses for certain significant applications. In a word, an entirely new[33] arithmetic lies hidden within, marvelously bountiful with theorems since in every kind of series the very notation of numbers operates by rules. In truth, it should not be used for ordinary calculations, but it leads to the resolution of difficulties where other known methods could not go.[34] But it would take too long to explain that here, and it is sufficient now before I stop that I make one more remarkable observation that is instantly visible in this notation, namely, why all numbers can be formed merely by combining numbers of the double or binary geometric progression, something that arithmeticians have already noted as an advantage of this progression. For example, 23 is $16 + 4 + 2 + 1$, that is $10000 + 100 + 10 + 1$, or all at once 10111,[35] and 113 is $64 + 32 + 16 + 1$, that is $1000000 + 100000 + 10000 + 1$ or 1110001, and so for others; and this is why assayers of money use small weights in a double progression. For in this way a few weights suffice for many numbers or weight calculations. For example, to form all numbers[36] from 1 to 63, you need only 6 numbers or weights, namely, 1, 2, 4, 8, 16, 32. And for 1 to 31 you need only five, namely,[37] 1, 2, 4, 8, 16 as the table shows, which shows at the same time the combination of weights [needed] to form the total weight corresponding to each number. If everyone had been minded to make coins in a double geometric progression too, we would have been able to make the most values with the fewest coins. And if instead of the decimal progression, the sedecimal progression were used, there would be a greater agreement between theory and practice than there is

---

31. 01 ▷ . I do not count the following [column] which has only 0s, therefore the next will be counted as the second, which has 0010 ◁ *deleted*. Note that this deletion is not recorded in transcriptions A or W.

32. calculation ▷ , whereas without that they need to be calculated ◁ *deleted*. Note that this deletion is not recorded in transcriptions A or W.

33. new ▷ geometry ◁ *deleted*.

34. go. ▷ But I will not dwell here on thoughts about a profound analysis ◁ *deleted*. Note that this deletion is not recorded in transcriptions A or W.

35. Leibniz mistakenly wrote 1011 here.

36. numbers ▷ from 1 to 31 ◁ *deleted*. Note that this deletion is not recorded in transcriptions A or W.

37. Leibniz mistakenly included 3 in this list, so it read "1, 2, 3, 4, 8, 16."

at present and we would have found uses there that the decimal system could not provide, and I say that just to remark upon it and not because I intend to reform public usage.

But my principal goal, my Reverend Father, has been to furnish you with a new confirmation of the Christian religion as regards the sublime doctrine of Creation, on a basis which in my opinion will hold great weight with the philosophers of China and perhaps even with the Emperor himself,[38] who loves and understands the science of numbers. To put it simply, all numbers are formed from combinations of unity and nothing, and that the nothing is enough to diversify them seems to me as credible as saying that God made all things from nothing, without using any primitive matter, and that there are only these two first principles, God and Nothing—God [being the principle] of perfections, and Nothing of imperfections or voids of essence. And if at the outset you set aside the origin of the invention of this calculus (which came from the analogy of the binary progression with the decimal) the thing will seem all the more admirable. Perhaps this great monarch will not be unhappy to learn that a European of your acquaintance, who is boundlessly interested in matters regarding China and its commerce in knowledge with Europe, made this discovery and wrote you about it expressly to dedicate it to his majesty. I have no doubt that you will lay stress upon this thing according to its importance, in order that its effect will be to the advantage of our religion; and perhaps it could induce this prince, in light of that, to instruct you to communicate wonderful Chinese knowledge to me personally, with some samples from this country, among others as regards the composition of paper that is extraordinary for its size and fineness, and likewise as regards some unusual experiments in physics or specific proven [remedies] in medicine. You may rightly conclude that I am not going to profit personally from them and that the public will draw all the benefit therefrom, but as I am head of a new society of sciences, as I have told you, I will not be the least bit unhappy to be useful to it by presenting something unusual to it, all the more as that will serve to further excite the prince who founded it, and that would contribute to the advancement of the sciences. This prince, whom the Muscovites already treated as Czar or King before his coronation, as also the Turks and Tartars, the King of Persia and all the other princes outside Europe, is important for China because from his territory alone comes the amber, to which the Chinese attach great value.[39] And it will be a wonderful opportunity for me to be the director of the scientific society of this prince, which you could make the most of. If this numerical invention has to be dressed up in order to present it to the Emperor and make him appreciate it more, you will know better than I what it takes to do this.

I come to the philosophy as it should be established and cultivated both for truth and for religion. I am delighted, my Reverend Father, that you are engaged with this as much as am I. You won't believe how far I have advanced in this; I have demonstrations in metaphysics whose like has never been seen before,[40] especially with regard to the cause, the effect, and the estimation of action.[41] The ablest people (right up to Father Malebranche, even though he is entirely Cartesian) now agree that the same quantity of motion is not conserved, as Descartes had thought, but they have thrown themselves into another extreme, for not yet understanding properly the true estimation of force, they now almost all believe, as is clear in the books published recently by Father Malebranche, Mr. de La Hire, and[42] others,[43] that absolute force[44] is not conserved and there is sometimes more and sometimes less of it in the world. This is because I have not yet unveiled to them my estimation by which I have demonstrated that the effect is always equivalent to the cause, and that not only is the same absolute force conserved in the universe or in bodies, which are thought to interact only with each other, but that even taking a certain time, for example, a quarter of an hour, there is as much

---

38. Kangxi.

39. In early modern Europe, Brandenburg-Prussia fiercely controlled a monopoly on amber deposits from the Baltic littoral.

40. before ▷ . But above all I show that the true rules of nature are infinitely different ◁ *deleted.*

41. See Leibniz (1695b). English translation: PPL 435–450.

42. and ▷ Mr. Parent ◁ *deleted.*

43. See Malebranche (1692), La Hire (1695), and Parent (1700).

44. force ▷ (and not just respective [force], as they think) ◁ *deleted.*

action in one quarter of an hour as in another quarter of an hour, either in the universe or in a system of bodies not interacting with anything outside. This shows that Descartes had seen something of the truth indistinctly, but had been caught out by a mistake of identity when imagining that his quantity of motion (which is obtained by multiplying the size by the speed and which is momentary) is the quantity of the action, and will give the estimation of what is conserved. One of the most renowned professors of Holland, Mr. De Volder,[45] who was a great defender of Descartes's philosophy, to the point that he wrote strongly against the criticism of the Bishop of Avranches,[46] surrendered entirely, after learning the basis of my reasons through various letters we exchanged.[47] Also, an English philosopher very famous for his considerable works has begun to abandon his own dogmas and his corpuscular philosophy, having recognized that there has to be something in corporeal substance which is different from size and impenetrability.[48] Also, I have demonstrated that if there were only that [in corporeal substance], there would be completely different laws and phenomena, unless God provided them by miracle.[49] That would be quite unreasonable, and unworthy of the author of things. Had I people who could assist me, I would give a metaphysics and some elements of physics truly demonstrated with rigor by means of a few axioms. But overwhelmed by a thousand tasks, of business of the court, of correspondence, of travel, without mentioning the History of the House of Brunswick, drawn from archives,[50] and of what I have done on the law of nations, in making public pieces that have not been printed concerning the treaties and negotiations of princes,[51] I will be obliged to let go of many things more important, in my opinion, than what concerns only certain times and certain men.

I recall having mentioned to you a plan for a quite extraordinary characteristic, a way of depicting not the word but thoughts, as algebra does in mathematics. In expressing the trains of thought in such characters, one would calculate and demonstrate by reasoning. I think a way could be found to combine this with the old characters of the Chinese that have already been the object of your meditation. This would serve marvelously to make them appreciate this invention, and this mysterious style of writing, which perhaps would be the greatest means that could be invented to establish the truth of religion by way of reason. It was to be hoped, as you mentioned in your letter,[52] that we might be able to discuss it in person, when Your Reverence was in Europe, but there being little chance of that now, we could perhaps accomplish something considerable, if you were to keep me fully informed of what is known or thought about about these old Chinese characters; and even more [could be accomplished] if, as a result of your representations sent to France, someone were assigned to help me in this great project of a characteristic that would transform their reasoning into calculation; and would even serve to determine the degree of probability in inferences that are only probable.

To return to your letter and to several supplementary matters, I beg you, my Reverend Father, to procure for me one day the *pater*, that is, the Lord's Prayer, with some samples of words often used in different languages of the Indies, and above all of the Tartars, Kalmyks or Mongols, Ölöts, and others

---

45. Burcher de Volder (1643–1709), a professor of philosophy and mathematics at Leiden University. He and Leibniz corresponded regularly between 1698 and 1706. See Leibniz (2013).

46. The Bishop of Avranches at the time was Pierre-Daniel Huet (1630–1721); Leibniz is referring here to Huet (1690); English translation: Huet (2003). Huet was criticized in De Volder (1695).

47. See Leibniz (2013, 169).

48. Leibniz is likely referring here to John Locke (1632–1704). For the changes he thought Locke had made in his views, see A VI 6, 29–34; English translation: Leibniz (1699).

49. See Leibniz (1695b). English translation: PPL 435–450.

50. In 1686, Leibniz was tasked with writing a history of the House of Brunswick to provide evidence of the connection between the Este family and the Guelf family, to which belonged Leibniz's employer at the time, namely, Ernst August, Duke of Brunswick-Lüneburg (1629–1698). Although initially intended to be a relatively short project, it gradually spiraled out of Leibniz's control as he unearthed more and more documents from various archives in Europe. When Leibniz died in 1716, thirty years after starting the project, it was still far from finished.

51. Leibniz is referring here to the collections of legal documents he published following his archival research; see Leibniz (1693) and Leibniz (1700).

52. Namely, in Bouvet's letter of 19 September 1699; see A I 17, 495.

who extend towards Persia, Russia, and the Eastern sea. And if there are in China itself languages different from the one that is common in the Empire, I would wish for the same about them. It would be good to have interlinear versions, word for word, of each *pater*, with the [written] characters of the [respective] peoples. I would also like this particularly in regard to Tibet, where the great Lama lives, since it is said that the Mongol Tartars heavily use the language and the characters of Tibet.

As this great conjunction of planets, observed by an old Chinese Emperor and reported I think by Martini,[53] provided material for many reflections on chronology, I ask you, my Reverend Father, to think, along with Reverend Father Grimaldi, if a few other select observations could also be found, to help us in Europe with regard to astronomy and chronology. I beg you also to pass on to him my very humble thanks for his remembrance, that I was extremely happy at the news of his health, and that I always hope to see something more from him. These observations would be principally from his office, since he is president of the Mathematics Tribunal.

I would also like to know whether the ancient Chinese knowledge of geometry was accompanied by some demonstrations, and in particular if they have long known that the square of the hypotenuse is equal to the squares of the two sides, or some other such proposition of uncommon geometry. Has anyone found that the art of distilling and other chemical operations are ancient in China?

I imagine also that Your Reverence will have found in Paris a small discourse by the famous Kepler, about which I had spoken with him [i.e., Grimaldi].[54] The discourse, written some time ago, was made on a letter sent from China by Father Terrentius of your company; I imagine your people will have given it some thought. In any case, I will speak about it again to Reverend Father Fontaney. What is your judgment, my Reverend Father, of the conjecture of Golius, the famous orientalist, who thought that the the the Chinese language, as well as the characters, was made by artifice?[55] Might it not be possible to obtain soon the grammatical analysis by Reverend Father Visdelou of the characters in some small Chinese book, to serve as a sample, likewise an essay on the grammar, so to speak, of the characters?[56] But as these things remain to be done, I hope to be able to obtain in the meanwhile things that are already done, namely, a dictionary of Chinese characters with definitions in some European language, and a dictionary that Reverend Father Grimaldi told me exists in China, where pictures of things are adjacent to the characters.[57] And likewise, if there are some other instructive Chinese books where the pictures aid the explanation. I would gladly and punctually pay for the cost of these books, and do as much as I could to be worthy of this favor besides.

You will remember, my Reverend Father, that Father Kochański had posed several curious questions, as had Mr. Schröck of Augsburg, for whom you were kind enough to forward a letter to Batavia.[58] I therefore beg you to think of them [i.e., Kochański's questions], or ask others to do so at your leisure, and to favor me by answers which I will receive as legatee of this good Father. In conclusion I wish you a good century and a happy New Year, for many years of the century we have just started, and in advance for 1702 in particular, especially since this wish will apparently not reach you until close to the coming year, and praying to God to often give us good reason to thank him for the graces he will bestow upon you and his Church, in the important mission in which you are involved, I am sincerely,

For Reverend Father Bouvet

---

53. See Martini (1655, 21–23), which reports that Zhuanxu (a mythical emperor said to have reigned 2513–2435 BCE and credited with inventing the Chinese calendar) observed the conjunction of five planets on the same day as he observed the conjunction of the Sun and the Moon.

54. Namely, Terrentius and Kepler (1630).

55. See Golius (1655).

56. Claude de Visdelou (1656–1737), one of the five Jesuits sent to China by Louis XIV in 1685.

57. characters. ▷ It also would be good to know what Father Verbiest had printed in Chinese and Latin for the Emperor ◁ *deleted*. Note that this deletion is not recorded in transcriptions A or W.

58. Kochański's questions were enclosed with Leibniz's letter to Bouvet of 2 December 1697; see A I 14, 788–793. Schröck's questions were in a letter to Andreas Cleyer of 16 January 1698, which Leibniz had sent to Bouvet with his letter of 30 January 1698. See A I 15, 247.

P.S. I am hesitant to leave a lot of empty space in a paper that has to travel so far. I beg you, my Reverend Father, to look over my past letters again, when you do me the honor of thinking about what is in this one. I am only too convinced of the great difficulties that you and your Apostolic companions must be experiencing at the gateway to this new world. For until now the stay you have made there can be considered only like that of novitiate. So I would be quite unreasonable if I did not understand that you should not be asked at the outset for information that perhaps others, or those like you, would never obtain, even if they had grown old there. So all these samples I wish for should be expected only in proportion to the time with which more important or necessary occupations leave you. However, as your company is increasing so considerably, as you have doubtless many locals to hand, and above all the Emperor who is worth a hundred million of them, you will be able to find the means to relieve yourself of many things and to have materials prepared for you. This great prince will not refuse this gratitude to Europe which is so generous with him in the most important communications. If the dictionary of the Chinese and Tartar languages is completed by order of the Emperor, I would very much like to be able to obtain a copy of it also, with your assistance. If Chinese characters are added, that would be so much the better. Would to God that some European language be added or rather that one were made specially for that, in which at the same time all the Chinese and Tartar words are expressed in their characters as well as ours. Perhaps your counsels will one day sway the Emperor. Could your portrait of this prince not be expanded in the form of short annals of his reign?[59] I was told that one of the princes, his children, learned some European language,[60] but I doubt it because you do not mention it. Mr. Mentzel has died.[61] Nothing has been found of Mr. Müller's key.[62] However, it seems that Chinese characters should at least have some interconnection. Are there not as it were root characters from which all the others are formed by certain grammatical rules of derivation and composition? These roots might have been hieroglyphs. I would very much like to hear your preliminary judgment about that, and also that of Reverend Father Visdelou. I hope that we may learn from you what should be concluded of the Strait of Anian,[63] but don't put yourself in danger for that. As I am curious about things related to mining and chemistry, are there in China metals and minerals unknown in Europe, and particular practices relating to them? To finish once and for all, wouldn't there be a way of sending a good number of Chinese to Europe to help with translations?

---

59. Leibniz is referring here to Bouvet (1697).

60. Namely, by Kochański in his letter to Leibniz of 28 May 1694; see A I 10, 405.

61. That is, Christian Mentzel (1622–1701), naturalist and physician to the Elector of Brandenburg.

62. Leibniz is referring here to Andreas Müller (1630–1694), provost at the Nikolaikirche in Berlin, and an orientalist and curator of the Chinese book collection of Frederick William, Elector of Brandenburg (1620–1688). Müller claimed to have found a key to Chinese characters but refused to publish it. Leibniz relates the story of Müller's key to Chinese in §18 of the preface to his *Novissima sinica* (1697). See WOC 56.

63. In early modern times, the Strait of Anian was believed to separate America from Asia and offer ice-free passage between the two. Confidence in the existence of the strait decreased markedly over the course the eighteenth century as a result of various voyages and expeditions that failed to find it.

23 Essay on a New Science of Numbers (26 February 1701)

Manuscripts:
M1: LH 35, 3 B 3 Bl. 1–2. Draft. French.
M2: Gotha A 449 Bl. 102–106 and 114–115. Copy of M1, corrected. French.
M3: Gotha A 449 Bl. 111. Partial copy of M2. French.
M4: Gotha A 449 Bl. 116–117. Partial copy of M2. French.
M5: Paris, procès-verbaux of the Académie royale des sciences 1701, pp. 144–150 and p. 155. Partial copy of no longer extant dispatched fair copy. French.

Z: Zacher 1973, 250–261 (following M2).

On 13 March 1700, Leibniz was elected as a foreign member of the Académie Royale des Sciences in Paris (see LH 41, 8). Writing to Leibniz on 8 December that year, the Académie's perpetual secretary, Bernard le Bovier de Fontenelle (1657–1757), informed him of plans to issue an annual proceedings of the Académie, the *Histoire de l'Académie Royale des Sciences* [*History of the Royal Academy of Sciences*], for which he solicited a contribution from Leibniz:

> We will put in this *Histoire* the most important pieces, as supplied [to us] by Academicians, and as this institution counts you as one of its principal members, Sir, when you wish to communicate some of your discoveries to it, we will be delighted to be able to adorn the *Histoire* with them. (A II 3, 665)

Fontenelle's letter was enclosed with one written on the same date by the Académie's president, Abbé Jean-Paul Bignon (1662–1743), in which he echoed Fontenelle's call for Leibniz to support the work of the Académie. Leibniz was only too happy to oblige. On the manuscript of Bignon's letter, he scrawled a number of possible topics for what would be his debut essay for the Académie, with the binary system at the top of a short list: "communicate 0 and 1. Ordinary infinitesimal analysis. Number of possible books. Anagogy." (A I 19, 293)[1] Opting for the first of these topics, Leibniz proceeded to put together the following essay.

Aside from a brief mention of the binary-creation analogy near the start, the essay is entirely focused on the mathematics of the binary system and is intended to capture his progress on binary to date. Among the topics he presents are (a) the representation of numbers in binary; (b) binary addition, including the distribution of carries into several columns when the sum of a column exceeds 3, a method Leibniz systematizes further in "Binary Addition" (see chapter 24); (c) the bit patterns occurring in tables of successive multiples of a common factor; (d) similarly, patterns occurring in tables

---

1. "Anagogy" may refer to Leibniz's essay "Tentamen Anagogicum: An Anagogical Essay in the Investigation of Causes" (GP VII, 270–279; English translation: PPL 477–484).

of successive squares, cubes, and other powers; and (e) what would now be called the boolean algebra of binary multiplication, a topic with which he had been experimenting as far back as 1679 (see
chapter 6).

Leibniz sent his completed essay to Fontenelle on 26 February 1701, noting in the cover letter that
he hoped his work on binary would be supplemented—by someone else—prior to publication:

> I enclose here an essay on a numerical science that is (would one believe?) brand new. Here
> in a few words is what it is about, which the enclosed paper will explain more. Using the
> binary progression by two instead of the decimal by ten, I write all numbers by 0 and 1,
> and I do this not for the sake of common practice but to make new discoveries. And it so
> happens that the sequences of numbers written according to this characteristic are wont to be
> periodical, and to carry their rule with them, making them continuable without calculation.
> By this means, we can reach knowledge that will be difficult to discover in another way. For,
> as this analysis of numbers is the most perfect and most thorough, and shows the most order,
> it should not be surprising if it leads further. Perhaps it would be good, before publishing
> it, to try to make a little more progress in it. If some astute person, well versed in numbers,
> were to take pleasure in that, he would find no reason to be sorry, and I would gladly assist
> him with my little advice. (A II 4, 15)

In a letter to another Académie member, the Marquis de L'Hospital, of 4 April 1701, Leibniz
stated explicitly that his essay was written in the hope of inspiring a member of the Académie to
develop the binary system further (see A III 8, 596–597), though as it turned out, his hopes in this
regard were to be unfulfilled.

Leibniz's essay was read to the Académie on 23 April 1701, aside from the opening two paragraphs, which the Académie's minutes indicate were omitted (see M5). A week later, Fontenelle
wrote to Leibniz to indicate that while the response to his essay was positive, there were some
concerns about the practical value of the binary system:

> I read to the Académie your new *Essay on Numerical Science*. The institution had some
> right over this work since you kindly wanted to be one of its members. It recognized therein
> that spirit of discovery characteristic of you, and impatiently awaits the applications you
> promised us of this new calculus. It has no doubt that the useful things you will draw from
> it will sufficiently compensate for the apparent inconvenience of the quantity of digits, and
> perhaps you have found some way of reducing this quantity. Anyway, a greater clarification
> on this matter will make us extremely happy. (A II 4, 19)

At the end of his letter, Fontenelle indicated that, despite Leibniz's stated uncertainty about the
merits of publishing the essay, the Académie fully intended to do so:

> Your *Essay on Numerical Science* will be in the *Histoire* of this year, 1701. If you want
> to add something to it, you will have enough time. The Académie invites you to enrich its
> annual *Histoire* as much as you can with some of your discoveries. (A II 4, 20–21)

Leibniz waited more than a year before responding on 5 May 1702. In the initial draft of his
response, Leibniz indicated a potential willingness to develop practical uses of binary if he had
time or assistance: "As for the binary calculus, I would promise applications if I had more time or
if someone had the desire to pursue it" (A II 4, 57). However, this remark was deleted and replaced
with a reminder of his belief that the value of the binary system lay in its potential to enable advances
in number theory rather than in the sort of practical applications the Académie was looking for:

> As for the objection which seems to have been raised, Sir, that the multitude of digits in
> binary calculus is inconvenient, this does not address the purpose I am proposing for it. I
> recommend this calculus not for common uses, but to find rules and methods, to which it is
> more suited than any other progression because of the simplicity of its elements. And what
> appears to be vast is abbreviated when one can thereby regulate the sequences which proceed
> to infinity. Also, once the rule has been found, there will be no need to write all these digits

expressly. If I had seen that someone had shared my idea, I would have enjoyed pursuing it and explaining its uses. (A II 4, 57)

Only a fragment remains of Fontenelle's reply of 17 June 1702,[2] and if any part of his reply did touch upon the binary system, that part is no longer extant. In his response of 12 July 1702, having received no indication that any member of the Académie was keen to take his work forward, Leibniz formally requested that the Académie not publish his essay:

> I dare to beg you, sir, as I had already done, not to publish my essay on binaries, because being very tractable, and easily putting myself in the shoes of others, I see from what you said about it that what I said about it is not enough to make someone want to push it forward, which was my aim. (A II 4, 64)

Fontenelle responded on 18 November 1702:

> In accordance with your request, I removed your binary calculus from the *Histoire* of 1701. I am beginning to see what it can be used for, but I cannot help, Sir, urging you to give some idea of it, if only by a single example, so that this discovery may be given to the public. As the common decimal progression produces by itself a very easy way to recognize all possible multiples of 9, which is only a trifle, it should be shown that your binary progression provides some other important knowledge. (A II 4, 96)

On 7 January 1703, Leibniz sent a further letter to Fontenelle, writing, "I thank you for having removed at my request my binary essay from the *Histoire de l'Académie Royale*. One must accommodate oneself to the public taste, which wants to see palpable uses. And from the outset I noted my wariness about publication when I sent it to you" (A II 4, 138). Within 3 months of sending this letter, Leibniz had written and submitted an entirely different essay on binary to the Académie (see chapter 30), and while Fontenelle did publish that essay in the *Histoire de l'Académie Royale*, serious doubts remained in his mind about the "uses" of binary (see chapter 30).

———————————∞○C🙰🙰○○———————————

---

2. Namely, a passage about aromatic oils that Leibniz quoted verbatim in a letter to Friedrich Hoffmann of 25 July 1702 (see A II 4, 62).

[M2]

Essay on a New Science of Numbers[3]

This project seems strange, to innovate in a science as well known and as cultivated from time immemorial as that of numbers. We will see, however, that it was not devised without reason. And what's more, there still remains a new geometry to devise, which I call *analysis situs*, which depicts situations without figures, and which gives a calculus of them very different from algebra, or from the calculus of magnitude.[4] It does not suppose any other elements, and will have very great uses. But let us dwell now on arithmetic.

Everyone agrees that the decimal progression is arbitrary, so others have sometimes been used. That made me think of the binary, or double geometric progression, which is the simplest and most natural. I first determined that it would have only two digits, 0 and 1, and that made me want to experiment with it, because being so unencumbered, it must have its variety only in the order. Consequently, I found that there is an admirable order everywhere and that the sequences carry their rule of progression with them in the very expression of the digits. This method should not be for the practice of ordinary calculation, but it can be very useful for the perfection of science and for more subtle practices. And as analysis cannot be pushed further, since here it goes down to the simplest elements, which are 0 and 1, I found that with this method we can overcome difficulties which are beyond other ones, as will be shown below. Some people have admired therein the wonderful analogy between the origin of all numbers from 1 and 0 and the origin of all things from God and nothing;[5] from God as the principle of perfections and from nothing as the principle of privations, or the voids of essence, without there needing to be besides that a matter independent of God. But let's get to the point.[6]

I merely suppose that there are only two digits, 0 and 1, and that consequently 10 sig-

$$\begin{array}{r} 11 \\ ..1 \\ \hline 100 \end{array}$$

nifies two, from which it follows that 100 will be four, 1000 will be eight, 10000 will be sixteen etc. This can be seen in the table [on the facing page], because 3, or $10 + 1$, is 11, and $11 + 1$ yields 100, since adding 1 to 1 gives 0 in the first column, with the remainder 1, marked by a dot in the next column, where being again added to 1, which was already there, it again gives 0, with the remainder 1 for the third column, which makes 100. This calculation can be continued forever by adding 1, as was done in the aforementioned table only to illustrate the foundation of the calculus. But as numbers, according to this expression, are arranged in their natural order, as is apparent in the table [on p. 142], we discover from the outset the method of writing them to infinity, because each column is periodic. The period of the first is 01 (i.e. 0101 0101 etc.), the period of the

---

3. In M5, before the title, Fontenelle recorded the reading of Leibniz's paper and the members present for it: "Saturday, 23 April 1701.
The meeting was attended by Honoraries: Abbé Bignon, Fathers Gouye, Seb[astian] Trouchet, Malebranche, Sauveur; Pensioners: Mery, Couplet, Varignon, Fontenelle, Jaugeon, Des Billettes, Lemery, Du Verney, De La Hire, Rolle, Marchant, Homberg, Du Hamel, Gallois, Le Fevre; Associates: Bourdelin, Littre, De La Hire, Morin; Pupils: Amontons, Lemery, Boulduc, Carré, Desenne, Pourpart, Thuillier, Du Verney, de Beauvilliers, Parent, Simon. A letter was read out that I received from Mr. Leibniz of Wolfenbüttel, dated 26 February 1701, in which is a new idea of arithmetic, not for ordinary use but, according to what the author claims, for numerical problems and research, which would be very difficult without it. The basis of it is the binary progression in place of the decimal." Fontenelle then gives the piece a slightly different title: "Essay on a New Science about Numbers." At the time, all members of the Académie were divided into one of four hierarchically ordered groups: Honoraries, Pensioners, Associates, and Pupils, the first composed of ten people appointed by the French king, the other three groups composed of twenty each.
4. For details, see De Risi (2007).
5. Leibniz means Duke Rudolph August; see chapters 14, 17, and 18.
6. M5 does not contain the opening two paragraphs. In the margin of M5 corresponding to the position of the sum shown here, the following binary sum appears instead: $\begin{array}{r} 11 \\ 111 \\ \hline 100 \end{array}$. The sum is incorrect (it is the equivalent of $3 + 7 = 8$); Fontenelle seems to have erred by replacing each dot in Leibniz's sum with a 1.

second column is 0011 (i.e. 0011 0011 etc.), of the third column 0000 1111, of the fourth column 0000 0000 1111 1111, and so on.

A simple glance at the table [on p. 142 below] reveals the reason for a remarkable truth, which arithmeticians had observed long ago, namely that by a privilege of the double geometric progression all numbers can be formed by the simple combination of numbers of this progression. The table shows how, for example, 23 is 10111, i.e. $10000 + 100 + 10 + 1$, or else $16 + 4 + 2 + 1$, and 53 is 110101, i.e. $100000 + 10000 + 100 + 1 = 32 + 16 + 4 + 1$. This is what led assayers to use weights in double progression, for in this way few weights are sufficient for many values. If coinage was ordered in this way, the fewest pieces would be sufficient for the most values, and as the small pieces have the smallest value, by their own constitution, they would be least needed. But that is said without any intention of correcting [common] usage, which has its reasons. This way of calculating has nothing to do with the Pythagorean table, as it is called, except the beginning, namely once one is one, which is an axiom. So this calculus does not assume anything known by memory, unlike the others, that is, everything therein is demonstrated by the operation itself. Multiplication is done simply by writing the number to multiply, and division is also always done without trial and error.[7] The whole difficulty is reduced to addition, the rule for which is this:[8]

1	1
1	
10	2
1	
11	3
·· 1	
100	4
1	
101	5
· 1	
110	6
1	
111	7
··· 1	
1000	8

One proceeds in effect as in ordinary addition, whose analogy permits that one begins with the columns furthest on the right, and that in the place of each number of the double progression contained in the column, beginning with the highest, we mark a 1 (by a dot if you like) on a column subsequent to the proposed one, or the one in question, by as many intervals as this number contains 0s. For example, on the first after the proposed one, if it is 10; on the second, if it's 100 etc. After which, depending on whether there is a 0 or a 1 left, a 0 or 1 will be written

---

7. Leibniz first recorded this observation in some remarks scribbled on the back of a letter he had received from Justus Christoph Böhmer in the second half of 1694: "All things being equal, it is the same operation for subtraction and division in binary as it is for decimal. This is important because in binary division there is no need for trial and error or for multiplication, since in fact it is nothing other than the repetition of simple subtraction" (LBr 82 Bl. 44r; English translation: Leibniz 1694).

8. this. ▷ {1} As long as the given column forms a number of the double progression, starting with the highest we mark a dot on a subsequent column as far away from the given one as the number contains dots; and when there is no such number left, if there is nothing at all left we will write 0 under the given column, and if 1 remains (that is, if what it makes overall is odd) we will write 1 there. {2} We discern the highest number of the double geometric progression in the given column and mark it with a dot on the next column as far away from the given one as this number contains zeros; after that, we proceed in the same way with what remains in the given column, still seeing where it goes the highest, and as long as there remains a number of the double progression, which we always mark on the next appropriate column, and if after all this there is nothing left, we write 0 under the given column; but if there is 1 left (that is, if what the column makes overall is odd) we write 1 there. ◁ M1 *deleted*. Parts of deleted passage {2} above are also found in M3.

M1 then continued as follows: "As long as the sum of the given column goes to 10 or 2, not including a higher number of the double progression, a dot (marking the carry 1) should be put on the next column along; as long as the given column goes to 100 (or 4), a dot should be put on the second column along; as long as 1000 (or 8), a dot should be added on the 3rd etc., in due time." Leibniz then continued the passage thus: time. ▷ {1} After having disposed of the highest number of the double progression in the column, we proceed in the same way with what remains there, seeing where it goes the highest and marking it under one of the following columns. And if after all this there remains nothing or 1, we will write 0 or 1 under the given column. If, besides the highest number of the double progression the column contains, it also contains another [number], as if besides 1000 it goes up to 10, that is, to 1010, we mark the highest number of the double progression the column contains with a dot on the column as far away from the given one as there are 0s in this number. {2} The highest number of the double progression that the given column contains should be marked by a dot on a subsequent column as far from the given one as there are 0s in this number, for example, 1000 on the 3rd column along, 100 on the 2nd column along, 10 on the first column along. And after that we will proceed in the same way with the highest number of the double progression contained in what remains; and if after all these numbers there is a 1 left, it will be written under the given column, otherwise only 0 will be written there. ◁ M1 *deleted*.

under the proposed. The whole thing will be clarified in the example below, where the small lines in the columns mark the numbers of the double progression, and the letters added both to these lines and to the dots which correspond to them, are not for practice, but only to make clear the order of the process in this example.

(I organized this method better)[9]

```
            m.   1 |   1
      r. p. 1    1 |   3
      ℓ.  1  0   0 |   4
          1  0   0  1 |   9
          1  0   1  1 |  11
      q.      1  1  1 |   7
      n.  1  1_q 1  1 |  15
          1  1  1  1  0 |  30
       1  0  1  1_r 1  1 |  47
  v.     1_t 1_s 0  1_n 1_ℓ |  27
  t._x 1_b 0  0  1  1  1 |  39
x.  s.  1  1  0  0  1_p 1_m |  51
    1  1  1  1  0  1  0  0 | 244
```

The tableau initially consists only of the addends, that is, the digits 0 and 1 left of the vertical line and above the horizontal line. The number of 1s in the rightmost column is $10_{10} = 1010_2$. The rightmost digit of this count becomes the rightmost digit of the result, namely 0, and there are carries into the columns one and three positions left of this rightmost column. Leibniz calls these two carries $m.$ and $\ell.$ respectively and writes them at the top of those two columns. The second-to-rightmost column, including the carry indicated by $m.$, also totals $10_{10} = 1010_2$; carries $p.$ and $n.$ are recorded in the third from rightmost and fifth from rightmost columns, and so on. (Leibniz develops the binary count of the 1s and carries in a column by starting from the top of the column and working down, underlining when the count reaches the largest power of 2 not exceeding the total for the column, underlining that bit, and repeating the process for the remainder of the column.)

But one of the great uses of this calculus consists in the order of sequences. It is a general rule that when the principles proceed in an orderly way, that which results from them does so as well. Now as the natural numbers have their order, as I have just said, and it consists in periods, their multiples, like ternaries, quinaries, septenaries, novenaries etc., also proceed by an order which is periodic. The ternaries [on the next page] have periods of the first column 01, of the second 0110, of the third 0010 1101 etc., the quinaries of the first 01, of the second 0011, of the third 0101 1010; the septenaries of the first 01, of the second 0110, of the third 0111 1000, and so on for the others. Here it is remarkable that in all the multiples one half of the period is always the complement of the other half, that is, that it has a 1 complementing a 0, or a 0 complementing a 1. For example, the period of the third column of ternaries is 0010 1101, i.e. $\frac{0010}{1101}$. Now it would be a matter of assigning to each column of each sort of multiple the period it has, and of ultimately finding the general rule. Reasoning can be joined to induction here, and as only half the period is needed, the search is reduced.

All the first columns have for the initial half only the digit 0. The second columns, having the initial half of only two digits, of which the first is always a 0, can have only 00 or 01 for this half; thus 00 for the unaries or natural numbers, 01 for ternaries, 00 for quinaries, 01 for septenaries. And an order will be found there. The third columns have the initial half in the unaries 0000, in the ternaries 0010, in the quinaries 0011, in the septenaries 0111 etc. But since the beginning is always a 0, the variations of the third columns can only be 0001, 0010, 0100, 0011, 0101, 0110, 0111, and therefore these kinds of periods at most must always return with a certain order in all the third columns of all multiples. There will likewise be a certain number of periods that will return in order in all the fourth columns of the multiples, and so on. And it will be the same for the columns of powers. Everything could be found by pure reasoning, without induction. But when using induction as well, the search is easier and more pleasant. Multiples will also give us new knowledge about aliquot parts and primes.

From multiples I pass to powers, where it is especially remarkable that powers have periods just as simple as multiples, although their nature is incomparably more compound. For the squares [see the table on p. 144 below] have their periods of the first column 01, the second has only 0s, the third therefore passes for the second and has the period 0010, the fourth has 0001 0100 etc. And in the

---

9. This sentence, which refers to the method developed in "Binary Addition" (see chapter 24), was added some time after the manuscript had been completed. It appears only in M2.

Ternaries

00 0000	0
00 0011	3
00 0110	6
00 1001	9
00 1100	12
00 1111	15
01 0010	18
01 0101	21
01 1000	24
01 1011	27
01 1110	30
10 0001	33
10 0100	36
10 0111	39
10 1010	42
10 1101	45
11 0000	48
11 0011	51
11 0110	54
11 1001	57
11 1100	60
11 1111	63

Periods   01
0110
0010 1101
0001 1100 1110 0011

Quinaries

000 0000	0
000 0101	5
000 1010	10
000 1111	15
001 0100	20
001 1001	25
001 1110	30
010 0011	35
010 1000	40
010 1101	45
011 0010	50
011 0111	55
011 1100	60
100 0001	65
100 0110	70
100 1011	75

Periods   01
0011
0101 1010
0011 0110 1100 1001

Septenaries

000 0000	0
000 0111	7
000 1110	14
001 0101	21
001 1100	28
010 0011	35
010 1010	42
011 0001	49
011 1000	56
011 1111	63
100 0110	70
100 1101	77
101 0100	84
101 1011	91
110 0010	98
110 1001	105

Periods   01
0110
0111 1000
0010 1010 1101 0101

cubes, the period of the first column is 01, of the second 0001, of the third 0000 0101 etc., so that if the periods of the powers were determined, the tables of the higher powers could be written almost without calculation. It follows that triangular numbers, pyramidal numbers, and an infinity of other sequences formed by the addition or subtraction of powers, will also have the same periods. What is more, the geometric progressions, which are of an even more compound nature than the powers since such a progression contains all the powers at once, are also subject to these laws, and even a complement of halves is noticeable there, as far as I can judge without having examined them.

If the laws of sequences of every kind in the *binary* progression were determined, we would go on to determine them also in the *ternary*, which has only 3 digits: 0, 1, 2. And then on to the *quaternary* etc., and finding the series of laws in this way, we would thereby discover, in the arrangement of the digits themselves, laws for every sort of sequence even in our ordinary or *decimal* progression, laws of which nobody would have been easily aware otherwise. But without going so far now, and sticking to the binary progression, I see extraordinary utility here for solving numerical problems which otherwise are very difficult; I even see uses for the most sublime geometry that would not be expected. The late Mr. Hardy told me that he and someone else had worked on logarithms following the double progression rather than the decimal, and that they had found considerable utility therein.[10] They would have found much more if the binary expression had also been known then.

After giving specimens of induction, I will add a key to reasoning for those who might like [to continue] this research. By the letters *a*, *b*, *c*, *d*, *e* etc. let us understand the positions (*sedes* in Latin)

---

10. Leibniz is referring here to Claude Hardy (1604–1678), a renowned mathematician and translator of Euclid and Erasmus. Leibniz and Hardy met during Leibniz's stay in Paris, 1672–1676.

in the natural numbers in the table, so that *a* signifies what is in the first position on the right, i.e. the last digit, which is 0 or 1 and again 0 or 1 etc., which it suffices to designate by 01, the period of the first column in the sequence of natural numbers, which the table lays out up to 64. And *b* likewise will signify the second position on the right, or the penultimate digit, i.e. 0 or 0 or 1 or 1, and again 0 or 0 or 1 or 1 etc., in a word 0011, period of the second column; and *c* will designate 0000 1111, period of the third column, and *d* 0000 0000 1111 1111, and so on. So multiplying *gfedcba* by 11 (i.e. by 3), to have some ternary [number], the product will be

$$g\ f\ e\ d\ c\ b\ a$$
$$g\ f\ e\ d\ c\ b\ a$$

Now *a* is 01, which will be the period of the first column of the ternaries [p. 141 above], the same as in the natural numbers. $b + a$ is thus:

$$b\ 0011$$
$$a\ 0101$$
$$\underline{0110}\ \text{period}$$
$$0001\ \text{carry,}$$

therefore 0110 will be the period of the second column of the ternaries, reserving a 1 for each fourth position in the next column, which makes 0001, because $1 + 1$ gives a 0 with the carry 1, indicated here by a dot above the first 0. So the third column will be $c + b + 0001$ or

$$c\ 00001111$$
$$b\ 00110011$$
$$\underline{\text{carry}\ 00010001}$$
$$\text{period}\ 0010110\dot{1}$$
$$\text{new carry}\ 00010011$$

Therefore 0010 1101 will be the period of the third column, with the new carry 0001 0011. For the fourth it will be $d + c + 0001\ 0011$:

$$d\ 0000\ 0000\ 1111\ 1111$$
$$c\ 0000\ 1111\ 0000\ 1111$$
$$\underline{\text{carry}\ 0001\ 0011\ 0001\ 0011}$$
$$\text{period}\ 0001\ 110\dot{0}\ 1110\ 00\dot{1}\dot{1}$$
$$\text{new carry}\ 0000\ 0011\ 0001\ 1111$$

which gives for the period of the fourth column 0001 1100 1110 0011, and for the next one the new carry 0000 0011 0001 1111. This calculus also reveals the reason for complementary halves, and by going further into it we will find laws of the periods. The same method is also used to find, without induction and by pure reasoning, the periods of powers. But before we get there, consider that here $a = aa = a^3 = a^4$ etc., and likewise $b = bb = b^3 = b^4$ etc., and so on. Moreover, it must be considered that when multiplying one position by another, like *a* by *b*, the period of *ab*, product of two positions, is as large as that of the posterior position, or as the greater period, which is *b*. But it has only 0s in the first half, and the period of the anterior position in the other half. Finally, when in some column there is $2ab$ or some other quantity multiplied by 2, one has only to put this quantity, like *ab*, under the next column, which also serves to place any other such quantity multiplied by a number. For example, if *ab* was multiplied by 3 (or 11), being $2ab + ab$, we would keep it in the column where it occurs and repeat it again in the next.

The following example will make this clear.

gfe	dcba	
000	0000	0
000	0001	1
000	0010	2
000	0011	3
000	0100	4
000	0101	5
000	0110	6
000	0111	7
000	1000	8
000	1001	9
000	1010	10
000	1011	11
000	1100	12
000	1101	13
000	1110	14
000	1111	15
001	0000	16
001	0001	17
001	0010	18
001	0011	19
001	0100	20
001	0101	21
001	0110	22
001	0111	23
001	1000	24
001	1001	25
001	1010	26
001	1011	27
001	1100	28
001	1101	29
001	1110	30
001	1111	31
010	0000	32
010	0001	33
010	0010	34
010	0011	35
010	0100	36
010	0101	37
010	0110	38
010	0111	39
010	1000	40
010	1001	41
010	1010	42
010	1011	43
010	1100	44
010	1101	45
010	1110	46
010	1111	47
011	0000	48
011	0001	49
011	0010	50
011	0011	51
011	0100	52
011	0101	53
011	0110	54
011	0111	55
011	1000	56
011	1001	57
011	1010	58
011	1011	59
011	1100	60
011	1101	61
011	1110	62
011	1111	63
100	0000	64

Periods	01	*a*
	0011	*b*
	0000 1111	*c*
0000 0000	1111 1111	*d*

$$
\begin{array}{llllcccccc}
\text{Multiplying} & & \ldots & \ldots & \ldots & \text{etc.} & e & d & c & b & a \\
\text{by} & & \ldots & \ldots & \ldots & \text{etc.} & e & d & c & b & a \\
& & 2ka & 2ha & 2ga & 2fa & 2ea & 2da & 2ca & 2ba & aa \\
& & 2hb & 2gb & 2fb & 2eb & 2db & 2cb & bb \\
\text{instead of etc.} & 2gc & 2fc & 2ec & 2dc & cc \\
& & 2fd & 2ed & dd \\
& & ee \\
\text{we will write} & ka & ha & ga & fa & ea & da & ca & ba & 0 & a \\
& & hb & gb & fb & eb & db & cb & & b \\
& & gc & fc & ec & dc & & c \\
& & fd & ed & & d \\
& & e
\end{array}
$$

Which shows that in the table of squares [on the next page], the first column has its period a or 01, the second column has only 0. But the third column is $ba + b$:

$$
\begin{array}{ll}
ba & 0001 \\
b & 0011 \\
\hline
& 001\dot{0} \ \text{period} \\
& 0001 \ \text{carry}
\end{array}
$$

Which shows that the period of the third column is 0010, with the carry 0001. The fourth column contains:

$$
\begin{array}{ll}
ca & 0000\,0101 \\
\text{carry} & 0001\,0001 \\
\hline
\text{period} & 0001\,010\dot{0} \\
\text{new carry} & 0000\,0001
\end{array}
$$

Therefore the period of the fourth column is 0001 0100, with the carry for the next column 0000 0001. We also see the beginning of periods in the table [on the next page], which is that of the cubes, and these periods can be found by reasoning as with the squares.

This may suffice to provide the inspiration and means to push this research forward by establishing over time the sequence of periods. However, this knowledge of periods is not the only thing that deserves to be looked for here. And there are still other important considerations that this new arithmetic can provide. But one should enter into that through the periods.

squares[11]

0 0000 0000	0
0 0000 000<u>1</u>	1
0 0000 0100	4
0 0000 1<u>001</u>	9
0 0001 0000	16
0 0001 100<u>1</u>	25
0 0010 0100	36
0 0011 <u>0001</u>	49
0 0100 0000	64
0 0101 000<u>1</u>	81
0 0110 0100	100
0 0111 1<u>001</u>	121
0 1001 0000	144
0 1010 100<u>1</u>	169
0 1100 0100	196
0 1110 <u>0001</u>	225
1 0000 0000	256

Periods   01
[0010]
0001 0100
[0000 1101 0101 1000]

Cubes

000 0000 0000	0
000 0000 000<u>1</u>	1
000 0000 1000	8
000 0001 10<u>11</u>	27
000 0100 0000	64
000 0111 1101	125
000 1101 1000	216
001 0101 0<u>111</u>	343
010 0000 0000	512
010 1101 100<u>1</u>	729
011 1110 1000	1000
101 0011 00<u>11</u>	1331

Periods   01
0001
0000 0101

---

11. In his original draft (M1), Leibniz also included a table of prime numbers from 1 to 67 expressed in binary, with the remark "primes lack periods." This table was not reproduced in M2 or M5. The second and fourth periods at the bottom of the table of squares are placed in square brackets to show that they have been corrected. In M5 (and also in M1 and M2), Leibniz wrote "0100" in place of the correct "0010" and "0001 1010 1011 0000" in place of the correct "0000 1101 0101 1000."

Manuscript:
M: LH 35, 3 B 3 Bl. 16. Draft. Latin.

Leibniz very quickly recognized that binary notation was too prolix to be useful. He first responded by proposing use of sedecimal (hexadecimal) notation, observing that "binary is for theory, sedecimal for practice" (chapter 9). Here he elaborates a different approach first shown in chapter 23, attempting to simplify the labor of adding up many binary numerals.

To see the basic idea, suppose that the total of the 1s and 0s in the rightmost (low-order) column of binary digits is 8. (That is, of course, just a count of the number of 1s.) A binary version of the grade-school addition algorithm would record 0 as a digit of the sum at the bottom of the column and carry 4 into the next column to the left, where it would be added to the total of the digits in that column. Leibniz observes that it would be simpler to record a carry of 1 in the *third* column to the left. If the total of the rightmost column had been decimal $10 = 1010_2$ rather than 8, then there would be carry bits of 1 into both the immediately adjacent column and into the third column over. In general, the pattern of 1s in the binary representation of the sum of a given column precisely describes which columns should be flagged with a carry bit.

To see that this strategy really does reduce the arithmetic overhead, imagine (for the sake of simplicity) summing a list of n numbers, each of which is n bits long. Since each numeral represents a number that could have magnitude almost 2^n, the sum can reach almost $n2^n$, requiring up to $n + \log_2 n$ bits to represent. Using the conventional algorithm, the carryout of the rightmost column could be as large as $\frac{n}{2}$, so the total of the neighboring column could be as large as $\frac{3n}{2}$ and the carryout of *that* column as large as $\frac{3n}{4}$, and so on. The sum of the leftmost column could approach $2n$, causing a carry of almost n out of the leftmost column, adding the extra $\log_2 n$ bits to the result. Using Leibniz's method, each column could have no more than $\log_2 n$ carry bits generated from adding up the bits in the immediately preceding $\log_2 n$ columns, so the mental arithmetic is at worst that needed to keep track of sums up to $n + \log_2 n$ rather than $2n$. The price to be paid is a bit of bookkeeping, for which Leibniz proposes a strategy.

There are two reasons for thinking that this manuscript was written shortly after "Essay on a New Science of Numbers" (chapter 23). First, the example of a long binary sum is the same as the one used in that essay, and second, Leibniz went back to that essay after it was completed on 26 February 1701 to add a short sentence referring to his new method of addition (see p. 140 above), suggesting that it was still to hand. Accordingly, this manuscript was likely written not long after that essay (i.e., in the spring or summer of 1701). This would make the new method for addition the first of two advances Leibniz made with binary in 1701, the other relating to working out the periodicity of columns (see chapters 25 and 27).

Binary Addition[1]

If there are any number of 1s in a column, the last 1 of the largest number of the double geometric progression in the column—as it occurs beginning from the top—should be marked by an inscribed numeral of the exponent. And in turn, the last [1] of the next largest [number of the double geometric progression], and so on, until no number of the double geometric progression remains, but only a 0 or a 1. Eight 1s, or 2^3, would be marked by the value 3; 4, or 2^2, by the value 2; 2^1, or 2, by the value 1. And the value can be represented by the same number of little lines: ‴, ″, ′, or else the common numeral itself can be used should too many other little lines be needed. Once this marking-up is done in a column, dots corresponding to the numerals of the present column should now be marked in the following columns. In other words, because of value 1, a dot should be marked in the first place of the first column from the present one. Because of value 2, a dot should be marked in the second place of the second column from the present one; because of value 3, a dot should be marked in the third place of the third column from the present one. And so on. And I prefer to count the places upwards, so that the first place for a dot is under the bottom number of the column, the second place under the second-to-bottom number of the column, that is, between the bottom one and the next one up; the third place under the third-from-bottom number of the column, that is, between the second-from-bottom number and the next one up.

```
                        1   |  1˙
                    1   1   |  3¨
                1   0   0   |  4ⁱᵛ
            1   0   0   1   |  9˙
            1   0   1   1   | 11ᵛ
                1   1   1   |  7¨
            1   1   1   1   | 15ᵛ⁺⁴
        1   1   1″  1   0   | 3̈0
    1   0   1   1   1   1   | 4̈7ᵛ
        1   1   0   1   1‴  | 27ᵛ
    ′ 1′  0″  0   1′  1‴  1 | 3̈9ⁱᵛ
  .   1   1   0‴  0   1′  1′| 51ᵛ
  ─────────────────────────
  1   1   1   1   0   1   0   0 |244
```

A calculation marked up in this way always verifies itself to an onlooker comparing the places of the dots in a column with the numerals of a far off preceding column according to the rules, or the numerals of a column with the 1s in the column itself and then the digits of the sum under the line with the remainder in the column below the numerals.

Rather than deviating too far from the literal translation by interpolating our interpretation, we annotate his example without the column of decimal numbers at the right, and then interpret it. (The significance of the dots and superscripts in the column of decimal equivalents is obscure.)

---

1. addition ▷ ¶ There is a binary column to be summed: count the 1s from the top up to the greatest number of the double progression contained in the 1s of the column. To the last 1 of this, affix the numeral which designates the exponent of a number of the double progression. ◁ *deleted*. Here and elsewhere, the "exponent" of a number of the double geometric progression is its base-2 logarithm, that is, for example, 2 if the number is $4 = 2^2$, 3 if the number is $8 = 2^3$, and so on.

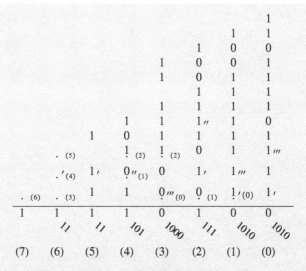

Here we have labeled the columns $(0), \ldots, (7)$ from low to high order, so the bits in column (i) contribute 2^i to the value of the result. Underneath each digit of the sum, we write the binary representation of the sum of the bits in that column. For example, the sum of the bits in the rightmost column is decimal 10 or 1010_2. That causes a sum bit of 0 (the rightmost bit of 1010) to be noted in column (0) and a carry of 1 into column (1) and a carry of 1 also into column (3). These carry bits are indicated by dots, and we have here marked those dots with the notation "(0)" to show that they resulted from a carryout of column (0). Of course, those carry bits themselves each contribute 1 to the sum of the column in which they appear, so the sum of column (1), which has nine 1s and one carry bit, is also decimal 10 or binary 1010_2 and causes carries into columns (2) and (4). Leibniz does not position the dots consistently, even though he implies that his rule is to insert dots starting from the bottom row and proceeding upward.

Leibniz uses the same method as he used in the example on p. 140 above (chapter 23) to achieve a binary count of the number of 1s and carries in a column. He determines the largest power of 2 in that count starting from the top ($8 = 2^3$ for the rightmost column in this example). Then, to record where to place a carry bit, he notes at that position in the column the exponent of 2 of that partial count, which will be the displacement from the current column of the column to receive a carry bit. He uses unary notation for that exponent, $'''$ for 3 in this case. He then repeats the counting process for the remainder of the column. The number of 1s below the $'''$ in the right column being $2 = 2^1$, he records $'$ at the position of the second and last 1 below the $'''$ in that column. In this way Leibniz completely avoids using decimal notation (or "ordinary numbers," in the language of p. 51, chapter 6 above). The annotated tableau has enough redundancy to "verify itself" in that (1) a unary count can be checked against the count of 1s and dots starting at that position and proceeding upward to the top or to the next unary count, (2) each dot can be correlated with a unary count in a column that may be "far off" to the right, and (3) each 0 or 1 digit below the line can be validated as the mod-2 sum ("remainder," after carries) of the 1s and dots above it.

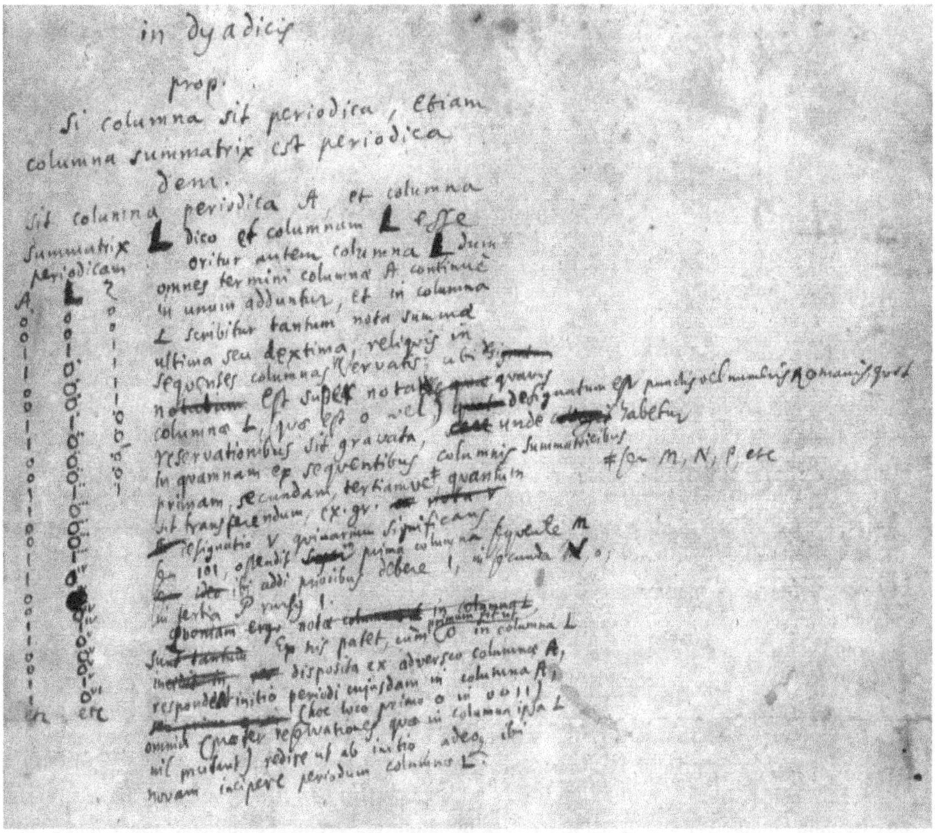
in dyadicis

prop.

Si columna sit periodica, etiam columna summatrix est periodica

dem.

Sit columna periodica A et columna summatrix L dico et columnam L esse periodicam

First manuscript page of “Periods in Binary” (LH 35, 3 B 5 Bl. 1r).

25 Periods in Binary (spring–fall 1701)

Manuscript:
M: LH 35, 3 B 5 Bl. 1. Draft. Latin.

Leibniz's enduring interest in the column periodicity of binary sequences peaked in 1701. In his letter to Naudé (chapter 21) and in the "Essay on a New Science of Numbers" (chapter 23), he began to advance from writing out sequences and inferring patterns inductively to using pure reasoning—treating binary digits as variables and manipulating them algebraically. And his attempts to interest others in this work had some success, with Pierre Dangicourt writing a paper on the column periodicity of natural numbers expressed in binary sometime between February and March 1701 (see A III 8, 534–545). This likely stimulated Leibniz to write the following manuscript, in which he revisits his interest in sequences such as the figurate numbers (see chapter 2), which are generated as the cumulative sums of initial segments of other sequences. He terms a sequence so generated the "summatrix" of the generating sequence, then focuses on the case in which the generating sequence consists of single-digit binary numbers; he shows that the rightmost bit of such a summatrix is periodic if the generating sequence is periodic.

The manuscript cannot be dated by its watermark, but its content shows it was written between Leibniz's receipt of Dangicourt's paper and the following October, when Leibniz further developed such ideas in several texts that culminated in November 1701 with "Demonstration That Columns of Sequences Exhibiting Powers of Arithmetic Progressions, or Numbers Composed from These, Are Periodic" (see chapter 27). Hence, we date it to spring–fall 1701.

In binary

Proposition:

If a column is periodic, the summatrix column is also periodic.

Demonstration:

Let there be a periodic column A, and a summatrix column L. I claim that column L is also periodic. Here column L is created when all terms of column A are continuously added into one, and in column L is written only the last or rightmost digit of the sum, the rest having been carried into subsequent columns; where[1] above any digit of column L (even a 0) it is marked by dots or Roman numerals what is bearing carries, there is had how much is to be transferred to which (first, second, or third, i.e., M, N, P) of the subsequent summatrix columns. For example,[2] the designation V, signifying five or 101, shows that in the first subsequent column M, a 1 must be added there to the earlier ones, a 0 in the second column N, and a 1 again in the third column P.

From these things it is evident that, when first it happens that a 0 in column L, laid out opposite to column A, corresponds to the beginning of some period in column A (in this case, to the first 0 in 0011), all things return as from the beginning (except for the carries, which change nothing in column L itself), and that therefore a new period of column L starts there.

A	L	Z
0	0	0
0	0	0
1	1	1
1	$0^{\cdot}$	1
0	$0^{\cdot}$	1
0	$0^{\cdot}$	1
1	$1^{\cdot}$	$^{\cdot}0$
1	$0^{\cdot\cdot}$	0
0	$0^{\cdot\cdot}$	$\overline{0}$
0	$0^{\cdot\cdot}$	$^{\cdot}0$
1	$1^{\cdot\cdot\cdot}$	1
1	$0^{\cdot\cdot\cdot}$	
0	$0^{\cdot\cdot\cdot}$	
0	$0^{\cdot\cdot\cdot}$	
1	$1^{\cdot}$	
1	0^{iv}	
0	0^{iv}	
0	0^{iv}	
1	1^{iv}	
1	0^{v}	
0	0^{v}	
0	0^{v}	
1	1^{v}	
1	0^{vi}	
etc.	etc.	

This proposition is true, but Leibniz's proof of it is incomplete and somewhat obscure. The "summatrix" for a list of numbers $A = a_1, a_2, \ldots$ is the list $L = \ell_1, \ell_2, \ldots$ that has in position n the mod-2 sum of the first n elements of A, that is, $\ell_n = (\sum_{i=1}^{n} a_i) \bmod 2$. Leibniz uses dot superscripts on ℓ_n for the number of carries out of the rightmost position, that is, $\frac{1}{2}((\sum_{i=1}^{n} a_i) - \ell_n)$, shifting from repeated dots to Roman numerals when the total is more than 3. Leibniz claims that if A is periodic, that is, if there is a p such that $a_{i+p} = a_i$ for all i, then L is periodic also. He works out an example with $p = 4$ and the repeating sequence $a_1 \ldots a_4 = 0011$, correctly showing that in this case, the summatrix is a repeating block $\ell_1 \ldots \ell_4 = 0010$. His argument is a bit troubled at this point, but it seems to consist in the correct observation that since the last element of the block $\ell_1 \ldots \ell_4$ is 0, when that is added to the first element of the next A block, namely, a_5, the result will be for the L block to repeat as from the beginning. All that is correct for the example that is worked out. Leibniz offers no explanation for the incomplete third column, to the right of the A and L columns, which seems to have Z as its header, but it appears to be a summatrix column based on column L. The period of the added column is 8 with block 0011 1100—as Leibniz seems to recognize, based on the small horizontal line drawn after the eighth entry of the Z column. The general rule, which Leibniz does not state, is that if the repeating block of A has an even number of 1s (as in 0011), then L is periodic with the same period as A, and if the repeating block of A has an odd number of 1s (as in 0010), then L is periodic with a period twice that of A, since after two repetitions of the digits in the block of A, an even number of 1s will have been accumulated and the entry in L corresponding to the last entry of the second block of A will be 0.

---

1. Omitting "est" here, probably part of an incomplete deletion.

2. To the left of this, Leibniz wrote: 0000 1111 1111 0000.

26 Periods and Powers (mid-to-late June 1701 [?])

Manuscript:
M: LH 35, 3 B 5 Bl. 25–26. Draft. Latin.

Leibniz's investigations into the column periodicity of binary numbers grew increasingly elaborate. In this manuscript, he fashions a series of notations for the columnar bit patterns seen in lists of sequential numbers and for operations on such patterns, especially bitwise multiplication (ANDing), which arises in some of the explorations seen in his "Essay on a New Science of Numbers" written a few months earlier (see chapter 23). In transcribing Leibniz's innovations, we have, where possible, employed typography suggestive of notation used in formal language theory today. For example, Leibniz writes $0 2^\ell 1 2^\ell$ for 2^ℓ 0s followed by 2^ℓ 1s, oversizing his handwritten 0 and 1 to distinguish these (as it were) object-language symbols from metalanguage letters and numbers being used in their ordinary algebraic senses. (The last page of the manuscript is reproduced near the end of this chapter.) Following the modern convention of treating such strings as a semigroup under concatenation, we write this instead as $0^{2^\ell} 1^{2^\ell}$, using exponentiation to denote repeated concatenation.[1]

Leibniz used similar notation in two other manuscripts (LH 35, 3 B 5 Bl. 87 and LH 35, 3 B 5 Bl. 97), on the back of one of which he wrote, "Draft a medal on the establishment of the English succession," which refers to the Act of Settlement passed in the English parliament on 12 June 1701.[2] Accordingly, it is likely that all of the manuscripts in which Leibniz experiments with such new notation were written around that time.[3]

The first, second, third, fourth, etc. column of any power of the natural numbers expressed in binary has a period of the digits [of length] 2, 4, 8, 16, etc. Hence [the period] can be known from that power of the numbers 0, 1, or 0, 1, 2, 3, or of the numbers 0, 1, 2, 3, 4, 5, 6, 7, 8, or of the

---

1. Leibniz's notation is less ambiguous than it might seem since he uses 0 and 1 only in the object language and other digits only in the metalanguage—with the exception of a couple of places where he needs a "1" digit to write "16" or "128" in the metalanguage.

2. Leibniz had a keen interest in this matter because he had lobbied for his friend and patron, Electress Sophie of Hanover (1630–1714), to be named rightful heir to the English throne should neither William III nor Anne bear further issue. The Act of Settlement secured the Hanoverian succession, as Leibniz wished, though he played no formal role in it.

3. In the first of these manuscripts, LH 35, 3 B 5 Bl. 87, Leibniz extends his grasp toward boolean algebra, defining the logical exclusive-or operator $\oplus$, which he writes as $\dot{\mp}$: "In binary, $a \dot{\mp} b$ to me signifies the digit which is produced from the digits $a + b$ added into one in the same column. For example, if a is 0 and b is 0, $a \dot{\mp} b$ will be 0. If a is 0 and b is 1, or the other way around, $a \dot{\mp} b$ will be 1. If a is 1 and b is 1, $a \dot{\mp} b$ will be 0, for $a + b$ is 10, where the 1 is carried into the next column. Hence $a + b = a \dot{\mp} b + 2ab$, for if a or b is 0, $2ab$ will be 0, but if either is 1, then $2ab = 2$. Hence $a \dot{\mp} b = a + b - 2ab$."

numbers 0, 1, 2, 3, ..., 16 etc.[4] And so the same numbers taken thence from 0 serve for that same column of any power:[5] 0 and 1 serve for the first column of any power and give the period 01.[6]

Leibniz scribbled several calculations in the right margin. The first calculates $3 \times 9 = 27$, $27 + 2 = 29$, $29 \times 3 = 87$, $87 \times 3 = 261$. The second gets the wrong result for 7^2 (should be 110001) through mishandling one of the carries. The third calculates $(5 \times 5) \times 5 = 125$. The fourth is 41×7 (41 being the erroneous result for 7^2). And the last is incomplete.

$$
\begin{array}{llllll}
\quad 1001 \\
\underline{\quad 1001} \\
\underline{\quad 11011} & & \quad 101 & \quad 101001 \\
\underline{\qquad 1} & \quad 111 & \quad 1010 & \quad 101001 & \quad 1100 \\
\underline{\quad 11101} & \quad 111 & \underline{\quad 11001} & \underline{\quad 101001} & \underline{\quad 111} \\
\quad 11101 & \underline{\quad 111} & \quad 11001 & \quad 100011111 & \quad 100100 \\
\underline{\quad 1010111} & \underline{101001} & \underline{1111101} & \quad 100011111 & \quad 100100 \\
\quad 1010111 & & & \underline{1000011111} \\
\underline{100000101}
\end{array}
$$

0, 1, 2, 3, or 0, 1, 10, 11, serve for the second column; 0 and 1 give a 0 for the second column, 10 gives 0 for the second column[7] of the square, and 11 also gives 0 for the second column of the square. Therefore the second column of the square has 0 everywhere.[8] Hence also [that] column of any even power will have 0 everywhere. For the cube, 0 and 1 give 0; 10 gives 0, 11 gives 1. Therefore, the period of the second column of the cube is 0001.[9] And generally in any power, 0, 1 and 10 give a 0 for the second column, but 11 in column 2 gives a 0 for even [powers] and a [1] for odd [powers]. Therefore in the second column we have 0000 for the period of even powers, and 0001 for the period of odd powers.

The third column should be investigated with the aid of the numbers 0, 1, 10, 11, [100,] 101, 110, 111. And 0 and 1 in that column give 0 everywhere. For square numbers, 10 gives 1, 11 gives 0, 100 gives 0, 101 gives 0, 110 gives 1, and 111 gives 0, making the period 0010 0010. And 0010 is sufficient for cubes: 10 gives 0, 11 gives 0, 100 gives 0, 101 gives 1, 110 gives 0, 111 gives 0, making the period 0000 0100.[10]

---

4. That is, establishing the repeating pattern in any power of x requires examining the same number of x values as does establishing the pattern in the corresponding column of x itself.

5. That is, in a table of of the binary representations of the powers of successive numbers, one per row, the bits in any column can be inferred from the first few.

6. Because $0^2 = 0$ and $1^2 = 1$, the rightmost column of any power of x alternates 01 just as the rightmost column of x itself does.

7. Leibniz here wrote "potentia" [power], but this is a mistake on his part; he is running through the four possible bit combinations for the last two bits of the base to determine the possibilities for last two bits of the square.

8. Here and below, Leibniz considers squaring or cubing a binary number whose last few bits we shall call c, b, a, each of which has the value 0 or 1. So the number represented is $2b + a$ if it has two bits or $4c + 2b + a$ if it has three. Then the last two bits of the square can be determined by evaluating $(2b + a)^2 = 4(b^2 + ab) + a^2$; thus the 2^1 position is identically 0 and the 2^0 position is $a^2 = a$.

9. In the expansion of $(4c + 2b + a)^3$, the only contributor to the 2^1 column is the term $2ba^2$, which is nonzero if and only if $a = b = 1$.

10. This is incorrect. $(4c + 2b + a)^3 = a^3 + 6a^2b + 12a^2c + 12ab^2 +$ (terms divisible by 8). Since $a, b, c \in \{0, 1\}$, $a^2b = ab^2 = ab$ and $a^2c = ac$. Thus the 2^1 term is ab with a carry of ab into the 2^2 position, and the 2^2 position is $3ac + 3ab$ plus a carry-in of ab, that is, ac plus carry-outs into higher order positions (since $3ac + 3ab + ab = ac \bmod 2$). So the 2^2 term of the binary representation is ac and the period is 0000 0101 as c, b, a assume the values $000, 001, \ldots, 111$. (The earlier "0010 is sufficient for cubes" is obscure.)

In a natural number, the first column has the period 01, the second [has the] period 0011, the third[11] $0^4 1^4$. We can represent this as follows:

8 4 2 1 instead of $0^8 1^8 \cdot 0^4 1^4 \cdot 0^2 1^2 \cdot 01$

1 gives 1, that is, 01 to 01 gives 01

$$\begin{array}{c} 0101 \\ \underline{0011} \\ 0001 \\ 0^2 01 \end{array}$$

$$\begin{array}{c} 0101\ 0101 \\ \underline{0000\ 1111} \\ 0^4\ (01)^2 \end{array}$$

$$\begin{array}{c} 0011\ 0011 \\ \underline{0000\ 1111} \\ 0^4\quad 0^2 1^2 \\ 0011\ 0011\ 0011\ 0011 \\ 0000\ 0000\ 0011\ 0011 \end{array}$$

16	8	4	2	1
16	8	4	2	1

$$\overline{0^{16}\,(01)^8}\ \ 0^8\,(01)^4\ \ 0^4\,(01)^2\ \ 0^2 01\ \ 1$$

$$1 \cdot 1 = 01$$
$$1 \cdot 2 = 0^2 (01)$$
$$1 \cdot 4 = 0^4 (01)^2$$
$$1 \cdot 8 = 0^8 (01)^4$$

$$2 \cdot 2 = 0^2 1^2$$
$$2 \cdot 4 = 0^4 \left(0^2 1^2\right)^1$$
$$2 \cdot 8 = 0^8 \left(0^2 1^2\right)^2$$
$$2 \cdot 16 = 0^{16} \left(0^2 1^2\right)^4$$

$$4 \cdot 4 = 0^4 1^4$$
$$4 \cdot 8 = 0^8 \left(0^4 1^4\right)$$
$$4 \cdot 16 = 0^{16} \left(0^4 1^4\right)^2$$
$$4 \cdot 32 = 0^{32} \left(0^4 1^4\right)^4$$

$$8 \cdot 8 = 0^8 1^8$$
$$8 \cdot 16 = 0^{16} \left(0^8 1^8\right)$$
$$8 \cdot 32 = 0^{32} \left(0^8 1^8\right)^2$$
$$8 \cdot 64 = 0^{64} \left(0^8 1^8\right)^4$$

Leibniz uses $1, 2, 4, 8, \ldots$ to denote the repeating block in the corresponding column of the sequence of binary representations of natural numbers and the $\cdot$ operator to denote bitwise multiplication (logical AND), with the further convention that if the two operands are different lengths, then the shorter is repeated to match the length of the longer operand. For example, the repeating block in the 2's column (second from the right) is $0011 = 0^2 1^2$, and the block for the 8's column (fourth from the right) is $0000\,0000\,1111\,1111 = 0^8 1^8$, so $2 \cdot 8$ is the bitwise product of $0011\,0011\,0011\,0011$ and $0000\,0000\,1111\,1111$, that is,

$0000\,0000\,0011\,0011$, or $0^8 \left(0^2 1^2\right)^2$.

Disregarding the even numbers, which here wander around outside their own column, as it were, let the remaining ones which proceed regularly be taken. Then the even numbers will be added in their own place.

---

11. Leibniz writes the immediately following expression as 0414 and similarly below. We adopt the modern convention of writing string repetitions using exponential notation.

		64	+32	+16	8	4	2	1
		64	+32	+16	8	4	2	1
				1·16	+1·8	+1·4	+1·2	+1·1
			2·16	+2·8	+2·4	+2·2	+2·1	
		4·16	+4·8	+4·4	4·2	+4·1		
	8·16	+8·8	+8·4	+8·2	+8·1			
16·16	+16·8	+16·4	+16·2	+16·1				

128	64	32	16	8	4	2	+0	1
1·128	1·64	1·32	1·16	1·8	1·4	1·2	+0	+1·1
	2·64	2·32	2·16	2·8	2·4			2·2
		4·32	4·16	4·8	4·4			
			8·16		8·8			
			16·16					

That is,

$0^{128}(01)^{64}$ $0^{64}(01)^{32}$ $0^{32}(01)^{16}$ $0^{16}(01)^8$ $0^8(01)^4$ $0^4(01)^2$ $+0^201$ $+0$ $+01$

$0^{64}(0^21^2)^{16}$ $0^{32}(0^21^2)^8$ $0^{16}(0^21^2)^4$ $0^8(0^21^2)^2$ $0^4(0^21^2)$ 0^21^2

$0^{32}(0^41^4)^4$ $0^{16}(0^41^4)^4$ $0^8(0^41^4)$ 0^41^4

$0^{16}(0^81^8)$ 0^81^8

$0^{16}1^{16}$

```
 11
 11
.. 1
────
101
```

$0^{15}10^710^5101$ $0^61^20^5101$ 0^71 0^{31}

$0^61^20^21^2011\dot{0}(01)^310(01)^20\dot{0}\dot{0}\dot{0}$ $0^41^201(01)^2100\dot{0}$ $0^2100\dot{0}$ $0+01$

0^31010^2

0^41^201

$(01)^201$

A new page begins here, and Leibniz devotes the rest of the manuscript to developing an algebra of column patterns. He introduces a new notation, writing "**1**3" for the sequence of 2^3 0s followed by 2^3 1s, and similarly for "**1**4," and so on. Where we have been able to identify that this convention is in use, we set the "**1**" in boldface, so "**1**2" denotes 0000 1111, for example.

$$\mathbf{1}4 + \mathbf{1}3 + \mathbf{1}2 + \mathbf{1}1 = N$$

$\mathbf{1}0 = 01$ $\mathbf{1}1 = 0011 = 0^21^2$ $\mathbf{1}2 = 0^41^4$ $\mathbf{1}3 = 0^81^8$, etc. $\mathbf{1}4 = 0^{16}1^{16}$

```
0101
0011
────
0001
```

$0^41^4 \cdot 0^{16}1^{16} = 0^{16}\left(0^41^4\right)^2$

$0^41^4 \cdot 0^{32}1^{32} = 0^{32}$

$\mathbf{1}\ell\mathbf{1}m = 0^{2^m}(\mathbf{1}\ell)(m-\ell)$

or $0^{2^\ell}1^{2^\ell} \cdot 0^{2^m}1^{2^m} = 0^{2^m}\left(0^{2^\ell}1^{2^\ell}\right)(m-\ell)$

The antepenultimate line to the left seems to be incomplete; to match the previous line, it would read

$$0^41^4 \cdot 0^{32}1^{32} = 0^{32}\left(0^41^4\right)^4.$$

(Concatenation takes precedence over the "·" operator.) Leibniz introduces the notation "(n)" to indicate 2^n repetitions of the immediately preceding string. However, in the last two lines to the left, $(m-\ell)$ should actually be $(m-\ell-1)$. If $\ell < m$, then $\mathbf{1}m = 0^{2^m}1^{2^m}$ has length 2^{m+1} and $\mathbf{1}\ell = 0^{2^\ell}1^{2^\ell}$ needs to be repeated $2^{m-\ell}$ times to match that length; the bitwise product is then 2^m 0s followed by $(0^{2^\ell}1^{2^\ell})$ repeated $2^{m-\ell-1}$ times.

Final manuscript page of "Periods and Powers" (LH 35 3 B 5 Bl 26r).

$$0^{2^m} 1^{2^m} + 0^{2^n} 1^{2^n} =$$

$$+0000\ 1111\ 0000\ 1111\ 0000\ 1111\ 0000\ 1111$$

$$+0000\ 0000\ 0000\ 0000\ 1111\ 1111\ 1111\ 1111$$

the same sum as the former   the opposite [sum] to the former $m + 1 - \ell + 1 - 1$

$$\left(0^{2^m} 1^{2^m}\right) (n - m) \left(1^{2^m} 0^{2^m}\right) (n - m)$$

$$0^{2^\ell} 1^{2^\ell} \cdot 0^{2^m} 1^{2^m} = 0^{2^m} \left(0^{2^\ell} 1^{2^\ell}\right)^{2^{m - \ell - 1}}$$

$$0^{2^{\ell+1}} 1^{2^{\ell+1}} \cdot 0^{2^{m-1}} 1^{2^{m-1}} = 0^{2^{m-1}} \left(0^{2^{\ell+1}} 1^{2^{\ell+1}}\right)^{2^{m - \ell - 2}}$$

$$0^{2^{m-1}}$$

$$0000\ 0000\ 0000\ 0000\ 0000\ 1111\ 0000\ 1111$$

$$0000\ 0000\ 0000\ 0000\ 1111\ 1111$$

$$0000\,1111\,0000\,1111\,0000\,1111\,0000\,1111 = 0^{2^{\ell}}1^{2^{\ell}} \qquad\qquad \ell = 2$$

$$0000\,0000\,0000\,0000\,1111\,1111\,1111\,1111 = 0^{2^{m}}1^{2^{m}} \qquad\qquad m = 4$$

$$0000\,0000\,0000\,0000\,0000\,1111\,0000\,1111 = 0^{2^{m}}\left(0^{2^{\ell}}1^{2^{\ell}}\right)^{2^{m-\ell-1}}$$

$$0000\,1111\,0000\,1111\,0000\,1111\,0000\,1111\,0000\,1111\,0000\,1111\,0000\,1111\,0000\,1111 = 0^{2^{2}}1^{2^{2}}$$

$$0000\,0000\,0000\,0000\,0000\,0000\,0000\,0000\,0000\,1111\,0000\,1111\,0000\,1111\,0000\,1111$$

$$= 0^{2^{5}}\left(0^{2^{2}}1^{2^{2}}\right)^{2^{5-2-1}}$$

$$0^{2^{\ell}}\left(0^{2^{\ell}}1^{2^{\ell}}\right)^{2^{p}}$$

$$\left(0^{2^{\ell}}\right)^{2^{p}} = 0^{2^{\ell+p}} \qquad \ell + p = m - 1 \quad \text{therefore } p = m - \ell - 1$$

$$0^{2^{\ell}}1^{2^{\ell}} \cdot 0^{2^{m}}1^{2^{m}} = 0^{2^{m}}\left(0^{2^{\ell}}1^{2^{\ell}}\right)^{2^{p}}$$

$$\left(0^{2^{\ell}}\right)^{2^{p}} = 0^{2^{\ell+p}} = 0^{2^{m-1}}$$

makes $\ell + p = m - 1$ or $p = m - \ell - 1$

Therefore $0^{2^{\ell}}1^{2^{\ell}} \cdot 0^{2^{m}}1^{2^{m}} = 0^{2^{m}}\left(0^{2^{\ell}}1^{2^{\ell}}\right)^{2^{m-\ell-1}}$

$$0^{2^{\ell+1}}1^{2^{\ell+1}} \cdot 0^{2^{m-1}}1^{2^{m-1}} = 0^{2^{m-1}}\left(0^{2^{\ell+1}}1^{2^{\ell+1}}\right)^{2^{m-\ell-1}}$$

$$+0^{2^{m}}\left(0^{2^{\ell}}1^{2^{\ell}}\right)^{2^{m-\ell-1}} = 0^{2^{\ell+n}}\left(0^{2^{\ell}}1^{2^{\ell}}\right)^{2^{n-\ell}} \quad \text{if } m = \ell + n$$

$$+0^{2^{m-1}}\left(0^{2^{\ell+1}}1^{2^{\ell+1}}\right)^{2^{m-\ell-1}} = 0^{2^{\ell+n-1}}\left(0^{2^{\ell+1}}1^{2^{\ell+1}}\right)^{2^{n-1}}$$

$$0^{2^{m-1}}0 \qquad\qquad\qquad 0^{2^{\ell+n-1}}$$

$$0^{2^{\ell+n-1}} \qquad\quad 0^{2}1^{2}$$

27 Demonstration That Columns of Sequences Exhibiting Powers of Arithmetic Progressions, or Numbers Composed from These, Are Periodic (November 1701)

Manuscripts:
M1:   LH 35, 3 B 4 Bl. 1. Draft. Latin.
M2:   LH 35, 4, 9 Bl. 1–2. Fair copy of M1. Latin.
M3:   LH 35, 4, 20 Bl. 7–12. Fair copy of M2, made in 1705, dispatched to César Caze. Latin.

GM: GM VII, 235–238 (following M1).

The following paper is a fuller exposition of, and a variation on, the ideas about column periodicity developed in chapter 25. The ideas had been worked on further in other manuscripts, most notably one that Leibniz had dated October 1701 and on the top of which had later added the following remark: "This is better in the paper in 4°, November 1701" (LH 35, 3 B 4 Bl. 2), a reference to the following paper, which was clearly a write-up of ideas developed in the months beforehand. At the top of the original draft of the paper, Leibniz indicated it had been written in November 1701 and shown to Pierre Dangicourt, the only mathematician he had been able to inspire to work on the same problem of column periodicity. Leibniz was sufficiently pleased with the paper that he not only had a fair copy made to show Dangicourt but, four years later, had another fair copy made to send to the mathematician César Caze (1641–1719), in both cases treating it as the culmination of his work on determining the rules for column periodicity.

The central claim of the paper is that when a periodic sequence is summed up, the cumulative sums are also periodic. Leibniz starts with some observations about the sequences of repeated differences, starting with the sequence of values of a polynomial $p(x)$ for $x = 1, 2, \ldots$. (That is, $p(1) - p(0)$, $p(2) - p(1)$, $\ldots$, and then the differences between successive members of *that* sequence, etc.) He then considers how the periodicity in a sequence $A = a_0, a_1, \ldots$ of 0s and 1s relates to the periodicities in each bit position of the sequence of binary representations in the "summatrix" $\sum_{i=0}^{n} a_i$ for $n = 0, 1, \ldots$. He expands the argument to establish the periodicities in the sum of two periodic sequences D and E, and finally to the periodicities in the sum of an arithmetic progression of three-bit numbers CBA, resulting in multiple carries into higher-order columns—he uses Roman letters for what comes from column A, Greek for column B, and Hebrew for column C. The argument might be seen today as involving a certain amount of handwaving but is fundamentally sound.

[M2]

Demonstration That Columns of Sequences Exhibiting Powers of Arithmetic Progressions, or Numbers Composed from These, Are Periodic[1]

<table>
<tr><td>1</td><td></td><td></td><td></td></tr>
<tr><td></td><td>15</td><td></td><td></td></tr>
<tr><td>16</td><td></td><td>50</td><td></td></tr>
<tr><td></td><td>65</td><td></td><td>60</td></tr>
<tr><td>81</td><td></td><td>110</td><td></td><td>24</td></tr>
<tr><td></td><td>175</td><td></td><td>84</td></tr>
<tr><td>256</td><td></td><td>194</td><td></td><td>24</td></tr>
<tr><td></td><td>369</td><td></td><td>108</td></tr>
<tr><td>625</td><td></td><td>302</td><td></td></tr>
<tr><td></td><td>671</td><td></td><td></td></tr>
<tr><td>1296</td><td></td><td></td><td></td></tr>
</table>

Every sequence of rational numbers, which are powers of arithmetic [progressions] of the same degree[2] (for example, the sequence of all x^4, where x is taken to be numbers of an arithmetic progression), when continued as far as is needed by differencing,[3] yields a consistent difference. (The ultimate difference of natural [numbers] is 1, of square [numbers] 2, of cubes 6, of quadrato quadrates 24, of sursolids 120 etc., that is, 1; 1.2; 1.2.3; 1.2.3.4; 1.2.3.4.5. Something similar happens if the arithmetic [progressions] are not the natural [numbers], but any other [arithmetic progressions], for example, the odd numbers set down in order, 1, 3, 5, 7 etc.)

It is the same if a sequence can be expressed by a complete analytical formula composed from powers, for example, $4x^4 + 6x^2 - 7x + 11$.

It is the same even if the formula has numbers in the denominator, but only if the terms are whole numbers. For when multiplied by a common denominator it yields, by differencing, a sequence of arithmetic [progressions]. Therefore even when not multiplied, for that does not change the uniformity of the difference. For example, the numbers $\frac{xx+x}{2}$ are always whole; therefore, a sequence of these has some consistent difference (in this place, the second), but the first differences are of an arithmetic progression.[4]

Hence all sequences of figurate [numbers], when continued by differencing, will yield a sequence of arithmetic [progressions] too.

And generally, every sequence of whole numbers expressed analytically by a formula, without a variable either in the denominator (as happens in a harmonic progression)[5] or in the exponent (as in geometric progressions), when the differences are continued as far as is needed, yields a sequence of arithmetic [progressions], or, what is the same, on the other hand, every sequence of this sort can be produced by the replicated summation of arithmetic [progressions]. For every sequence is composed from its own differences, no matter the degree, with the summations continued.

From this I now proceed to the expression of numbers by geometrical proportions, whether the binary [expression] or decimal or any other. And since all sequences of arithmetic [progressions] have periodic columns, this being especially evident in those expressed in binary, it will follow that all sequences of powers or of figurate numbers or of the others mentioned above have periodic columns, whereby it will have been shown that every summatrix sequence of a periodic [column] is periodic. For thus also a summatrix of a summatrix will be periodic. Indeed, there follows from this what we had presupposed, namely that an arithmetic sequence is periodic, since it is a summatrix of a sequence of consistent [periods], which is of course periodic.

To show that *a summatrix of a periodic sequence is periodic*, which is lemma 1, let us consider separately to begin with any column of a summand sequence, as if this column itself were a summand. I say: if the summand sequence of one column should be periodic, a summatrix sequence would

---

1. In M1, at the top of the page, Leibniz wrote, "Berlin, November 1701. I have demonstrated this to Mr. Dangicourt."

2. That is, a sequence $\{x_i^k\}$ for a fixed k, where successive x_i differ by a constant.

3. For clarity, we use the word "differencing" to describe the operation of taking differences between consecutive members of a sequence of integers, though it translates the same Latin word as Leibniz uses elsewhere for differentiation in the infinitesimal calculus.

4. $\frac{x^2+x}{2}$ assumes the values 1, 3, 6, 10, ... for $x = 1, 2, \ldots$. The first differences are 2, 3, 4, ..., and the second differences are 1, 1, 1, 1, ...

5. The nth element of the harmonic progression (series) is the sum of the reciprocals of the first n positive integers.

have all columns periodic, for example,[6] a summatrix sequence composed of columns LMN etc. of a consistent summand sequence from column A, I say that because A is periodic, LMN are periodic too. In other words, where it happens that a 0 corresponds to a (simple or replicated) period of column A in the summatrices columns, all things in those columns return as before. For columns L and M, that happens in place ⊙, where the simple period of column A ends. But for columns L, M, N, that happens in place 𝔻, where the duplicated period of column A ends, which also constitutes an entire period in itself. And it can be foreseen by the replication of the period of column A how many 0s occur in the summatrix. For it is evident how many 1s a simple period of summand A has, for example, in this place 4. Now one times 4 gives 100 in binary, and thus a 0 in the first and second summatrix column; but the column of the two periods of column A, that is, twice four, is 1000 in binary, that is, it gives a 0 in the third summatrix column too etc. If a simple period were to have three 1s, then it would

$$\overbrace{\text{once 3}} \qquad \overbrace{\text{twice 3}} \qquad \overbrace{\text{thrice 3}}$$

be $1 \times 3 = 11$, and $2 \times 3 = 110$, and $3 \times 3 = 1001$ and $4 \times 3 = 1100$ etc. And thus the period of the first summatrix column would be twice as long as the period of the summand column, the period of the second summatrix column would be four times as long as the period of the summand column, that of the third [summatrix column] would be eight times as long etc., and that happens whenever the sum of period A itself is an odd number.

I now take up the second lemma: *when two or more periodic columns are added into one, so that a term is added to a corresponding term, they yield a periodic sequence.* For if there are two columns to be added, DE, either 0 and 0 will always be added, or 0 and 1, or 1 and 1. In the first case it produces 0 in the first column F of the united sequence FG; in the second case, 1, in the third, 0 , or 10; that is, 0 with a carry of 1 to the next column G. And from the two united [columns] DE are produced only the two columns F and G, and each of these follows the period of the column endowed with the longest period from the first two, which here is E. For the first elements always return after this. Indeed I suppose, as is always the case in our binary, that a longer period always contains a shorter one, being twice or four times or eight times as long etc. (although if one did not accommodate the other, the common period would still ultimately be formed from periodic numbers, so that if one period were of 3 terms and the other one of 4, a common period would be of 12 terms). From this in turn it follows that even three or more other columns united into one, in such a way that the corresponding terms are added into one, yield a periodic sequence, since the third periodic column is to be added to the one composed of the first two, which as I have already shown consists of periodic columns.

We have hitherto spoken about summand column A alone; now we should consider that a given summand sequence can be composed from several columns, A, B, C etc. and however many summatrices as they have, as A has L, M, N etc. B has λ, μ, ν etc.; and the summatrix sequence itself has columns X, Y, Z etc. I say that X will be L, but Y will be formed from λ and M, and Z will be formed from $\flat$, μ, N, and so on. Or, to express matters more clearly, columns L, M, N, summatrices of column A, have terms denoted by little circles; columns λ, μ, ν, summatrices of column B, have [terms] denoted by triangles, and the summatrices of C, namely), ▢, $\flat$ have terms denoted by little squares. It is evident that X is just composed of column L's little circles, Y is composed of column M's little circles and column λ's triangles, and lastly Z is composed of column N's little circles, column μ's triangles, and column $\flat$'s little squares, and so on. Or if you express it

	N	M	L	A	
0	0	0	0	0	
0	0	0	1	1	
0	0	1	0	1	
0	0	1	0	0	
0	0	1	0	0	
0	0	1	1	1	
0	0	1	1	0	
0	1	0	0	1	⊙
0	1	0	0	0	
0	1	0	1	1	
0	1	1	0	1	
0	1	1	0	0	
0	1	1	0	0	
0	1	1	1	1	
0	1	1	1	0	
1	0	0	0	1	𝔻

G	F		E		D
	0		0		0
	1		1		0
1	0		1		1
1	0		1		1
	1		1		0
	1		1		0
	1		0		1
1	0		1		1

---

6. In the example, column A consists of repetitions of the sequence 0110 0101, and in row n. the other columns show the binary representation of the sum of the bits in column A up to and including row n.

in words: *X*, the first column of the summatrix sequence *XYZ*, is composed from *L* (first column of the first summatrix column of summand *A*), but *Y*, the second column of the summatrix sequence, is composed from *M* (the second column of the first summatrix column of summand *A*) and from λ (the first column of the second summatrix column of summand *B*), and *Z*, the third column of the summatrix sequence, is composed from *N* (the third column of the first summatrix column of summand *A*) and from μ (the second column of the second summatrix column of summand *B*) and lastly from $\flat$ (the first column of the third summatrix column of summand *C*), and so on.

But since (by lemma 1) the individual columns being united are periodic, a united [column] will also be periodic by the preceding demonstration, that is, by lemma 2, and although when they are united something is again transferred into the following columns, it is nevertheless clear that it is also periodic itself, and therefore when adding it to other periodic [columns] it must ultimately form periodic columns. Which was to be demonstrated.

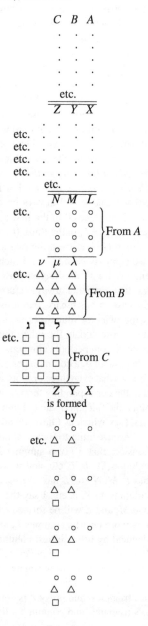

Manuscripts:
M1: LBr 105 Bl. 21–28. Fair copy, dispatched. French.
M2: Paris, BiBl. Nat. ms. Fr. 17240 pp. 75–88. Copy of M1. French.

Transcriptions:
H: Bouvet (1704) (following M2 [?] but with variants).
A: A I 20, 533–555 (following M1).
W: W 330–374 (following M1).

Bouvet's response to Leibniz's letter of 15 February 1701 (see chapter 22), and in particular to the outline of the binary system therein, was enthusiastic, as the following letter attests. Its centerpiece is Bouvet's hypothesis of a "wonderful correlation" between Leibniz's binary numeration and the sixty-four hexagrams of the Yijing, or *Book of Changes*, an ancient Chinese divinatory text.[1] Each of the hexagrams consists of six stacked horizontal lines, each line being either unbroken — (representing yang) or broken -- (representing yin). From these two elements, arranged in all possible groupings of six, emerge the sixty-four hexagrams. To demonstrate the alleged correlation between the hexagrams and binary, Bouvet enclosed a diagram (reproduced near the end of this chapter) in which the sixty-four hexagrams are arranged in a square and again in a large circle around the square.[2] Unhelpfully, Bouvet offered little explanation of how he had read either the hexagrams themselves or the circular and square arrangements of them in order to construe the arrangements as equivalent to the ordered sequence of the decimal numbers 0–63 in binary, advising only that the broken lines of each hexagram should be read as 0 and the unbroken lines as 1. But using the diagram Bouvet supplied, it is easy to determine his thinking. Bouvet read the individual hexagrams from the bottom line upward, such that he took the bottom line to represent the leftmost digit (32's place) of a six-digit binary number, the next line up to represent the leftmost-but-one digit (16's place) and so on, such that the hexagram ䷁ would be equivalent to 10 0000 in binary (32 in decimal).[3] As regards the square arrangement of hexagrams, Bouvet started at the top left and then read horizontally, from left to right, working from top to bottom in this way, just as one would read books written in European languages. Thus in the square arrangement, Bouvet considered the top row of hexagrams to represent the decimal numbers 0–7 in binary, left to right, the second row to represent the decimal numbers 8–15, and so on. As for the circular arrangement, Bouvet saw it as two semicircles divided by a vertical line running from top to bottom through the center; he started with the hexagram ䷁ (for him denoting 00 0000) at the bottom of the right-hand semicircle, then proceeded counterclockwise up the semicircle until he reached the top, at which point he continued the count from the hexagram ䷁

---

1. For a modern English edition, see Wilhelm and Baynes (1950).

2. Note that the diagram Bouvet sent is not itself part of the Yijing, even though the component hexagrams are.

3. Alternatively, one can ascertain the binary number Bouvet thought each hexagram represented by rotating the hexagram 90° clockwise and reading off the lines from left to right.

($10\,0000_2 = 32$) at the bottom of the left-hand semicircle, proceeding clockwise up the semicircle until he reached the top. (Despite the lack of explicit instructions, Leibniz clearly discerned Bouvet's method, since he added the numbers 0–63 to each of the hexagrams in the circular and square arrangements in the diagram Bouvet sent him.)

While Bouvet's reading of the individual hexagrams (from bottom to top) accords with how the Chinese have traditionally read them, his reading of the two arrangements of hexagrams in the circle and square does not. For example, the traditional reading order of the square is to start with the hexagram on the bottom right and then proceed right-to-left, bottom-to-top. Whether Bouvet's European-style left-to-right, top-to-bottom reading was a deliberate inversion of the traditional reading, or a straightforward mistake on his part, is unclear,[4] though the fact that he had been a student of the Yijing since 1693 suggests he would have known the traditional Chinese reading order.

Nevertheless, there were gaps in Bouvet's knowledge. For example, he indicated to Leibniz that the circular and square arrangements of hexagrams were the work of the mythical Chinese philosopher-emperor Fuxi, said to have lived in the third millennium BCE. However, this is not the case; the two arrangements of hexagrams in the diagram, known as the Prior to Heaven sequence, were not a more-than-4500-year-old product of the legendary Fuxi, as Bouvet believed, but rather a much more modern rearrangement, by the neo-Confucian Shao Yong (1011–1077), of the classical "King Wen" sequence, or Later than Heaven sequence. The King Wen sequence, in which there is no prospect of identifying a representation of a sequence of binary numbers, was not suitable for Bouvet's purposes, and he did not mention it to Leibniz (though Leibniz had come across it before in Intorcetta et al. 1687, xlii and xliv).[5]

Bouvet's interest in and interpretation of the Yijing had been a topic of his side of the correspondence with Leibniz for several years. In a previous letter to Leibniz, of 28 February 1698, he indicated that he was working on solving the riddle of the Yijing, in particular focusing on the "small whole and broken horizontal lines" (A I 15, 355), to which he thought he had found the key. And in a letter to Charles Le Gobien of 8 November 1700 (which Bouvet asked Le Gobien to forward to Leibniz), he stated his belief that numbers were the basis of the Yijing, which was "nothing more than a numerary metaphysics" (A I 20, 573) which, once properly understood, could explain all that was still unknown.[6] Bouvet made this claim about the Yijing as a whole, rather than just the hexagrams (the Yijing comprises not just the hexagrams, which are individually named, but also a short, enigmatic essay on each one), but clearly his prior convictions about the nature of the Yijing made it relatively easy for him to see a correlation between the hexagrams and binary numeration once he learned of the latter.

Despite suggesting a correlation between binary and the hexagrams, Bouvet made no further efforts to investigate it or incorporate it into his own studies. While he remained a student of the Yijing

---

4. While some scholars have supposed that Bouvet simply erred in his "European" reading direction of the hexagrams (see, for example, Zacher 1973, 112; Swetz 2003, 288–289), others have claimed that it was a deliberate decision, made possible—at least in Bouvet's mind—by his belief that the circular and square arrangements allowed for flexibility of interpretation (see, for example, Mungello 1977, 51).

5. For further details on why Bouvet erred in thinking that the hexagrams were intended as a binary system, see Sypniewski (2005) and Zhonglian (2000).

6. "Judging by the analysis I made of it, it is nothing more than a numerary metaphysics, or a very perfect general method of the sciences, and drawn up not only according to the rules of three kinds of number progressions, but also according to those of the figures and proportions of geometry and the laws of statics, all equally necessary to build a system as simple and as fitting as this one to give order to all the sciences, and to explain all that we admire in the works of the Creator who, according to the testimony of Holy Scripture, observed there all these rules and has made everything *in accordance with number, weight and measure*. In order to show in passing to those who have knowledge of the philosophy of numbers (which is properly the science of this system) that the ideas I am putting forward are not simple conjectures, I will tell you that it consists of a double sequence of numbers, planes and solids, so linked together by all the consonances of music, and by a perpetual harmony, that, conforming exactly with the 64 characters and the 384 small lines of the figure, they represent the periods with all the harmony of the celestial movements, and besides that all the principles necessary to explain the nature and the properties of all things, the causes of their generation and their corruption" (A I 20, 573).

for the rest of his life, Bouvet's focus was on Hermetic and Kabbalistic readings of the Yijing itself rather than in interpreting its hexagrams as an encoded number system. In a later work, "Idea generalis doctrinae libri ye kim" [General Idea of the Doctrine of the Book of Yijing] from 1712, Bouvet portrays the Yijing as a hieroglyphic or symbolic philosophy about the creation, corruption, and restoration of the world, with no mention of Leibniz or binary.[7] And on the rare occasion Bouvet did mention Leibniz and binary (for example, in a letter of 3 December 1707 to some fellow Jesuits in China), he made no connection at all to the hexagrams of the Yijing.[8]

While Bouvet's hypothesis may not have ultimately had much significance for him, it certainly did for Leibniz. Leibniz received Bouvet's letter on 1 April 1703 and was immediately inspired to write a new paper for the Académie in Paris, to replace the one he had withdrawn (see chapter 23). The new paper, "Explanation of Binary Arithmetic" (see chapter 30), drew heavily on Bouvet's hypothesis. Further, Leibniz was sufficiently thrilled at the thought that his binary system was the key to deciphering an ancient Chinese enigma that, upon receiving Bouvet's letter, he wrote to a number of correspondents to let them know about it (see, for example, A I 22, 321–326, and Aiton 1981), even informing Pope Clement XI (see A I 23, 353).

Most of Bouvet's letter to Leibniz—including all of the material on the correlation between binary and the hexagrams of the Yijing—was published in the French journal *Memoires pour l'histoire des sciences et des beaux arts* in January 1704 and therefore appeared in print before Leibniz's only work on binary that was published in his lifetime, namely the aforementioned "Explanation of Binary Arithmetic," which was written in April 1703 but not published until 1705.

⸻⸻ ∞◦€ઽᴅ◦€◦ ⸻⸻

[M1]

Sir,   At Peking, 4 November 1701

You doubtless did not expect to receive such an early response to the very kind letter you did me the honor of writing on 15 February of this year. The speed of the ship that brought it to me, along with that of the individual to whom it was given on arrival for safekeeping,[9] was such that I still had the time needed to convey my very humble gratitude to you by the same ship. I was quite concerned to learn of the inconvenience that obliged you to make the trip to Bohemia.[10] I was also not a little touched by the death of Father Kochański which you took the trouble to tell me. But there was at the same time a special joy for me to learn that His Most Serene Majesty the King of Prussia has set up a new Society of Sciences and has given you its leadership.[11] The enthusiasm you have long shown for everything having to do with China, and in particular for the advancement of the Gospel, surely merits that China alone make as much of this news as all of Europe, which undoubtedly is making a great deal. The one sorrow I have in that is to see that the great talents heaven has given you for more important matters, such as the most profound sciences, draw you towards occupations that deprive the commonwealth of sciences of a good part of the great benefits it could expect from your learned meditations.

What you briefly mention regarding new discoveries made by you and others on the principles of your new analysis of infinitesimals has greatly piqued my curiosity. If the captain who commands the ship on which Father de Fontaney returned from Port Louis to Canton in five months and several days

---

7. The text is transcribed in Collani (2007, 278–283).

8. See note 3 in chapter 17.

9. Namely, the French merchant ship *Amphitrite*, which arrived in Canton (Guangzhou) on 9 September 1701. Leibniz's letter to Bouvet of 15 February 1701 was carried in the luggage of the missionary Jean de Fontenay, who was returning to China after having been in Europe since 1699. He passed on Leibniz's letter to Bouvet. See Madrolle (1901, 54 and 83n1).

10. See note 5 in chapter 22.

11. See note 3 in chapter 22. Bouvet here follows Leibniz's incorrect styling of Frederick William, whose official title was King in Prussia rather than King of Prussia.

of navigation had followed the very sage advice you have given to others, to observe the barometer to anticipate storms, he would have been only too pleased to benefit from it and would not have lost all his masts, which an unforeseen typhoon tore off, putting the ship a hairsbreadth from ruin a few days before making port.

We have notified the Emperor of China of the news you took the trouble to convey to me of the complete defeat of the Czar's army by that of the King of Sweden.[12] The Emperor was amazed that Sweden, with such unequal forces, had won such a great victory over the Russians. But ultimately the news does not seem to have troubled him at all. I hope that the Czar's suspicions do not wreck the projects you are pursuing on behalf of China by way of Russia.

Although[13] you had not told me in your previous letters about your new numerical calculus, Sir, what I heard about it in general from one of my friends, who had himself heard about it,[14] gave me a great desire to learn something of it directly from you. But what you were kind enough to tell me about it in your last letter exceeded my expectations, and has excited in me a real passion to be able to learn the entire system from you, not just for the sacred use you identify for it on behalf of religion,[15] which is the principal reason it ought to be held in high regard, especially by all the people of my profession, but also because of the wonderful correlation I found between your principles and those on which I imagine is founded the ancient Chinese science of numbers and the other sciences whose knowledge they have lost. Among others, [these include] physics, or the science that teaches the principles and causes of the generation and corruption of all things, in which the ancient sages of China found exactly the same analogy as in numbers, in which the whole science was founded on a system in no way different from your numerical table which you establish as the basis of your numerical calculus, from which you pass, as do the Chinese, from the generation of numbers to the production of things, while maintaining the same analogy in the explanation of both.

Besides, I speak here only of the numerical table to which you add the double geometric progression, which the Chinese chose as the simplest of all and the one which contains the most perfect harmony. And to make you see, Sir, that this table is—without changing anything—the same thing as the system of gua,[16] or small lines of the prince of Chinese philosophers, by whom I mean Fuxi,[17] I ask you three things that you will surely grant me without inconvenience: First, I beg you to extend your numerical table (which in your letter is not continued beyond the 5th degree of the double geometric progression, that is, up to the number 32), to extend, I say, your table up to the 6th degree of this same progression, that is, up to the number 64 or rather only up to the number 63, since 63 with the zero which is at the top of the progression makes 64. Once your table has been extended by half[18] by the addition of this degree of the double progression, I will ask you, second, to suppose that all the zeroes that are like images of nothingness, or imperfections, are changed to as many broken | as

you see in this parenthesis (¦) and that your entire table is composed just of little whole and broken lines, like that of Fuxi. With this supposition made, I beg you, third, to cut your table into two halves, each of which will contain 32 rows each of six little whole or broken lines, and to arrange them next

---

12. See note 19 in chapter 22.

13. Transcription H starts here.

14. Probably the Jesuit missionary Claudio Filippo Grimaldi, to whom Leibniz had outlined his invention of the binary system in a letter written in 1697 (included in chapter 19).

15. Namely, the idea that the binary system may make it easier for non-Christians to understand the doctrine of creation of all things from nothing by God.

16. Bouvet is here referring to the sixty-four hexagrams of the Yijing. Each hexagram is composed of unbroken lines (symbolizing yang) and broken lines (symbolizing yin).

17. A mythological figure and cultural hero, said to be the first emperor of China, and responsible for developing Chinese civilization through inventions such as fishing, agriculture, and writing. A seventeenth-century work on Chinese history by the Jesuit missionary Martino Martini (Martini 1658, 11–13), with which Bouvet would have been familiar, reported *inter alia* that Fuxi's reign began in 2952 BCE and that he was skilled in astrology, instituted marriage between men and women, and invented the hexagrams of the Yijing, the claim that Bouvet makes here.

18. Doubled, actually.

to each another as two columns, in such a way that the two ends, which touched each other before the bisection, are opposite each other. This makes these two numerical columns curve toward each other in a semicircle, in such a way that these two upper ends are united together, as well as the two lower ends. Then compare this figure with the circular Chinese figure I am sending you,[19] and see if you notice any difference and if you don't discover all this wonderful harmony which is found in your table.

This figure is one of the two figures of Fuxi which, by the admirable art of a perfected science (to which you seem to be taking the right path to have men recover what had been lost to them for more than 30 or 40 centuries), was known to contain, as it were under two general and magical symbols, the principles of all sciences and true wisdom.[20] And this great philosopher, whose features have nothing Chinese, though this nation regards him as the author of sciences and the founder of the monarchy, had built this system with its circular figure, it seems, to calculate and know exactly all the periods and movements of the celestial bodies, and to provide clear understanding of all the changes which happen continually and successively in nature through their influence.

The second figure you see enclosed in the first and which is a kind of magic square[21] is the same as the first in terms of its constituent parts, but very different no doubt in terms of the situation of its parts and in terms of the uses to which Fuxi, its author, put it. These uses are not and have not been known to anyone, any more than those of the first figure, perhaps since the beginning of this ancient monarchy. The various commentaries written on it at different times, even before Confucius, have served only to make its meaning more obscure.[22]

But if your numerical table is absolutely the same as the one that Fuxi used to form his system (which is considered by all scholars as a system of numbers), as you yourself will recognize as soon as you have compared them, you will perhaps also recognize at the same time, Sir, that the science— in which you rightly find that the natural and harmonious order of numbers of this table, continued according to the degrees of the double geometric progression, affords great insights—should not be considered a new science, at least in China, since the system of small lines attributed to the founder of the most ancient monarchy in the world seems to be none other than this science whose discovery should not fail to bestow upon you in Europe the same glory among the learned as if no one in the world had ever thought of it.

As for me, while thinking for quite a long time about the system of little lines, which for a very long time has passed here for an inexplicable enigma, I thought I glimpsed the nature, beauty, and scope of this science, as you will have been able to discern from a letter I wrote last year to France on the subject and which I asked Father Gobien to share with you.[23] I will confess to you that the conformity of your discovery with this ancient system, which is the thing I value the most in the world in the realm of science, has greatly increased in my mind the great esteem I already had for you, and makes me hope that, in opening these new paths to our scholars, we will one day be able to recover the knowledge of the true system of nature and of all the other sciences, more or less as the first patriarchs of the world once had, whether they obtained it as a gift from heaven or acquired it over time by dint of their genius and their assiduous and long labors.

Sharing the same principles as you, Sir, I am not at all surprised at the design for a characteristic you propose, in order to depict thoughts in such a way that the same characters serve both for calculation and for demonstrating by reasoning etc. For this type of writing seems to me to encompass the true idea of the ancient hieroglyphs and of the Hebrew Kabbalah, as well as the characters of Fuxi, who is regarded in China as the first author of this nation's letters or hieroglyphs, for the formation of which it is commonly said that he used the 64 combinations of his system's whole and broken

---

19. The diagram was included on a separate sheet. A facsimile appears at the end of this chapter.

20. The circular arrangement of the hexagrams is attributed to the Confucian philosopher and cosmologist Shao Yong (1011–1077).

21. A square arrangement of integers in which all the rows, columns, and main diagonals sum to the same value.

22. Bouvet is here referring to *The Ten Wings*, a collection of ten ancient commentaries on the Yijing. Although traditionally ascribed to Confucius, his authorship has been questioned for almost a millennium.

23. Namely, Bouvet's letter to Charles Le Gobien of 8 November 1700. See A I 20, 571–576.

lines, which are, as it were, as many different expressions of numbers. But as the Chinese are completely ignorant of the value of the numbers hidden under these expressions or algebraic symbols, and consequently are completely unaware of the correspondence between their characters and these primordial symbols, they were so altered and corrupted by common usage, to which they have been subjected for such a long time, that there remain very few which have been preserved whole enough to be able to be referred to their true and first origin. Accordingly, it seems to me that the shortest way of re-establishing this ancient characteristic, the idea of which seems to me rather akin to yours, would be first of all to re-establish Fuxi's system and use the double geometric progression, as he did, to yield a natural metaphysics, and to reduce in a very simple and very natural order all the ideas to their genera and species arranged by more or less universal degrees as in a genealogical tree, of which all the divisions and subdivisions of the branches from the trunk to the last twigs would be made by following this double progression. So that, first, the dot (.), which is the simplest sign one can imagine for representing unity, be the characteristic sign of the first principle and the transcendent being and serve to represent the simplest, the most perfect, and the most fecund of all beings. Second, the two genera that issue immediately from the first principle, namely the perfect and the imperfect,[24] be represented by the binary and ternary number signs (..) (...) or by two small lines, one whole — and a sign of the perfect and odd number three, and consequently very suitable for representing the universal genus of perfect things or ideas; and the other broken - - and sign of the even and imperfect number and suitable for representing the genus of imperfect things or ideas. Third, since according to the double geometric progression, the second degree of generation is four, 1 producing 2 and 2 producing 4, the two genera of the first degree, namely the perfect genus and the imperfect genus, must produce four subaltern genera, two subordinate to each of the two higher genera, namely the most perfect genus and the least perfect genus along with the more imperfect genus and the less imperfect genus, which can be represented simply and naturally by double lines in this fashion ☰ ☱ ☲ ☳, or else by the symbols of your table 11, 10, 01, 00, which amount to the same thing. Fourth, as the third degree of generation in the double geometric progression is 8, double of 4, the third degree of generation of genera of things or ideas must contain 8, namely 2 subordinate to each of the 4 superior genera, and differentiate between them by the simple ideas of more and less. For example, to make the matter clear in the subject of colors, of which light and shade are like the two universal genera or principles, the luminous genera will produce the white genus in a more intense degree, and the genus of the color yellow in a less intense degree; and shadow, on the other hand, will produce black in a more intense degree and blue in a less intense degree; these are, as it were, the four elementary colors. From these 4 colors, which are like 4 genera through the sole difference of more and of less, there emerge eight others, which can be distinguished by names, which describe their genera and their differences as more and less white, more and less yellow, more and less blue, and more and less black. And the idea of these eight genera, or sorts of colors, can be very well represented, and in a manner equally simple and natural, by the 8 combinations of 3 lines, to mark the third degree of generation in this way:

☰ ☱   ☲ ☳   ☴ ☵   ☶ ☷
white   yellow   blue   black

These are the 8 characters of the fundamental figure of the Fuxi system, called *ba gua tu* (figure of the 8 gua), which all the learned have always regarded as the foundation of all sciences, without anyone having yet been able to understand the real meaning, which I nonetheless think comes down roughly to what I have just said on this subject. Fifth, if one goes from this third degree to the fourth: because 8 produces 16 in the double progression according to this analogy, the 8 genera represented by the 8 trilinear characters of the figure of Fuxi produce 16 others, namely each one produces two, which keep between them the same analogy as their generators, differing only in the more or less, and they can be represented very naturally by 16 characters, each composed of four broken and whole

---

24. Bouvet's marginal comment: "By 'imperfect' I mean here not the absolute imperfect but the respective imperfect, like rest in relation to motion, cold in relation to hot etc."

lines and combined according to the same analogy. Sixth, from the fourth degree of generation, the fifth is reached in the same manner. It contains 32 specific genera, or generic species, representable by 32 characters each composed of the same two sorts of lines combined together in varying ways in fives, to mark the fifth degree of generation. Lastly, from this fifth degree the sixth degree is reached, containing 64 species of things, or ideas, which Fuxi doubtless wished to represent by the 64 sextilinear characters of his figure. They are all natural symbols of the ideas they represent, which together in this attractive order in which he arranged them, in his double square and circular figure, provides a very perfect method for treating the principles of every kind of science in depth and explaining them very clearly and without hindrance or difficulty, and by the same terms and analogies.

Here there is a reflection to make, namely that to get from the third degree of generation to the sixth, the shortest way would be to multiply by eight each of the 8 terms of the third degree. By this multiplication, we would get the same number of 64 and would thereby rid ourselves of two degrees of subaltern genera, the multitude of which can cause only difficulty and burden in a science, like this kind of metaphysics, which should be both simple and universal. Besides that, this multiplication would happily yield the 8 degrees of difference that Aristotle gives to his qualities.[25] So to apply this to the same subject, which we considered above as an example, I mean about colors, one will distinguish in this way with admirable clarity 64 different colors under eight principal genera, each of which contains 8 different colors. There is something admirable in this, namely that when these 8 genera of colors are arranged according to the order we have already noted above, with each distinguished in its 8 degrees, which form eight different shades, these eight shades united together will make but a single[26] universal shade of 64 kinds of colors, each different from its neighbors in only one degree. What I have said of colors can and must be said in the same way of every other subject[27] to which one might wish to apply this new, or rather this ancient metaphysics, which perhaps has some similarity, Sir, to the one for which the learned are hopeful that you will provide demonstrations, and which for this reason perhaps deserves that you deign to give it some attention.

According to this system, when it is said about some subject that it is "white as four," "black as eight," "hot as two," "cold as six," "dry as five," "wet as seven, or in the 7th degree" etc., we will have an idea as clear and precise of what is being said as when it is said that a degree is the 360th part of a circle, or that a celestial sign is the 12th part of the zodiac, etc. And it is very likely that Aristotle, and the other great philosophers who talked as he did, had these same ideas when they talked like this. And if posterity has found some obscurity in their expressions, it's only for lack of having understood their thoughts and having conceived aright their true system which, when it has been well examined, will perhaps prove to be running entirely on the same principles as that of Fuxi. In my opinion, as much for his antiquity as for his clarity, simplicity, solidity, and universality, he deserves with very good reason to be regarded as the prince of all philosophers, a view I think I can advance without thereby claiming to cede to China scientific superiority over all other nations of the world. (God forbid that I commit such a dreadful insult against Europe, my dear homeland, which is as superior to China in the high sciences as China is distant from Europe!) For though this nation considers this great man as its first master and the founder of its monarchy, there are good reasons, endorsed by very able scholars (I'm talking about the majority), who argue that Fuxi never set foot in China. And indeed, most of what is reported, both from the time he was alive and about what

---

25. Aristotle identified eight qualities in his *Categories*: habits, dispositions, natural capabilities, natural incapabilities, affective qualities, affections, shape, and external form. See Aristotle (1984, I, 14–16: 8b25–10a16).

26. Bouvet's marginal comment: "To see this entire shade of colors represented by the harmony of characters in the circular figure, the two pieces have to be put together so that one is viewed from the other side."

27. Bouvet's marginal comment: "With just two principles, or two kinds of matter, one fiery and mobile in nature, the other earthly and without any action by its nature, both produced by the creator, by following the rules and analogies of this system the production of all the kinds of things there are in the universe are explained without difficulty etc." In H, this sentence is placed in the body of the text, at the end of the paragraph, and is slightly different from the marginal comment in M1: "Thus with just two principles, or two kinds of matter, by following the rules and analogies of this system the multiplication of all kinds of purely corporeal beings is explained." H 143.

he did, is such that it is easy to judge, by the almost complete conformity of all that with what our ancient authors, and those of the east, reported of Zoroaster,[28] of Hermes Trismegistus,[29] or even of Enoch,[30] that Fuxi was none other than one of those great persons. Even the two hieroglyphs of his name give some basis for this conjecture. For the first hieroglyph, *Fu*, is made up of two other characters, namely *ren* (man) and *quan* (dog), which means something like "man-dog, or [possessor of] the canine instinct for hunting and seeking the causes and principles of all things." The same is true of Hermes Trismegistus, whom the authors of emblematic hieroglyphs have represented with a dog's head on a human body.[31] The second character, *xi*, means victims, a term that suggests Fuxi was a Sacrificer or high priest, and that it was he who organized sacrifices and religious worship. Aside from his common name, in books he is also given the title of *Taihao*,[32] which means very great, even three times great, or Trismegistus.

Returning to his system, on which I should further note: First, that the 8 genera of the third order of generation, each with its eight species, are represented in the square figure in as many horizontal lines containing 8 sextilinear characters. Assuming the generation of 64 species, made by the eightfold multiplication of each of the 8 genera of the third degree of generation, these characters should be considered as produced immediately by the combination of each of the 8 trilinear or simple characters, both with themselves and with each of the seven others, in such a way that the simple combined character forms the lower half, and the simple character added by the combination forms the upper half of the double character, as can be seen in the square figure.

In the second place, this system, which was devised more than 4600 years ago, if Chinese chronology is to be believed, is therefore the most ancient monument that exists on earth. It was devised by an infinitely enlightened man, one who could not have been ignorant of the age of the world and the order that the creator had maintained in creation, since through this system he claimed above all else to teach posterity the causes and true principles of the production of the universe and of all the parts that compose it. I should remark, then, that this philosopher had to have his doctrine received to have regard for the order of the creation, knowledge of which was still very distinct and universal in his time, when the children of Noah were still alive or had only recently died.

So it is for this reason, it seems, and to represent in a natural way the sequence of productions of all creatures in the six days reported by Moses,[33] that this philosopher, who must have lived before him, established six degrees of generation in his system. It is also doubtless to preserve the memory of the 7th day, which God sanctified by resting after the work of six days and which he required man to observe religiously to preserve the memory of the infinite blessings of its author, that Fuxi consecrated the seventh number in the system of his circular figure, in which all revolutions end and begin at the seventh. This is a mystery that the Chinese never understood, which we hope one day to unravel with the special help of Heaven.

To make evident the conformity of your beautiful demonstration of the existence of a God, Creator of all things, based on numbers, with that contained in Fuxi's system, I should remark in the third

---

28. Zoroaster was a Persian priest who founded the monotheistic religion that bears his name, Zoroastrianism, following a series of visions in which he questioned Ahura Mazda (God) and other divine beings.

29. Hermes Trismegistus, traditionally thought to be an initiate in esoteric wisdom and long identified with the Egyptian deity Thoth, was purportedly the author of the Hermetic corpus, a series of mystical texts concerning magic, astrology, alchemy, and philosophy. In some accounts, Hermes was said to be a contemporary of Moses and even Moses's teacher.

30. In Jewish and Christian tradition, Enoch was the great grandfather of Noah (see Genesis 5). In the apocryphal books attributed to Enoch, he was the recipient of secret knowledge from God.

31. In fact, traditionally, Hermes Trismegistus has been depicted with the head of an ibis on a human body.

32. Tai means "highest" or "supreme," and hao "limitless" or "vast sky." In Chinese mythology, Taihao was originally one of the five gods of the directions, being god of the east (as well as that of wood and of spring), but during the Han dynasty (second century BCE–second century CE) was gradually identified with Fuxi, who inherited Taihao's reign over the east and took on Taihao as an honorary name.

33. That is, the creation story in Genesis 1.

place that, since his system turns on the double geometric progression—1, 2, 4, 8, 16, 32, 64—continued to the 7th term, of which 1 or rather (.) is the principle or the 1st term and as it were the generator of all the others, 2 is the first degree of generation, 4 the second degree, 8 the third degree etc., and since he represents in this system the six last terms of this progression as symbols or images of the six degrees of generation of all natural things, it follows that as the second term, or first degree of generation, 2, which is the exhibitor of the characteristics of heaven and earth; it follows, I say, that as 2 is the first degree of generation and that it has 1 as principle of its generation, in the same way heaven and earth must be, according to Fuxi, the first degree of natural productions, and have a single principle of their production, represented by 1, first of the seven terms of the double progression, 1, 2, 4, 8, 16, 32, 64. A truth which, moreover, is clearly established in the classical books of China, since we find this passage in the *Book of Rites: Taiyi fen er wei tian di*, that is, *the great or three-times great unity*, or to put it better yet, *the Triune unity* (since the two hieroglyphs *Tai-ji* encompass all three meanings) is *the principle of Heaven and Earth.*[34]

But to show that this Taiji,[35] or this great unity, spoken of in the *Book of Rites*, is a real, intelligent, all-powerful unity, and the same thing as *Shangdi*,[36] supreme Lord of Heaven and Earth, it is enough to say that it is written in the *Shiji*, which is an ancient book of annals, much revered and written before the birth of Jesus Christ, that *in ancient times* victims of the first order were sacrificed and offered to Taiji, that is, The Greatest One, in the *jiao*, the place intended for sacrifices to Shangdi.[37] To which I might add that it is reported in a ritual almost as ancient that sacrifices were once made in the same way to *Yi-San*, that is, *to the Triune Unity*.[38,39] Which shows, Sir, that the ancient Chinese had knowledge not only of God as creator and as principle of all natural things, but even of the venerable mystery of the very Holy Trinity. If I were to add that their ancient books contain a hundred passages which can make it reasonable to conclude that they have had knowledge of the sin and punishment of the evil angels, of the long life of the first men, of the corruption of human nature by sin, of the flood, of the future incarnation of the Divine Word, and of the principle circumstances of the mystery of our redemption, matters which are indicated by such clear and elaborate features that they seem to me comparable or even preferable to the Sibylline oracles,[40] yet I would have to be resigned to seeing these propositions contradicted even by the majority of the same missionaries who recognize, as I do, as a matter beyond doubt, that the Chinese once had a quite perfect knowledge of the deity. But that would not deprive me of the hope of perhaps one day confirming at least the greater part of my opinion, which is certainly very well founded, as I hope to be able to demonstrate.

If you want to be soon convinced of this truth,[41] as well as we have been, Sir, for a long time, and for very solid reasons, despite what has been published against it these past few years throughout

---

34. Bouvet misquotes and misinterprets this passage from book 4 of the Liji [Book of Rites]: "From all this it follows that rules of ceremony must be traced to their origin in the Grand Unity. This separated and became heaven and earth" (Legge 1882, 386–387). Whereas Bouvet takes "the Grand Unity" to be God, Legge notes that classical Chinese commentators took it to refer to "the original vapoury matter of chaos, before the separation of heaven of earth" (Legge 1882, 387n2).

35. Taiji (literally "supreme [ridge]pole") is variously translated as "grand unity" (Legge 1882, 386), "supreme ultimate" (Zhang 2002, 179), or "great primal beginning" (Wilhelm and Baynes 1950, 86, 616).

36. The term used for the high God during the Shang dynasty (c. 1550 BCE–1027 BCE).

37. See Qian (1993, 26).

38. Transcription H omits all of the material from the end of this sentence to the sentence that begins, "You mentioned Golius's hypothesis …" three paragraphs hence (and Golius's name is omitted there also).

39. *Yi-San* is an inversion of "San-yi" (literally "three one"), a triad of independent gods—the heavenly unity, the earthly unity, and the great unity—recognized in Daoism during the Han dynasty. See Qian (1993, 27). The inversion was common among Jesuit missionaries of the time, who used it to argue—as Bouvet goes on to do—that the ancient Chinese had recognized a triune deity.

40. A reference to a collection of oracular prophecies written in Greek purportedly by sibyls, divinely influenced prophetic priestesses. In some of the oracles, the sibyls supposedly prophesied various events recorded in Jewish and Christian scriptures, including the future Messiah. However, the extant oracles were written by Jewish, Christian, and Pagan authors in the second to the sixth centuries CE.

41. In the margin, Bouvet added here: "I mean that the Chinese have had knowledge of the divinity."

Europe,[42] take the trouble to browse a short report in Latin we have just printed here in Peking, a copy of which I am sending you.[43] As you read the article about ancient tradition and all that relates to temples and sacrifices, as well as to maxims and proverbs, if you find any passage for which you would like clarification, be kind enough, Sir, to let me know, for as I worked on this part, perhaps it will not be difficult for me to satisfy you in this regard. In the meantime, if I can send to France this year the translation of another small Chinese work I prepared for new missionaries and for the Chinese who were curious to understand their classic books,[44] you will be able to find there something to fully satisfy your curiosity on this matter. This work is a complete collection of all the finest sentiments about the deity found in classical Chinese books, arranged in the form of a catechism suitable for instructing and showing, without any dispute or argument, that the ancient Chinese were not the least bit inferior in this matter to those who had the most enlightened knowledge of the deity.

If I were to follow my inclination I would say more on this subject, but I am aware that such a long letter, written with such haste, so little order, and so many crossings out, must be very tiresome for you. Nonetheless, before I finish I must respond to a few other points in your letter, while promising you to do my best to get you versions of the Lord's Prayer in every language I can etc.

Father Grimaldi, who has become quite infirm, was deeply touched by the honor of your remembrance. I asked him, on your behalf, to find some old and selected observations on astronomy that you desire. For our part, we will also do what we can to satisfy you in this regard. The Chinese used to have (I doubt, however, if this applies to anyone other than Fuxi, whose system seems to me founded on a perfect understanding of the science of numbers and of geometry) an understanding of numbers and geometry and in particular of the wonderful properties of the sides and the hypotenuse of right-angled triangles. I said "I doubt" because I have not examined whether what is now found in their books on this matter, for which they claim authorship, is really theirs, or actually came to them from the Arabs. You mentioned Golius's hypothesis concerning the characters and language of the Chinese;[45] this strikes me as true, as I believe just as much as he does not only that there is something artificial in their construction, but also that it was the result of a perfected science and wisdom of the first patriarchs of the world, whom I believe to be the authors and not the Chinese, whose hieroglyphs, corrupted as they are, seem to me to be the same as those of the Egyptians.[46] And their [Chinese] words, which have undergone a much more far-reaching corruption, in terms of sound and meaning still have many similarities with other languages, and especially to the simplest and most ancient languages, such as the holy language[47] and the other eastern languages. If I have the time I will one day be able to make some interesting observations on this subject, which seems to me to be very fruitful.

Since we embarked upon this mission,[48] the work of the French Jesuits has been subject to so many distractions and thwarted by so much difficulty that we have not yet been able to add to the ordinary work of the mission—which as you know must be our principal task—the systematic plan we had drawn up for working to clarify Chinese matters and to impart knowledge of this nation's sciences to

---

42. For example, Anon (1700) and Alexandre (1700).

43. Namely, Thomas et al. (1701). The work was later published in Europe. Judging from a remark Leibniz made on a letter dated 13 March 1703 that he received from another Jesuit missionary, Charles Le Gobien ("Father Bouvet says on p8 [of his letter of 4 November 1701] that he is sending me a report in Latin, printed in Peking"), it seems Leibniz did not receive the work. See A I 22, 278.

44. Probably Bouvet's "Tianxue benyi." See Collani (1988).

45. See Golius (1655).

46. In line with common Jesuit thinking of the time, Bouvet assumed that the Chinese were descended from Shem, one of Noah's three sons (see Genesis 10), and that their writing was derived from the Egyptian hieroglyphs.

47. Namely, Hebrew.

48. In 1685, the year Louix XIV sent Bouvet and four other Jesuit missionaries to China.

Europe.[49] But now Heaven puts us in a position to start working on it seriously—thanks to the rest we are going to enjoy with God's help in the midst of our work, and the number of our companions, which will increase considerably every year, as well as the number of our establishments—we hope henceforth to be able to send information to Europe every year which will give real pleasure to the learned and in particular to those who, like you, are interested in and contribute in such an extraordinary way to the perfection of the arts and sciences.

For our part, too, we presume and hope from the great zeal you show, Sir, for the propagation of the faith in this great empire, that you will kindly communicate your own discoveries to us, and those of this new and illustrious academy which is beginning to work in your example and under your auspices. These will be the new arms which the missionaries will use here to gradually destroy the empire of the devil and firmly establish that of Jesus Christ in its place.

I wish I had some dictionary translated into some European language to send you, but I don't. Father Visdelou is in the process of starting one, which will leave nothing to be desired on this subject, except perhaps for the analysis of the characters you want. For myself, I was strongly inclined to work on this analysis, but I think for that I should not distract myself from some other studies, which seem to me more important for the good of religion and even for our sciences. The translation of the Chinese dictionary into Tartar[50] which the Emperor had done by very able men has not yet been completed.[51] As soon as it is, one of us will translate it into Latin or French to give Europe the knowledge of three foreign languages at the same time. For the Chinese language and its characters are like two different languages, one speaks to the ears and the other to the eyes.

In the meantime, I shall look for the book you mentioned, I mean the one in which pictures of things are added to the words, along with another which contains pictures of ancient sacrificial vessels and other similar things with their inscriptions in ancient hieroglyphs. This work is here of the same style as our books of ancient medals, the explanation of the inscriptions is adjoined in hieroglyphs or modern characters. If they can be found in time, I will send them to you this year along with the sample of paper you requested, and then the secret of making this excellent paper, if I can obtain it.

For your wonderful discovery of numbers, along with the demonstration you draw from it to establish the dogma of creation, it will have to await some favorable opportunity to discuss it with the emperor, who for several years having been satisfied with what he wanted to know of the theory of our sciences, does not give us the same opportunity as before to talk to him about these sorts of things. Besides, when we talk to him about something, we have to be prepared for all the questions he might ask on that subject, and as you have not yet done me the favor of telling me everything that might be desired on this matter, I may need to wait for some new knowledge in this regard.

Since I returned from Europe,[52] the emperor has occupied me with just one task in his service, that of working with two of my companions,[53] whom I brought to him, and with Father Thomas, on a topographical map, 20 square feet, of a canton of this province 15 leagues from this town toward the south. We made it for him last winter. But this task has given me no opportunity to make any discovery I think worth writing you about, except that when making the map of the *haizi* pleasure house where the emperor goes every year to enjoy several days of hunting,[54] we saw among the wild animals, of which there are several species and in very large numbers, a kind of mule no different from the domestic ones, except that this species breeds in Tartary, where they are caught like other

---

49. The distractions and difficulty Bouvet mentions here likely refer to the rites controversy, which concerned the religiosity (or otherwise) of Confucianism and Chinese rituals. While many Jesuits had held that the rites carried out by the Chinese to their ancestors, and in some cases to Confucius, were essentially civil and secular rather than religious, others argued otherwise, such as the theological faculty of the Sorbonne in Paris. Pope Clement XI eventually banned the rites in 1704. Leibniz sided with the Jesuits on the rites controversy; see WOC 61–65.

50. That is, into Manchu.

51. This is the Qingwen jian [Mirror of the Manchu Language].

52. Bouvet ended his stay in Europe on 6 March 1698.

53. Namely, Jean-Baptiste Regis (1663–1738) and Dominique Parrenin (1665–1741).

54. Bouvet is referring to the Nanyuan or Nanhaizi Palace Garden to the southeast of Beijing.

animals. Perhaps what makes the domestic ones sterile is this constrained and managed rearing, just like the elephants we saw in Siam, where those who have the stewardship of the king's elephants, as well as their grooms, all assured us that the elephants became sterile as soon as they were tamed and raised in a stable, whereas when they were allowed the freedom to go into the woods, where the king of Siam had several studs, they breed as well as the wild ones, with which the female elephants of the royal studs mate. This year the emperor, beside his usual hunt in Tartary which lasted nearly four months, went on a special one. It's elk or moose hunting. He killed more than twenty of them by his own hand. He pursued this fine hunt in his own lands. The moose, and the beaver, which are found in those regions as in North America, is a new reason to speculate that there is some connection of these two continents.

I am keeping Mr. Schröck's and Father Kochański's questions to answer myself or with some of my companions.[55] Those who said that the crown prince of the empire was learning or had learned some European language are ill informed.[56] The emperor, as well as the princes, his children, has indeed sometimes had the curiosity to make us write the letters of our alphabet in front of him, and to make us read from our books, but never with the intention of learning them. It is true that his majesty and some of his children, following his example, having found the characters of our numerals simpler and more convenient than theirs for calculation, became so accustomed to them that they use our tables of sines etc., and logarithms, as we do. I had several copies neatly bound in France for this reason, which they are now using.

You have very good reason to think, Sir, that Chinese characters must be interconnected, since indeed they are, and so closely that if one fully knows this interconnection and the real basis for the construction of their characters, which has nothing to do with caprice or chance, it would be extremely useful for learning how to understand them and memorize them without fear of forgetting them. If this interconnection were understood and dictionaries were compiled according to a method based on this interconnection then it would be a real pleasure to study these characters, but for want of that it is the most difficult and disagreeable study there is in the world. In all dictionaries there is at the beginning a list of root characters from which all the others are composed,[57] but because in these [Chinese] dictionaries only the number of strokes is considered for the order of their arrangement, in order to be able to find the ones you are looking for more easily, hardly any other advantages can be drawn from these dictionaries.

Before finishing this letter, which I find twice too long for what it contains, you will perhaps not be sorry to see the analysis of several of the hieroglyphs which have been used for over 4,500 years by philosophers of the east to indicate the supreme being, which they have known at least as clearly as other philosophers of all nations other than the Christians and the Jews. These hieroglyphs will be a proof which could be taken for a demonstration by individuals capable, like you, of assessing these sorts of things.

As the dot and unity are the principles of the science of continuous magnitude and of discrete magnitude, or rather, of arithmetic and geometry, the two bases in which all Chinese science is anchored, it should not be surprising that ancient sages used the dot and unity signs to express the supreme being, the principle of all others and the simplest of all. For this purpose they used these two characters (.) and (—), the two simplest it is possible to imagine, and as it were two first principles of all the others. For this reason, first, in ancient times they used a single dot (.) to express the idea of master, of lord, of sovereign, a sign they pronounced *zhu* (lord), as evidenced by all or almost all dictionaries. Second, in the classical book *Shujing*, the character for unity (—) is also used to denote the unique and supreme being, *wei yi*.[58]

---

55. See note 58 in chapter 22.
56. See note 60 in chapter 22.
57. In the margin, Bouvet added here: "and under which they are distributed in so many classes,"
58. "The only one." Bouvet's information here is incorrect; the *Shujing* contains no uses of the character — to denote wei yi.

In the text of the book *Liji* quoted above, this same character serves to indicate the deity, as I said. *(Taiyi) fen er wei tian di*, that is, thrice greatest unity, or one thrice greatest is the principle of heaven and earth, for the term tai means "the greatest," yi "one," and these two terms together include the ternary number, or a trinity, as the Chinese usually say following their dictionary, without now understanding what they are saying, *Taiyi han san*, that is, the greatest one contains the Trinity (namely of persons).

It should be noted here that the character Tai, meaning "the greatest," is composed of two others, namely da, meaning "great," and the dot (.), meaning "lord." So Tai-yi when analyzed means precisely "one great lord," a title which suits only God, as well as "one greatest triune."

Hence, to denote this same idea with a single word they joined the character of unity (—) to the character 大 da, meaning "great," and so formed the character 天 Tian, "heaven," which signifies the heaven archetype rather than the material sky, for which it is also used. Add also to that the great similarity of the word Tian, both conceptually and phonetically, to the Greek word θεόν, θεός,[59] both being equally aspirated in pronunciation. A saying they often use is *yi da wei tian*, that is, the character – yi, "the one," and the character — da, "great," form the character tien, "heaven," or in the sense of the proverb, "the great one is heaven," something that is said more naturally of the Lord of heaven than of heaven itself.

The hieroglyph di, "emperor," Shangdi "supreme emperor," is the most common symbol used to express the supreme lord of all things. This hieroglyph, I say, is composed of two characters, namely li 立, which means "first to establish," and jin 巾, "banner or standard," which is the mark of command, which suggests that it is God who has supreme authority in the world, that it is he who commands all things and can demand obedience to the slightest beckoning of his will.

The hieroglyph di 帝 is commonly explained, even by the less orthodox philosophers on the subject of the knowledge and existence of a God, by these two other characters, zhu zai 主宰, that is, Lord who governs (Lord Governor). We will do an analysis of each. First, the hieroglyph zhu, "lord," is composed of two others, namely wang 王 (king) and the dot (.) Lord. So according to the analysis, zhu actually means Lord King. Second, the hieroglyph zai 宰 is composed of three parts: the first or upper 宀 is set up to signify cover in the form of a roof and denotes the sky; the second, which is shi 十, is the character of the number 10 among the Chinese as among the Romans and the Egyptians, and which is taken by the former for the totality of things enclosed by heaven, just as in Pythagoras, who represented all things by his famous quaternary 1, 2, 3, 4, whose sum is ten; the second part of this character, I say, which is shi, represents the totality of things enclosed by heaven. Finally, the third part of this hieroglyph, namely li 立 which is between the two preceding ones, and which means to erect or build, when joined to the two others marks in a very appropriate way the idea of the creator of heaven and all things, founder of heaven and of all things within its compass.[60]

The character huang 皇, which at first was attributed only to Shangdi, and which the emperors adopted for themselves (ever since that famous emperor usurped that magnificent title by burning all the books of the empire after having subjugated all the kings who shared China between them at the time), this character, I say, which means "august," is composed of these two others, bai 百姓, "hundred," and wang 王, "king, kings." It very clearly designates supreme power and shows that the sages of Chinese antiquity had recognized Shangdi by this title, as the Sovereign of Sovereigns, King of Kings and Lord of Lords. It should be noted that the hieroglyph bai, "hundred," is a universal

59. "God."

60. compass. ▷ In the Book of Odes called *Shijing*, there is another hieroglyph composed of this one, zai, and another, jun, that is, monarch or absolute ruler. These two characters joined together make that of bi, which consequently, by virtue of this composition, must mean the creator and supreme lord of heaven and all enclosed therein. Thus this sentence of *Shijing*, dangdang shangdi xiamin zhi bi, contains an idea of Shangdi quite worthy of God, that is, that the supreme emperor Shangdi, creator and absolute master of all peoples as well as of heaven and all that heaven contains, is great, o powerful is he! ◁ *deleted*. The passage Bouvet cites here is from *Shijing* III.1.1. A nineteenth-century English edition translates it as: "How vast is God, The ruler of men below"; see Legge (1871, IV: 505).

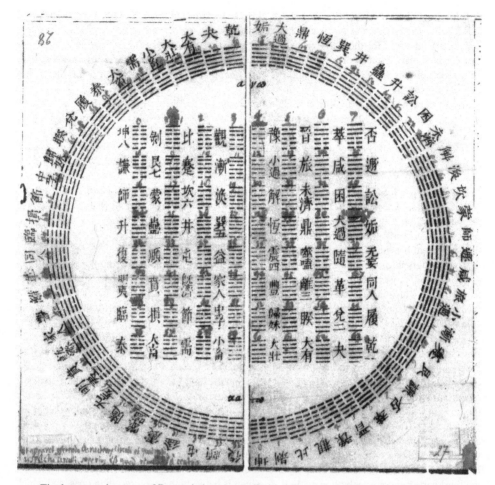

Final manuscript page of Bouvet's letter to Leibniz, 4 November 1701 (LBr 105 Bl. 27r).

collective term. For example, baixing, "a hundred surnames," is used to mean all people, while baiguo, "a hundred fruit," generally to mean all the fruits of the earth.

I could also add here the analysis and explanation of several other hieroglyphs specific to Shangdi which the Chinese use to express his infinite perfections. But what I have related about it seems to me sufficient,[61] by the force of the hieroglyphs, to convince people harder to persuade on this matter than you, namely that the ancient Chinese truly recognized a God, and adored something other than the material sky. And consequently, all those who deny this knowledge little understand the power, energy, and true meaning of their characters. I beg your pardon, Sir, if by the length and the poor organization of this letter I have transgressed the boundaries of profound respect with which I beg you to believe that I am, Sir, your very humble and very obedient servant J. Bouvet J.[62]

As you have knowledge of various books which I do not have here and which could be useful to me in the type of study to which you see I apply myself, kindly indicate them to Father Le Gobien, who will gladly take on the responsibility of looking for them on my behalf. But above all else, do not deny me, I beg you, your own insights, which I will value more than any other possible assistance.

---

61. Transcription H ends here.
62. Presumably, the second "J" stands for "Jesuit."

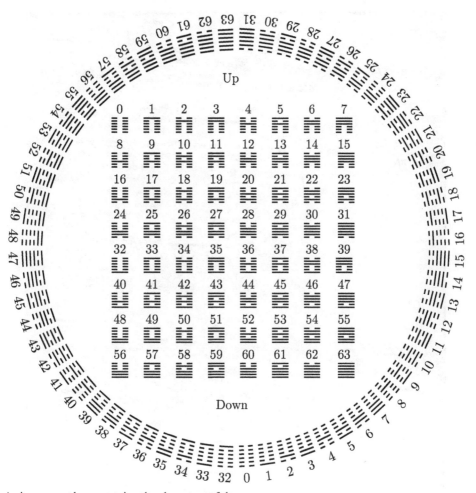

As is apparent by comparing the characters of the circle and the square, with respect to the circle the top [of a hexagram] is that which is furthest from the center.[63]

In the figure above, the words "Up" and "Down" are written in Greek in Bouvet's hand, and the numbers around the circumference of the circle and within the square array are in Leibniz's hand. The method Bouvet and Leibniz used to convert the hexagrams in these diagrams to numbers in an ordered sequence treats the unbroken lines as 1 and the broken lines as 0. They read the lines in each hexagram within the square from the bottom upward to identify the binary representation of the number, working through the square from left to right and top to bottom. With the circle, they read each hexagram from the innermost line toward the outermost. The order begins with the bottommost hexagram on the right-hand side and proceeds counterclockwise upward to the topmost hexagram on the right, then continues from the bottommost hexagram on the left-hand side, proceeding clockwise upwards to the topmost hexagram on the left.

---

63. Leibniz added this remark to the bottom left of the figure.

Second manuscript page of Leibniz's letter to Bouvet, April [?] 1703 (LBr 105 Bl. 30r).

29 Leibniz to Bouvet (early April [?] 1703)

Manuscript:
M: LBr 105 Bl. 30–35. Draft. French.

Transcriptions:
A: A I 22, 348–368.
W: W 396–434.

Leibniz received Bouvet's letter of 19 November 1701 only on 1 April 1703. His excitement at Bouvet's suggestion that the binary system might unlock the secrets of the Yijing is clear in the following lengthy response, not least in his claim that it was a matter of providence that he should have outlined his binary system to Bouvet at the very time when Bouvet was trying to discern the meaning of the Yijing, thus making possible the association between the two. Leibniz also speculates that the hexagrams of the Yijing, as an embodiment of the binary system, may even have been constructed by the ancient Chinese to symbolize creation from nothing by God, the very symbolization Leibniz himself had seen in the binary system and the one that led him to present the system to Bouvet (and other missionaries) in the first place. This fed his conviction, clear in various writings from the last decade of his life, that the ancient Chinese were not only advanced in mathematics but also had a fully worked-out natural theology.[1] The remainder of Leibniz's letter concerns matters relating to the Chinese missions, the deciphering of Chinese characters, and his ongoing efforts to learn more about Chinese science and culture.

Only Leibniz's undated draft of the letter is still extant. Two different datings have been proposed. The first, by Widmaier,[2] is based on a letter from Jean de Fontaney to Leibniz of 13 June 1704, in which Fontaney thanks Leibniz for his letter of 18 May 1703.[3] Since Leibniz indicates in this chapter's letter that he was enclosing a letter for Bouvet to pass on to Fontaney,[4] Widmaier concludes that the following letter was written around the same time and thus gives the date as 18 May 1703. However, this dating is ruled out by the existence of a short text dated April 1703 in which Leibniz indicates that he has already replied to Bouvet's letter.[5] The second suggested dating, by the editors of the ongoing critical edition of Leibniz's writings, is based on a letter to Leibniz from Christophe Brosseau in Paris of 30 April 1703,[6] in which Brosseau thanks Leibniz for his (now lost) letter of 17 April, enclosed with which were other letters for him to forward on, which the editors suggest were one for Jean de Fontaney[7] and the following one for Bouvet.[8] This would date both letters

---

1. See, for example, his "Dialogue on Natural Theology of the Chinese" (1716), WOC 75–138, especially 133–134.
2. See W 733.
3. See A I 23, 417.
4. See below, p. 188.
5. See Leibniz (1903, 327). The text is translated below; see footnote 45 in this chapter.
6. See A I 22, 393.
7. See A I 22, 372–373. English translation: Leibniz (1703b).
8. See A I 22, 347–348.

to 17 April at the latest. However, given that the evidence for this dating is suggestive rather than conclusive, and that in the lengthy postscript to the following letter Leibniz details research he had carried out at the royal library in Berlin after the main part of the letter had been written, indicating that the postscript was written some time after the main letter, we think it a mistake to affix a specific date of composition to a letter that was clearly not written on a single day. There are grounds for thinking the letter was written, or at least started, very early in April 1703, since it does not mention "Explanation of Binary Arithmetic," the paper Leibniz wrote on 7 April 1703 for publication in *Histoire de l'Académie Royale des Sciences*; if the following letter to Bouvet was written later than this, one might reasonably expect it to contain some mention of that paper, as a matter of courtesy, since it concerns and was inspired by Bouvet's hypothesis about the connection between binary arithmetic and the Yijing. We have hence tentatively dated the letter to early April 1703. The question of dating notwithstanding, Leibniz received no reply to this letter, suggesting that Bouvet likely did not receive it.

⸺⋘◌⋙⸺

My dear Reverend Father,

The letter from Your Reverence dated Peking, 4 November 1701, was given to me on 1 April 1703, through the mediation of the Very Reverend Father Gobien. I am thrilled to learn that Your Reverence is well, and that the affairs of the Europeans are going well in China, since it concerns the progress of the Christian religion. But I cannot avoid exhorting you and your companions to take advantage of favorable times a bit more than has been done up to now, in order to obtain the fine knowledge of the Chinese, especially in the arts, in exchange for ours. For one part of your letter made me conclude that the goodwill there towards foreigners will very much cool off some day, when it is thought it will no longer be needed so much. It's the passage in which you say, my Reverend Father, that for several years, the Emperor, having been satisfied with what he wanted to know of the theory of our sciences, no longer gives Europeans the same opportunity to talk to him about these matters.[9] So it's extremely important to work on the history of the arts of China without delay, as much as the means allow, all the more since it is easier for them to learn our sciences and knowledge than for us to learn theirs, both because ours consist more in reasoning and theirs more in experience, and because ours are public for the most part whereas the Chinese sciences are hardly known except by those who practice them, and are passed on there as if by tradition. I think you could adroitly interest even the Emperor in this research on the arts, getting him to understand that the way to perfect the arts of his country is to provide descriptions of them, and in this way it will be possible to see what the Chinese lack so as to provide it with the help of Europeans. Likewise, that it's important to have a complete dictionary, or rather several which make up a complete one, which contains all the characters specific to each profession, and that it's important to the Tartars that the Chinese no longer hide anything from them. Now, on the matter of terms, when it comes to explaining them, there will be provided figures of natural and artificial things and descriptions of practices or secrets. In particular, my Very Reverend Father, I beg you to obtain for me as clear a description as possible of the manufacture of all kinds of Chinese paper and the ways of embellishing it. The pieces of type they use for printing their characters, what are they made of? Mr. van Helmont, the son,[10] once told me that they were made of a lacquer or varnish in which these characters were easily engraved, while it was still soft, but which then hardened like wood. That would be convenient. I would also like to learn if they have good ways of working leather or some fabric so that air and water can't penetrate, particularly air, since I have not yet found anyone in Europe who has succeeded at this enough for my liking to make big bags or cushions that could be inflated and slept on.

---

9. See above, p. 171.

10. Namely, Francis Mercury van Helmont (1614–1698), alchemist, philosopher, and student of the Kabbalah. Leibniz met him twice in Hanover in 1696 and twice before that.

As for affairs in Europe, they are in a state to make us envy the Chinese. The death of the King of Spain has sparked a terrible war,[11] which will decide the fate of Europe. The war between the Czar and the King of Sweden is still going on;[12] but as the latter, instead of pushing back the Russians, turned his forces against Poland and entered the heart of that kingdom,[13] the Russians regained strength and are harassing the Swedish provinces that are exposed to them, having even taken Nöteburg,[14] a fairly important place. They are also trying to expand out to the coast of the Black Sea, building a new fortress,[15] and if the Sultan had not absolutely wanted to preserve the peace after a war in which the Ottoman Empire suffered so much,[16] and if he had wanted to consent to the solicitations of the Tartars of Crimea, he would have perhaps already restarted the war with the Russians, whose aim seems to have been to have some port on the Baltic Sea again, but they will find it difficult. Till now, the King of Sweden has seemed very far from a peace with the King of Poland, whom he really wanted to have removed from his Kingdom, having effectively on his side a large faction of Poles and Lithuanians who leaned that way. But matters seem to have turned around, and it seems that the King of Sweden will finally be able to bring himself to deal with the King and the Republic, with whom he would like to make an alliance against the Russians. However, I don't think the King of Poland will consent to this, since the Czar has faithfully stood by him, and as a perpetual peace was concluded between the Poles and the Russians already some years ago,[17] the Poles have no reason to break it as far as I know.

But let's leave aside these worldly affairs, which are hardly of interest to China, and return to the sciences. The analysis of infinitesimals has given us a way, both to me and to those who have advanced my methods,[18] to resolve difficulties where geometry has not succeeded heretofore. For example, no one had determined the shape of the catenary that Galileo thought was a parabola.[19] However, I have shown that by means of a very fine chain, like those of watches, suspended at both ends, one will have ready-made logarithms without calculation as precisely as the size, subtlety, and flexibility of the chain permit.[20] I also don't know if I spoke to you about my science of dynamics or forces, where I found the true way of estimating them, drawn *a priori* from sublime principles of real metaphysics and then perfectly conforming to experience.[21] Here I have found not only that it is false that the quantity of motion is conserved, as the Cartesians believed, but that absolute force (properly understood) is conserved, whereas Mr. Mariotte,[22] Mr. de La Hire,[23] and others recently have only a conservation of something respective. I also found that there is as much action in one hour as in another, whether in the universe or in bodies conceived as communicating only with

---

11. Charles II of Spain (1661–1700) died on 1 November 1700, sparking the War of Spanish Succession (1701–1714).

12. That is, the Great Northern War (1700–1721).

13. Leibniz is here referring to the Battle of Kliszów (July 1702), in which the numerically inferior army of the Swedish king Karl XII (1682–1718) defeated the Polish-Saxon army of August II The Strong (1670–1733).

14. Leibniz is here referring to the Siege of Nöteburg, one of the first sieges of the Great Northern War. The Swedish fortress of Nöteburg was surrendered to the Russian forces on 22 October 1702. The town was subsequently renamed Shlisselburg, a name it retains to this day.

15. Probably the Peter and Paul Fortress on Hare Island in the city now known as St. Petersburg. Construction of the fortress started on 27 May 1703.

16. Mustafa II (1664–1703) was sultan of the Ottoman Empire at the time. On 11 September 1697, Mustafa's army was defeated at the Battle of Zenta, bringing an end to the Great Turkish War of 1683–1697. This led to the Treaty of Karlowitz that ended hostilities between the Ottoman Empire and the Holy League (consisting of Austria, Poland, Venice, and Russia). The treaty was signed on 26 January 1699.

17. Namely, the Treaty of Perpetual Peace between Russia and Poland, signed in 1686.

18. Namely, Jacob Bernoulli (1655–1705), Johann Bernoulli (1667–1748), and Guillaume de l'Hospital (1661–1704), all of whom adopted and developed Leibniz's infinitesimal calculus.

19. See Galilei (1638, 146). English translation: Galilei (1914, 149).

20. See Leibniz (1691b). English translation: Leibniz (2001).

21. See Leibniz (1695b). English translation: PPL 435–450.

22. Mariotte (1673).

23. La Hire (1695).

each other. But the quantity of motive action is quite different from what the Cartesians call the quantity of motion. I further show that force is of the essence of corporeal substance, and that it is the entelechy of the ancients, although it needs to be determined by the interaction of bodies or by God's original arrangement. So it is by considering force that I justify the ancients against the overly material philosophy of the moderns to whom I grant that everything happens mechanically in bodies. But I show them that they do not adequately consider that the very principles of mechanism come from something higher than what is merely material or subject to the imagination.

But let's come to what is one of the main subjects of your letter—the connection between my binary arithmetic and the figure of Fuxi, who is considered one of the most ancient kings and philosophers known in the world, and as it were founder of the Chinese empire and sciences.[24] This figure is therefore one of the most ancient monuments of science found in the universe today; it seems to be more than 4000 years old and may not have been understood for thousands of years. It is quite surprising that it agrees perfectly with my new style of arithmetic and that I should have written to you about it at precisely the time I did, that is, while you were devoting yourself to deciphering these lines. I confess to you that had I not established my binary system, I might stare for a long time at this system of gua or lines of Fuxi without understanding its purpose. It has been well over 20 years since I had in my head this arithmetic of 0 and 1, from which I saw wonderful consequences for advancing the science of numbers to a perfection surpassing everything that came before, but I held back revealing myself about it until I was in a position to demonstrate its great uses at the same time. But as a thousand other matters and considerations prevented me from working on it, I finally opened up about it (though still not in any printed book) so that an idea of this importance would not perish.[25] And I am delighted that this was at precisely the time when you were in a position to make such a wonderful use of it as that of deciphering this ancient monument of China. In truth, there seems to be some influence of providence in this. And I do not doubt that when this arithmetic is further advanced, as it soon would be if there were a few people capable and inclined to work on it, this would be a new way of reviving the attention of the Emperor and the most eminent men of China, although your deciphering of the gua of Fuxi should already be enough to surprise them.

I find that Fuxi's square is the same thing as his circle, and even serves to explain it. For the order is a bit confused in the circle. I begin in this circle at the bottom of the right side, that is with $\blacksquare\blacksquare$ and $\blacksquare\blacksquare$, etc., or indeed with 00 0000 and 00 0001. For one must look at the figure as if the eye were placed at its center and begin with the lines near the center. This is also shown by the characters, since they appear as they are if they are looked at from the center, or if the circle is turned so that the part of the character furthest from the center is topmost, which then shows the correspondence with the characters of the square, which thus appear the same. So that with regard to the circle there is no up or down, any more than with regard to the globe of the earth, where the furthest from the center is taken as the highest, and perhaps Fuxi even took this into consideration. But you were nevertheless right, my Very Reverend Father, to write "up" and "down" there in relation to the square inscribed in this circle.

But to return to the gua or linear figures of the circle, starting with 0 and 1 at the bottom of the right side we climb up to the highest point of the right side, that is, up to $\blacksquare$ (the line closest to the center always considered as the first), that is, 01 1111, or 31. After that (instead of continuing in the circle and going to the top of the left side and descending from there), we start again at the bottom of the left side to go back up; now on the left side at the bottom is $\blacksquare\blacksquare$, that is, 10 0000 or 32, apparently because 0 and $32+0$ or else 1 and $32+1$, or 2 and $32+2$ (which, when expressed in binary, differ only in that the linear figure of one begins with $--$ or 0, and that of the other by $-$ or 1),[26] are

---

24. sciences. ▷ I confess to you that this connection surprised me, and seemed to me ◁ *deleted*.

25. Leibniz likely means here his decision in the late 1690s to start advising his mathematician correspondents of his invention of binary arithmetic, which culminated in his writing "Essay on a New Science of Numbers" (see chapter 23) for the Académie Royale des Sciences in Paris.

26. In a one-page draft about the association between binary and the Yijing that was likely written very soon after receiving Bouvet's letter (i.e., very soon after 1 April 1703), Leibniz considered whether it was appropriate to designate 1 by $-$ and 0 by $--$ rather than the other way around, writing "even if one could suppose, by

opposite each other when going from bottom to top on both sides. Perhaps also this was intended to be a bit confusing and to show some apparent difference between the circle and the square. Finally, going from bottom to top on the left side you finish with ▦ , that is 11 1111 or 63.

As for the inscribed square, its order is perfectly natural in beginning with:

䷀	䷀	䷀	䷀	Leibniz presents these hexagrams upside down by
00 0000	00 0001	00 0010	00 0011 etc.	comparison with their appearance in the square on
0	1	2	3	Bouvet's diagram (end of chapter 28).

where it is worth noting that the order of numbers goes from left to right in each line or row of numbers, and these lines or rows follow each other from top to bottom. Both follow the order we use in Europe, contrary to the custom of the Orientals and Chinese.[27] It is true, however, that in each number the lines have to be counted from bottom to top, perhaps imitating the correspondence to the circle, where we go from bottom to top, while beginning with what is closest to the center. This correspondence of such an ancient monument to European writing reminds me of having read that very ancient Oriental monuments also seem to have been read in the European way. If I am not mistaken, this is the case in the extraordinary writing found in the ruins of Persepolis or Chehel Minar, where the writing consists of lines and triangles and may well predate the Persian monarchy.[28] This ancient monument is mentioned in the report of Figueroa, Spain's ambassador in Persia,[29] and in what is found among the relations of voyages of the late Mr. Thévenot,[30] and elsewhere.[31]

But let's return to the binary calculus which uses only nothing and unity. You have sensed very well one of its main uses for religion, which is that it is an admirable symbol of creation, that is, of the origin of all things from God alone and from nothingness, without any pre-existing matter; and that nothingness here is not absolute but relative, that is, its limitation is found essentially in creatures in proportion to their imperfection, this limitation being nothing but a negation of the further progress of pure reality or pure actuality, as when a circle is bounded by its circumference, which gives it its *non plus ultra*.[32] And I think that when scholars of China immerse themselves in this idea and see above all the artifice of Fuxi conforming to ours, they will be sufficiently inclined to believe that this great man had also wanted to represent God, the author of things, and the [act of] creation by which he drew them out of nothingness. And so this may be one of the most important articles of your catechism drawn from the classical authors of China and worthy of being explained to the Emperor himself.

As for the use of the binary expression of numbers, that is, by 1 and 0, for the perfection of numerical science, I have demonstrations that can convince us that it is a way to lead this science far beyond its current state.[33] For I have found that squares, cubes, and other powers of natural numbers

---

turning it all around, that — stands for 0 and - - for 1, I think it is more fitting to understand the unbroken and complete [line] as unity and the broken [line] as nothing, that is, imperfection; for negation or limitation mixed with positives and absolutes forms the imperfections of creatures" (LBr 105 Bl. 93r; English translation: Leibniz 1703a).

27. Leibniz gets this completely wrong: the chart of hexagrams he received from Bouvet (end of chapter 28) is traditionally read right to left from the bottom upward, rather than left to right from the top downward, as Leibniz supposes here. (Though that too would be in numerical order, if the interpretation of solid and broken lines as binary digits were reversed, with a solid line representing 0, a broken line 1, and the lines in each hexagram read from most significant at bottom to least significant at top!)

28. Leibniz means Old Persian cuneiform script, which was deciphered only in the nineteenth century.

29. See Silva y Figueroa (1667).

30. Leibniz is here referring to the short essay, Thévenot (1663).

31. elsewhere. ▷ Now, this resemblance between the line of Fuxi and my numerary table being beyond doubt ◁ *deleted*.

32. Literally "no more beyond" or "no further beyond," a term often used in early modernity to mark an outer limit.

33. Leibniz is referring here to his "Demonstration That Columns of Sequences Exhibiting Powers of Arithmetic Progressions, or Numbers Composed from These, Are Periodic" (see chapter 27).

also have periods in their columns like the natural numbers, which provides a wonderful opportunity for determining them, drawing up tables of them with ease, and discovering previously unknown patterns. I even find that this will have great effects for expressing incommensurable geometric quantities by sequences of whole numbers which approach infinity,[34] whereas what is lacking in Ludolph van Ceulen's formula for a circle (for example) is that there is no rule for continuing further.[35] But as I do not have time to advance this research, in the event I do not find anyone with the capacity or desire to help me, I will leave to posterity the benefit that will be derived from it, greater than can be imagined by those who look only at the surface of things.

As for the use of the double geometric progression in the arrangement of ideas, of which Your Reverence also speaks,[36] experience has taught me how useful dichotomies are for the formation of concepts. It is true that dichotomies can be made in different ways, but they lead to the same goal, that is, to the same species by different subaltern genera, and this is what makes it possible for the same terms to have different definitions, which are ultimately shown to be consistent by analysis. I think that few people have worked more on the systemization of concepts and have a greater apparatus for it [than I]. But in the midst of occupations and distractions, I cannot unscramble my chaos without the assistance of individuals who have more energy and more free time than I do. However, if God gives me a few more years of life, I shall make efforts to bring some order to this work, which seems to me one of the most important that can be undertaken, since it would give reason a *filum Ariadnes*,[37] that is, a tangible method in the form of a calculus to guide us in both judging or inventing.

I suspect that Fuxi assigned the 64 numbers (either simple or doubled to 128 or beyond) to terms, which he conceived as the most basic, and that he gave to each of these terms its character which also designated its number or rank, and that then from these simple and fundamental terms and characters he formed others by adding small lines. But in the course of time these characters became altered, both by the nature of their popular use, which gradually changes their features (as can be seen by comparing ancient and modern writing of some language), and by those who, no longer knowing the rationale or method of the characters, accommodated them to their whims often based on metaphors or some other more tenuous connections. If we were familiar with Chinese literary history, which when writing to you on another occasion I wished to see established according to [the standards of] a good critical methodology, in order to discern the ancient and the modern,[38] we could judge it better. It is also likely that different princes or philosophers have made different revisions of the characters, in order to improve them, but having not always followed the same laws of grammar or etymology, as it were, the origins finally became completely obscured, even more than in the broken lines of Fuxi, whose meaning had been lost but which we have recovered. I would be delighted to know the meaning of the Chinese characters adjoined to each number in the figure of Fuxi, and beg Your Reverence to send me the explanation, provided it doesn't cause you too much trouble.[39] Apparently these are modern explanations or accounts devised without knowing the real usage of the characters, that is, calculation by 0 and 1.

If you find that the Chinese of today have no knowledge of this calculus, Your Reverence will be able to confidently claim both to the Emperor himself and to the principal scholars of China to have personally deciphered the real meaning of Fuxi's figure and its broken lines with the help of a new discovery from Europe of the way of calculating with 0 and 1. In my opinion, this should not seem a matter of little consequence and should increase among the Chinese the esteem for European sciences and consequently for our religion. That will also give them a great expectation regarding mysteries

---

34. Leibniz is thinking here of the decimal expansion of π, as will become clear.

35. The German mathematician Ludolph van Ceulen (1540–1610) calculated π to twenty decimal places in Ceulen (1615). He later calculated π to thirty-five decimal places. Leibniz's complaint is that van Ceulen did not provide an algebraic formula that would enable the calculation of π with even greater precision.

36. See above, pp. 166–167.

37. "Ariadne's thread," which she gave to Theseus so he could retrace his steps out of the Labyrinth of the Minotaur.

38. Leibniz is referring here to his letter to Bouvet of 2 December 1697. See A I 14, 826–835.

39. trouble. ▷ The way I think I will use to apply numbers to expressions of ideas or terms is not yet ◁ *deleted*.

still hidden that remain to be revealed, and will give us free rein to invent a new characteristic,[40] which will appear a continuation of Fuxi's and will provide the starting point of the analysis of ideas and of this wonderful calculus of reason I have planned.[41] This secret and sacred characteristic would give us also a way of introducing the Chinese to the most important truths of philosophy and natural theology in order to smooth the path to revealed theology. And entirely new or different as it will be from theirs, it will be acceptable thanks to Fuxi and will eventually become, as it were, a special language of the highest class of the most enlightened scholars and the ones most attached to Fuxi, even worthy of its own separate bodies or academies. So this discovery could have great consequences for the whole Chinese Empire, if one knew how to profit from it in your country or rather in Europe.[42]

For now that we have found the key to these lines, and that the Chinese can't doubt our success, we will have an incomparably higher reputation with them. By relating ideas to numbers, the characteristic we can build on Fuxi's lines will at the same time have the advantage of subjecting ideas to calculation like numbers, which is beyond anything that could be hoped for in these matters, and which many people will believe not unreasonably to be impossible. Also, to do that there has to be a form of expression which is not easy to figure out but which carries its demonstration with it. Judge for yourself, my Very Reverend Father, if that should not awaken those interested in the perfection of the functions of the human mind, and above all in the progress of virtue and the true religion. But the present European war, and perhaps the importance and very grandeur of this discovery of the general characteristic, greatly diminishes my hope of seeing it carried through with the assistance that would be necessary for it, since one wonders whether providence might want to reserve it for other times, when men are more apt to win the graces of heaven.

You see from this, my Very Reverend Father, that it would not be enough to form some divisions and subdivisions to come to the species, as that would not provide an analysis capable of showing the properties of species by the characters that would have been attributed to them by virtue of these divisions. There is still a big distinction to be drawn here, between confused and distinct ideas.[43] For the characters of distinct concepts can be perfect from the outset, but those of confused ideas can be no more than provisional and proportionate to what has been recognized as distinct, leaving it for posterity to perfect them following new experiments. And this observation is important, as much for not promising what is beyond the present power of men and making us suspected of vanity as for knowing what actually needs to be done. As for colors, experiments with the prism show that besides white and black, the principal colors are yellow, blue, and red, which makes purple together with blue, and that green is just a combination of yellow and blue.[44] This is determined by the bending of light rays in refraction, and not just by mixing white and black together. But that is said in passing, and you have put forward colors only as an *example of clarification*, where the rule holds good that

---

40. That is, a philosophical notation that would allow one to calculate with ideas much as one does with numbers.

41. Leibniz wrote numerous pieces to support and develop his plan for a universal characteristic. See, for example, those in A VI 4, 1–1002.

42. Europe. ▷ It has been a long time since I saw or read something imperfect on the broken lines of Fuxi (because I do not remember seeing the whole circle of them, until you sent it to me) and that I knew that the mystery was believed lost. I thought it could be used for the new characteristic of ideas I am planning, and to give them a meaning and thereby make it acceptable in China, without knowing then what these lines meant. ◁ *deleted*. Leibniz here alludes to Intorcetta et al. (1687, xlii and xliv), who depict the eight trigrams and sixty-four hexagrams of the Yijing in rows and columns arranged according to the classical King Wen sequence.

43. For Leibniz, a distinct idea is one in which it is possible to identify the elements or features that distinguish it from other ideas. Such is, he claims, the idea that assayers have of gold, which enables them to distinguish gold from everything else. A confused idea, on the other hand, is one in which the distinguishing elements or features are not separately identifiable. A VI 4, 585–592, at 586–587; English translation: PPL 291–294, at 292.

44. Leibniz is likely thinking here of Isaac Newton's prism experiments, which showed that all colors were present in white light and so were not formed by mixtures of white and black, as had previously been believed. See Newton (1671/2, 3075–3087).

says *the truth of examples is not required.*[45] I would also not want anyone jumping from the third degree to the sixth, since either the method is not well enough founded or it should not allow these sorts of jumps while remaining intact. The eight degrees the Scholastics give to qualities (for I don't recall if Aristotle used them)[46] came from imagining that two of the four elements always have a common quality, so they wanted to attribute four degrees to each of the elements, for example, four degrees of heat in a fire (as chemists do) and four others in air. A certain Englishman by the name of Suisset, called The Calculator par excellence, who lived more than 300 years ago, refined intension and remission to an extreme degree, that is, refined the degrees of qualities.[47] But this way of calculating did not seem to suit our characters, as things as well as concepts do not differ only in degree, but in many other ways.

It seems, however, that the eight gua, or eight linear figures considered fundamental by the Chinese, could lead us to think that Fuxi even had creation in mind, in making everything come from one and from nothingness, and that he even pushed the connection to the story of Genesis. For 0 can signify the void that precedes the creation of heaven and earth, then follow the seven days, each of which indicates what existed and was made when that day began. At the beginning of the first day there was 1, that is, God. At the beginning of the second day, 2, there was heaven and earth, as they were created during the first. Finally, at the beginning of the seventh day the whole already existed, which is why the last is the most perfect and the Sabbath, since everything is then created and fulfilled, so 7 is written 111 without 0. And it is only in this manner of writing by 0 and 1 that shows the perfection of the number seven, which is considered sacred, where it is even notable that its character [i.e., 111] has some relation to the Trinity.

I think I once told you about a theologian from Berlin who died a few years ago called Andreas Müller,[48] who was Pomeranien by birth and had a profound knowledge of oriental literature and of languages in general, and wrote some excellent notes on Polo's *Cathay.*[49] He thought he had found some key to Chinese characters, according to what he made public in a small printed paper.[50] But it is known that he demanded a considerable sum of money to reveal it. I mentioned him to Very Reverend Father Grimaldi, who wished to talk with him on his way from Vienna to Poland. But Mr. Müller missed the meeting. He had much merit, but he was equally capricious. The late Elector of Brandenburg[51] seemed disposed to give him what he wanted, but he was not at all accommodating and withdrew himself from Berlin to his homeland. He threatened to burn his papers before he died, and it is said he kept his word.[52] It does not appear that he knew the calculus hidden under the gua of Fuxi, as he was not enough of a mathematician to be aware of such an arithmetic. Nonetheless, I cannot easily persuade myself that a man of such merit wanted to boast of a secret without having some probability of it, at least from his perspective. So I suspect that he glimpsed some relation between

---

45. Leibniz expanded on this point in a short text written shortly after this letter to Bouvet. The text reads in full: "In examples which are declaratory, not comprobatory, the common rule holds good which says the truth of examples is not required. I responded with this distinction to Reverend Father Bouvet, who used the example of colors, which he composes from white and black in order to construct eight [colors] to elucidate the eight linear characters of Fuxi. I responded that, even if colors do not in fact originate from the mixing of white and black, the example is sufficient for illustration, and when there is no other aim the truth of the examples is not required. April 1703." (Leibniz 1903, 327)

46. See note 25 in chapter 28.

47. Leibniz is referring here to Richard Swineshead (or Suisset), an English mathematician who flourished in the fourteenth century. He acquired the nickname The Calculator from his work *Liber calculationum* (c. 1350), in which he discussed the Scholastic problem of intension and remission, that is, how a subject may admit more or less of a quality during qualitative changes such as heating and cooling. Leibniz had read Swineshead's work in late 1689; see A II 2, 306.

48. Leibniz mentioned Müller in his letter to Bouvet of 15 February 1701. See above, p. 134.

49. Namely, Müller (1671).

50. Leibniz is referring to the three-page pamphlet by Müller (1674), which contains Müller's reasons for not publishing his key. The pamphlet was reprinted in Bayer (1730, 182–187).

51. That is, Frederick William, Elector of Brandenburg.

52. See note 62 in chapter 22.

ordinary numerical characters of the Chinese and their root character for things, and perhaps also made some observations on the small additions which vary the root characters to form derivatives, or perhaps even composites. But this is what Your Reverence and Reverend Father Visdelou will judge better than I could. For I would have to have good dictionaries of Chinese characters along with a European explanation in order to think about it.

Your Reverence tells me that such dictionaries are not found where you are.[53] But they must be hidden from you. For it is certain that there are some, even printed ones, as I recall. My evidence comes from the *Catalog* of oriental manuscripts of the incomparable Golius that were auctioned off in Holland about six years ago to my great astonishment and regret.[54] For they were a public treasure worthy of the greatest king, and which the University of Leiden, aided by the States General or by the Society of the East Indies, ought to have held on to. Now I recall that in this *Catalog* there were two Chinese dictionaries,[55] one in Chinese and Spanish, printed I believe in the Philippines, and I don't recall if the other was in Portuguese or Flemish. I hope I have kept this *Catalog*, and when I return to Hanover I shall try to find it, in order to clearly indicate everything to you. One of these two certainly explained the characters, but I cannot say if both did or if one was entirely in Chinese. I am told that most of these items wound up in England.

I don't know what to say about the hieroglyphs of the Egyptians, and I have trouble believing that they have any affinity with those of the Chinese. For it seems to me that the Egyptian characters are more popular and rest too much on their resemblance to things that can be perceived, such as animals and others, and consequently on allegories; whereas Chinese characters are perhaps more philosophical and appear to be based on more intellectual considerations, such as are provided by numbers, order, and relations. So [in Egyptian hieroglyphs] there are only isolated features that bear no resemblance to any kind of[56] body. I know many have believed the Chinese were a colony of the Egyptians, based on the supposed similarity of the characters,[57] but that is implausible. There is an ancient book on Egyptian hieroglyphs by one Horapollo;[58] one can also find in Ammianus Marcellinus, if I am not mistaken, the explanation of the characters of an obelisk that still stands in Rome.[59] Pierio Valeriano,[60] Father Kircher in his *Oedipus Aegyptiacus*,[61] *Obeliscus Pamphilius*,[62] and other works, and Lorenzo Pignoria in *Mensa Isiaca explicata* (which is an old piece full of Egyptian hieroglyphs, which I learned can be found at present in the cabinet of the Duke of Savoy) have tried to shed light on these hieroglyphs.[63] I would advise you to obtain and use these books if I thought they could in any way assist the understanding the characters of China. The late Mr. Thévenot, the king's librarian, had published a book of characters the Mexicans once used,[64] but I believe they are unrelated either to the Chinese or to the Egyptians.

As for Very Reverend Father Grimaldi, who I have just mentioned in connection with the Chinese key of the late Mr. Andreas Müller, I am delighted that he remembers me favorably, but upset that his health is not the most robust. I beg Your Reverence to tell him how much I honor him and how

---

53. In Bouvet's letter of 4 November 1701; see above, p. 172.

54. Leibniz is referring here to Golius (1696). Golius's collection of manuscripts was auctioned in Leiden on 16 October 1696.

55. These were *Thesaurus rarissimus, in quo explicatur ultra 10000 Characteres Chinensium lingua Hispan* and *Vocabularium Hispanico Sinense, cum annotat. J. Golii item libellus Hispanicus de pronuntiatione charact. Chinensium.* See Golius (1696, 27).

56. of ▷ corporeal nature ◁ *deleted*.

57. For example, Kircher (1652–1654, III: 8-2).

58. Horapollo Nilous, *Hieroglyphica* (late fifth century). English translation: Nilous (1840).

59. See Marcellinus (1935, I: 327–331).

60. See Valeriano (1556).

61. Kircher (1652–1654).

62. Kircher (1650).

63. Pignoria (1669). Victor Amadeus II (1666–1732) was Duke of Savoy from 1675–1730.

64. Thévenot was the translator of this work rather than its author. See Gage (1696).

concerned I am for his health.[65] I have also been asked to send greetings to him on behalf of Very Reverend Father Vota,[66] confessor of the King of Poland, who recently left Berlin, where he stayed for a short time as well as in Hanover where he had followed the Queen of Prussia.[67] He has now returned to the king, his master, to assist him during Holy Week. He told me that the account of the voyage Father Grimaldi had sent him was lost last year along with other papers and luggage when the army of the King of Poland was put in disarray by the army of the King of Sweden, with both kings being present, by a panic that broke out among the Poles and then the Saxons. For there is nothing so contagious as fear. He would very much like to have this account [again], as would I. So if the very Reverend Father Grimaldi will favor us with it, I beg Your Reverence to send it to me as well as the astronomical observations I hope can be obtained from him and other friends.

What Your Reverence told me in the last two pages of your letter about the signs of a worship of a supreme God and the traces of the true revealed religion among the Chinese found in their characters and in their classic books, strikes me as important. I have always had a tendency to believe that the ancient Chinese, like the ancient Arabs (think of the book of Job), and perhaps the ancient Celts (that is, Germans and Gauls) were far from idolatry and were rather worshippers of the supreme principle. In his *Germania antiqua*, Clüver, an excellent man no doubt, went so far as to believe that the old Germans had some knowledge of the Trinity.[68] I am not of his sentiment, although I don't see why he should be condemned for it. And I find it strange that there has been so much commotion against your Fathers for writing that the ancient Chinese had the true religion. What harm is there in that? Even if it is wrong, is it an error that gives rise to dangerous consequences? Not at all.[69] Would to God that the Chinese had no other errors.

Besides, I beg you, my very Reverend Father, to remember the Lord's Prayer in different languages; if it were possible to add a small list of other frequently used words, so much the better. Likewise, [I beg you] to remember the geography of Tartary, with remarks on Mr. Witsen's map that you are sure to have with you.[70] Likewise, to remind Father Grimaldi and others to obtain for me some old Chinese observations, and even some good modern observations, especially if you have some of the eclipse of the year . . . .[71] which would be a good way to establish the meridian of Peking. I would also like to know if it is possible to learn something more specific about this ancient conjunction of all the planets, with which an ancient emperor began the cycle the Chinese use in their chronology.[72] I also repeat my requests about the manufacture of paper in China, on the method of trapping air in

---

65. health ▷ , although he has not been in a position to favor me as he had given me to hope ◁ *deleted*.

66. Carlo Maurizio Vota (1629–1715) was an Italian Jesuit and advisor and confessor of two Polish kings, John III Sobieski (1629–1696), as well as his successor, Augustus II (1670–1733).

67. That is, Queen Sophie Charlotte (1668–1705).

68. See Clüver (1663, 212–218).

69. all. ▷ Would to God that it were true, and that the Chinese believed it. Leaving passions aside, that [assessment] could come only from the prejudices of a few good people who are infatuated with themselves and the special grace they conveniently believe God has granted them, quite similar to the young lady from Lüneburg (otherwise wise but who looked upon the world condescendingly because she believed herself the wife of Jesus Christ, who appeared to her), and lack sufficient charity, that is, universal benevolence. These people are the polar opposite of Celio Secondo Curione, who wrote *De amplitudine regni coelestis*, and are ready to damn even infants who die unbaptized. This shows the strange idea they must have of God; if only they would sufficiently consider the consequences of what they attribute to him. ◁ *deleted*. The "young lady from Lüneburg" is Rosamund Juliane von der Asseburg (1672–1727?), who from an early age had claimed to have visions of Christ and also that he dictated things for her to write down, some of which were prophetic in nature. Leibniz discussed her case at some length in his correspondence with Electress Sophie of Hanover and Queen Sophie Charlotte of Prussia in 1691–1692; see LTS 69–96. The book Leibniz mentions in this deleted passage is Curione (1617), in which it is argued that the number of the saved is incomparably greater than the number of the damned, the view that Leibniz compares favorably here to the traditional belief that God would not even save unbaptized infants.

70. Namely, Witsen (1687).

71. The ellipsis is Leibniz's own, and presumably functions as a placeholder for the desired year, which would have been added to the final, dispatched letter.

72. See note 53 in chapter 22.

leather or in some other fabric so that even when pressed very hard it does not escape through pores. Likewise, the Hebrews who are said to have been in China from time immemorial, about which I think I wrote to you in order to know if it is possible they have the entire Hebrew Old Testament or just the Pentateuch and Joshua, as do the Samaritans;[73] and what writing they use, whether it is Samaritan, which is the older, or the one that Jews still use today, which is believed to have been brought from Babylon. Likewise, whether they have vowel points and accents or not. And if I have asked either Your Reverence or Father Grimaldi for something else, I repeat it and beg you to think about it at your leisure with your friends, as also about the questions of Mr. Schröck or the late Father Kochański. And not to neglect even the question I think I asked about diseases,[74] among others if gout and stones[75] are as common in China as in Europe, likewise smallpox, and even the other pox.[76] For all these accounts are important. Likewise, if the Chinese have remedies against certain ailments as reliable as quinine is against fever and ipecac is against dysentery. I also remember having read in books that speak of China, and if I am not mistaken in Father Martini's Atlas, that the Chinese have the means to make silk twice a year, in the same place and from the same mulberry trees.[77] It would be good to know if it is true and what it consists of. Finally, I pray to God that he keep you for a long time and that he give you always great success in your excellent endeavors, being sincerely,

My very Reverend Father Yours, etc.

P.S. I beg Your Reverence's pardon for being helped by another hand because I am rather distracted and am not too well.[78] I looked at (after writing and having the above copied) the characters of Fuxi in the *Confucius* of Paris printed according to the accounts of Reverend Father Couplet, where I found only the four and the eight gua properly arranged.[79] The 64 are given in a scrambled order.[80] But adjoined there are the meanings, apparently presumed, such as air, water, etc., which these linear figures are or were thought to have. Could it be that the characters around the printed circle you sent me also mean what is indicated in the *Confucius* of Paris? I beg you to examine it, my Very Reverend Father, and let me know. For I don't doubt that you have in Peking this book printed in Paris.

I will say a bit more to Your Reverence about the Chinese dictionaries. Since writing my letter, I think I have found here in the library of the King of Prussia one of the two dictionaries mentioned in Golius's *Catalog* of manuscripts, namely the one in Chinese and Spanish. I see that this dictionary, which is in manuscript, actually explains the Chinese language and not the characters, though to each word of the language is adjoined its Chinese character. I was once told of a large dictionary of characters called *Hai pin*,[81] or Ocean, which is doubtless to be found in China, but is not explained in any European language. However, this other dictionary or, rather, vocabulary with pictures of things, about which I spoke in my preceding letter, and for which Your Reverence made me hope, seems to me interesting and convenient.

If Your Reverences kindly want to send me something either about physics, or mathematics, the arts, languages, or geography, I beg Your Reverence either to have it addressed to me or to indicate

---

73. Samaritans recognize as canonical only the Samaritan Pentateuch, consisting of the first five books of the Hebrew Bible written in the Samaritan alphabet. There also exists a Samaritan Book of Joshua, which is considered noncanonical (and should not be confused with the Book of Joshua in the Hebrew Bible).

74. The question appears in Leibniz's letter to Jean de Fontaney from mid-February 1701; see A I 19, 419; English translation: Leibniz (1701).

75. That is, kidney stones.

76. That is, syphilis.

77. Martini (1655, 4b).

78. Leibniz is apologizing for having an amanuensis write out the (now lost) fair copy of the letter that would have been dispatched rather than doing it himself.

79. See note 42 in this chapter.

80. In fact, in that book, the hexagrams are arranged according to the classical King Wen sequence rather than the Fuxi (or Shao Yong) sequence that Bouvet had used in his letter to Leibniz of 4 November 1701 (see note 20 in chapter 28).

81. Leibniz was informed of this dictionary in 1698. See the letter from Hiob Ludolf to Leibniz of 12/22 March 1698; A I 15, 424.

very clearly to your friends in Paris that they should pass it on to me, otherwise I will have difficulty in benefiting from it. I have not yet been able to learn anything in detail or to have passed on to me everything that Very Reverend Father Fontaney had brought [from China], and about which I am again inquiring. Perhaps these things have already passed into other hands.

As this Father wrote me a short and very obliging letter from Canton after his return [to China],[82] in reply to the one I had the honor to write him while he was in Paris,[83] I replied to him with the enclosed, which I beg Your Reverence to have conveyed to him.[84]

The title of the Chinese-Spanish dictionary I just mentioned is: *Vocabulario de letra China con la explicacion Castellana hecho con gran propriedad y abundancia de palabras por el Padre Don Francisco Diaz de la Orden de Predicadores ministro incansable en esto Reyno de China.*[85]

About the mules that breed in Tartary, someone pointed out to me a passage from Strabo which says the same about mules of Cappadocia.[86]

Does the monument of the Christians who lived in China more than a thousand years ago, which Father Kircher published, still exist?[87] I would like a well-supported account with Chinese testimonies themselves, taken from their registers of that time, to shut the mouths of people who dare to say that the modern Christians have made it up. That does not seem credible to me, when I consider the inscription.[88]

Some time ago, my very Reverend Father, you sent me a Lord's Prayer in the language of the Chinese Tartars, but if I recall correctly you told me you had composed it,[89] so if the one used among the Christian Tartars is different from that one, I beg you to send it to me along with the others.

Since writing this, I found in the library of the King of Prussia another small Latin dictionary with Chinese characters and words, printed in China itself apparently at the instigation of the Reverend Father Jesuits,[90] but this small dictionary contains only the most essential words.

Lastly, I beg you to give the enclosed to the Very Reverend Father Fontaney.[91] It responds to the letter he did me the honor of writing from Canton on 15 September 1701 which I received several months before yours. I am sending him a list of questions,[92] asking him to think at his convenience to enlighten me about them, and I beg you to assist him as if it were for you too, as it is in fact.

---

82. Leibniz is referring here to Jean de Fontaney's letter of 15 September 1701. See A I 20, 469–470. Fontenay had arrived back in China in August 1701. See Fontaney's letter to Leibniz of 13 June 1704; A I 23, 417.

83. Namely, Leibniz's letter to Jean de Fontaney of mid-February 1701. See A I 19, 417–419. English translation: Leibniz (1701).

84. Leibniz is referring here to his letter to Jean de Fontaney for which the editors of the ongoing critical edition of Leibniz's writings have suggested a date of 17 April 1703. See A I 22, 372–373. English translation: Leibniz (1703b).

85. That is, *Chinese Letter Vocabulary with the Castilian Explanation Made with Great Propriety and Abundance of Words by Father Don Francisco Diaz of the Order of Preachers, Tireless Minister in This Kingdom of China.* Diaz's manuscript was written between 1640 and 1643 and has been digitized (Diaz 1640–1643).

86. See Strabo (1928, V: 381).

87. Leibniz is referring to the Xi'an Stele (or Nestorian Stele) erected in 781. The inscription on the Stele documents 150 years of Christian communities in northern China. It was buried in 845 and not rediscovered until 1625. The stele is discussed in Kircher (1667, 1–10).

88. For the stele's inscription, see Kircher (1667, 10–35).

89. See Bouvet's letter to Leibniz of 28 February 1698 (A I 15, 356–357), in which he provides a Manchu translation of the Lord's prayer and indicates that he had composed it "in haste" and that it was not entirely accurate because he had forgotten some of the words.

90. Jesuits, ▷ in which the Latin letters which express the Latin and Chinese words have been engraved with Chinese characters, and the whole printed together ◁ *deleted*.

91. Namely, Leibniz's letter to Jean de Fontaney for which the editors of the ongoing critical edition have suggested a date of 17 April 1703. See A I 22, 372–373. English translation: Leibniz (1703b).

92. These must have been separate from the letter to Fontaney, which contains no such list of questions. They have since been lost.

30 Explanation of Binary Arithmetic, Which Uses Only the Digits 0 and 1, with Some Remarks on Its Usefulness, and on the Light It Throws on the Ancient Chinese Figures of Fuxi (7 April 1703)

Manuscripts:
M1:   LBr 68 Bl. 21. Draft. French.
M2:   LBr 68 Bl. 126–127. Draft, revised and expanded version of M1. French.
M3:   Paris, procès-verbaux of the Académie royale des sciences 1703, pp. 135–137.
        Partial copy of no longer extant dispatched fair copy. French.
M4:   Gotha 449 A Bl. 95–98. Copy of E, made in 1734. French.

Transcriptions:
E: Leibniz (1705a) (following no longer extant fair copy of M2).
GM: GM VII, 223–227 (following E).
Z1:   Zacher 1973, 293–296 (following M1).
Z2:   Zacher 1973, 296–301 (following E).

Leibniz's reticence to go public with the binary system was finally overcome by Joachim Bouvet's suggestion, in his letter of 4 November 1701 (see chapter 28), that binary was the key to deciphering the mysterious hexagrams of the Yijing. Leibniz received Bouvet's letter on 1 April 1703 and almost immediately began writing a paper for publication in *Histoire de l'Académie Royale des Sciences*, the same periodical for which the earlier "Essay on a New Science of Numbers" (see chapter 23) had been intended. What follows is a first draft of the paper, followed by the final version.

The draft opens with an overview of the mathematics of binary before turning about halfway through to the parallel between binary and the hexagrams of the Yijing. The outline of binary is quite minimalistic, with Leibniz offering a brief explanation of binary notation along with a table of binary numbers up to 1 0000 (16) and two examples of addition sums before noting the periodicity in the columns of natural numbers expressed in binary. The remainder of the draft is devoted to outlining Bouvet's hypothesis that binary is the key to deciphering the hexagrams of the Yijing. Whereas the earlier, more mathematically sophisticated "Essay on a New Science of Numbers" (see chapter 23) was designed to stimulate interest among mathematicians to undertake further research on binary, the intention behind "Explanation of Binary Arithmetic" was quite different, with Leibniz content to provide just enough information about binary to enable a reader to see the connection between binary and the hexagrams.[1]

Having sketched out his initial draft, Leibniz quickly produced an expanded version. While Bouvet's hypothesis about the parallel between binary and the hexagrams of the Yijing remained the focus of the expanded version of the paper, Leibniz also sought to deepen his presentation of the binary system to a certain extent. Among other changes made to the initial draft, Leibniz expanded the table of binary numbers up to 10 0000 (32), discussed column periodicity in more detail, included examples of all four arithmetic operations (albeit without any explanation as to the methods involved), and added a remark about the practical advantage of using either duodecimal or his own sedecimal system in place of decimal (which would make this the first published reference to base 16).

On 7 April 1703, Leibniz sent the finished paper to the Académie's president, Jean-Paul Bignon, accompanied by the following cover letter:

---

1. The same is true for a couple of letters Leibniz wrote in April 1703 to alert correspondents to binary in light of Bouvet's hypothesis. See Leibniz's letters to Carlo Maurizio Vota of 4 April 1703 (A I 22, 321–326) and Hans Sloane of 17 April 1703 (English translation in Aiton 1981).

I have waited for a good reason to write to you so as not to abuse the honor you have done me by responding. A letter from China provided me with it.[2] I communicated to Reverend Father Bouvet my way of calculating by 0 and 1.[3] And he responded by saying that he discovered straightaway that it is precisely the meaning of the figures of Fuxi, Chinese king and philosopher thought to have lived more than 4,000 years ago. It is the oldest known monument of science, the Chinese having lost its true explanation and having offered a great deal of chimerical ones, and yet now here we are, its deciphering by means of Europeans. This confluence is curious and could be of consequence for that country. I have put it all in the attached paper to be inserted, if it is deemed appropriate, in the *Memoires* of the Académie that are printed from time to time. This is instead of the earlier paper I sent about this same arithmetic,[4] which is not so appropriate to be printed. (A I 22, 332)

On 18 April, Leibniz wrote to Bernard le Bovier de Fontenelle, editor of the Académie's annual proceedings, the *Histoire de l'Académie Royale des Sciences*, alerting him to his new paper:

I wrote to Father Bignon in the last post, and time did not permit me to write to you also, Sir, and yet having [since] received your last [letter],[5] I will not hesitate to tell you that the occasion of the agreement of my binary arithmetic with the ancient figures of Fuxi means that I believe we can now speak of this arithmetic in the *Memoires* of the Académie. But I implore you, Sir, to insert there not my old paper, but the one I have just sent to Father Bignon that is shorter and talks about this agreement Reverend Father Bouvet mentioned to me. He was the one who deciphered the enigma of Fuxi with the help of my binaries. I had feared the public would despise a computation whose fruit is not immediately apparent, but this new remark on the Chinese figures perhaps renders it acceptable. (A II 4, 161–162)

Leibniz's paper was read at the Académie on 5 May 1703 under the title "On the Binary Calculus." The official minutes of the Académie indicate that less than half of the paper was actually read, with Fontenelle focusing solely on the part dealing with the hexagrams of the Yijing (see M2); the outline of binary in the first half of the paper was ignored, perhaps because members of the Académie were already familiar with the ideas from Leibniz's "Essay on a New Science of Numbers" (see chapter 23), which had been read there two years earlier. On 6 July 1703, a little over two months after the reading of Leibniz's "Explanation of Binary Arithmetic," Fontenelle replied to Leibniz, informing him, "I read to the Académie your explanation of the mysterious lines of Fuxi by binary calculus, and it came across as very ingenious and quite excellent" (A II 4, 171). Fontenelle also stated that Leibniz's paper would be published in the *Histoire de l'Académie Royale des Sciences* for 1703, which it was, though delays meant that the volume for 1703 did not appear until 1705.[6]

Each volume of the *Histoire* was divided into two parts: the first part (the *Histoire* proper) contained Fontenelle's remarks on each of the papers that had been read at the Académie that year, while the second part (the *Memoires*) contained the papers themselves. In the case of Leibniz's paper, Fontenelle's remarks are contained in an essay entitled "New Binary Arithmetic," which was almost as long as Leibniz's paper (see Fontenelle 1705b; English translation: Fontenelle 1705a). Fontenelle's essay begins with a straightforward outline of binary arithmetic but halfway through turns increasingly mischievous in tone. First, Fontenelle (mis)informed his readers that

it was in 1702 that he [Leibniz] communicated to the Académie this binary arithmetic, announcing only that it would have great uses for the sciences but not revealing them. He did

---

2. Namely, Joachim Bouvet's letter to Leibniz of 4 November 1701 (see chapter 28).

3. Namely, Leibniz's letter to Bouvet of 15 February 1701 (see chapter 22).

4. Leibniz is referring here to his "Essay on a New Science of Numbers" (see chapter 23).

5. That is, Fontenelle's letter of 24 March 1703 (see A II 4, 151–153).

6. At just five printed pages, Leibniz's paper may seem short for a scientific paper, but it was not unusually so for the time. The volume in which Leibniz's paper appeared contained 55 papers in all, half of which were shorter than Leibniz's, including 5 papers that were only a single page.

not want it to be mentioned in the *Histoire* until this new invention could be accompanied by its uses. (Fontenelle 1705b, 60)

Leibniz had in fact communicated binary arithmetic to the Académie in 1701, in his "Essay on a New Science of Numbers" (see chapter 23), and he had subsequently asked Fontenelle not to publish the essay until he could provide "better examples" (A II 4, 64) rather than "uses" as Fontenelle claimed. Nevertheless, having raised the issue of Leibniz's inability to identify uses of binary arithmetic, Fontenelle (1705b, 60) proceeded to claim that "in the present year, it was found that it had one, which Mr. Leibniz himself had not expected." Fontenelle identified this "use" as the role of binary in deciphering the hexagrams of the Yijing, a suggestion owing more to Leibniz's description of this as a "fruit" in his letter to Fontenelle of 7 April 1703 (see above) than anything written in "Explanation of Binary Arithmetic," where Leibniz identifies binary's use as facilitating advances in number theory (or "the perfection of the science of numbers," as he put it; see below). To compound matters, Fontenelle (1705b, 61) then all but dismissed Leibniz's/Bouvet's suggestion that Fuxi had binary arithmetic in mind when devising the hexagrams, before identifying someone else who had allegedly made inroads in binary, namely, Thomas Fantet de Lagny (1660–1734), a professor of hydrography at Rochefort: "If Mr. Leibniz did not have the same thoughts on binary arithmetic as Emperor Fuxi, at least Mr. de Lagny did have the same thoughts as Mr. Leibniz on this very subject." Fontenelle then explained that de Lagny had devised binary arithmetic during his efforts to improve logarithms, a clear example of a "use" that, according to Fontenelle, Leibniz himself had been unable to provide.[7] After devoting almost a third of his essay (which was notionally on Leibniz's paper) to de Lagny, and sensing that he was on the verge of sparking a new priority dispute, Fontenelle (1705b, 62–63) closed his remarks by stressing that de Lagny had developed binary arithmetic independently of Leibniz, or rather that he *appeared* to have done so:

> As the greatest mathematicians can very legitimately be jealous of the glory of having had the same thoughts as Mr. Leibniz, without having followed him, we owe this testimony here to Mr. de Lagny, that having always been at Rochefort, he does not appear to have had any knowledge of what Mr. Leibniz had sent to the Académie on the binary calculus.

It is likely that Leibniz would have been unhappy with the tone and content of Fontenelle's essay, though as it happened, he never got to see it, or even learn of its existence, or ever find out that his "Explanation of Binary Arithmetic" had been published. His enquiries regarding the fate of his paper received no reply, and in 1706 and in 1711, he had to resort to asking other correspondents if they were able to find out whether his paper had been published or not, again with no success.[8] In 1715–1716, in a lengthy essay on the natural theology of the Chinese, at the end of which is discussed the parallel between binary and the hexagrams of the Yijing, he wrote (WOC 134) that mention of the binary system in printed works could be found only in an essay published by Wilhelm Tentzel, in 1705 (namely, Tentzel 1705) and in another published by Pierre Dangicourt (namely, Dangicourt

---

7. In "New Binary Arithmetic," Fontenelle (1705b, 62) claims that de Lagny had already published his work on logarithms using binary, in 1703, though he does not mention the title. In a later essay in the same volume of the *Histoire*, Fontenelle (1705c, 64) tells a different story: "It was said above that Mr. de Lagny is working on a new trigonometry. He will call it *Trigonometrie Françoise or Reformée*.... In this new trigonometry, Mr. de Lagny puts in place of the old logarithms, which he finds arbitrary and defective, the natural logarithms of binary arithmetic. He also has new views on the tables of sines, tangents, and secants, and he has given to the Académie a small sample of his work on tangents and secants and an assurance of his promises." Here it is clear that de Lagny's book was merely planned, not already published, as Fontenelle had earlier claimed. As it happens, de Lagny published nothing in 1703 and never did publish a book entitled *Trigonometrie Françoise or Reformée*, though it was clearly something he intended to do, as in an essay from 1725, de Lagny (1725, 291–292) proposes "to divide the quadrant into 30 degrees, one degree into 32 minutes, one minute into 32 seconds etc." and that "this is the subject of a preliminary dissertation of my *Trigonometrie Françoise or Reformée*." However, no such work was ever published, and if it was ever written, it is no longer extant. While it is not uncommon to find claims that de Lagny had worked on binary arithmetic, these all derive from the unsubstantiated claims made in Fontenelle's essay "New Binary Arithmetic."

8. See Zacher (1973, 138–139).

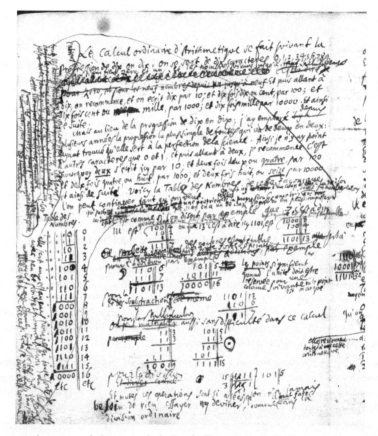

Second manuscript page of the second draft of "Explanation of Binary Arithmetic" (LBr 68 Bl. 126v).

1710), indicating that even in the last year of his life, Leibniz was unaware that his "Explanation of Binary Arithmetic" had been published.[9]

Despite Fontenelle's uncollegial and incautious framing of Leibniz's invention on its debut in the world of letters, "Explanation of Binary Arithmetic" would be frequently republished in the eighteenth century (see, for example, Nolte 1734, 17–24; Leibniz 1768, III, 390–394) and thereafter. Moreover, the association between binary and the hexagrams of the Yijing would be endorsed often, sometimes enthusiastically (see, for example, Tentzel 1705), sometimes cautiously (see, for example, Bilfinger 1724, 359–360), and could be found repeated in numerous European and American encyclopedias and other reference works until the middle of the nineteenth century (see, for example, Croker, Williams, and Clark 1766, n.p.; Brewster 1832, II: 382–385; Anon 1857, III: 591).

---

9. Quite possibly the reason for Fontenelle's half-truths, misleading statements, and inappropriate allusions was that he did not think "Explanation of Binary Arithmetic," with its focus on the Yijing, was fitting subject matter for the proceedings of a scientific academy. It is also likely that Fontenelle did not think much of the binary system; when writing a lengthy eulogy for Leibniz in 1716, Fontenelle outlined many of Leibniz's innovations and discoveries but did not mention binary even once. See Fontenelle (1718, 103–183).

[M1]

Explanation of Binary Arithmetic, which Uses Only the Digits 0 and 1, with Some Remarks on Its Usefulness, and on the Light It Throws on the Ancient Chinese Figures of Fuxi, by M. D. L.

Remark on Binary Arithmetic by 0 and 1 and Its Usefulness and on the Explanation It Affords of the Ancient Chinese Figures of Fuxi, by Mr. de Leibniz[10]

The ordinary reckoning of arithmetic is done according to the progression of tens. Ten digits are used, which are 0, 1, 2, 3, 4, 5, 6, 7, 8, 9, that is for zero and the nine numbers; and then, when reaching ten, one starts again, writing ten by 10, ten times ten, or a hundred, by 100, ten times a hundred, or a thousand, by 1000, ten times a thousand by 10000, and so on.

But instead of the progression of tens, I have for many years used the simplest progression of all, which proceeds by twos, having found it useful for the perfection of science. Thus I use no other digits in it bar 0 and 1, and when reaching two, I start again. This is why *two* is here expressed by 10, and two times two, that is four, by 100, two times four, that is eight, by 1000, two times eight, that is sixteen, by 10000, and so on. Here is the Table of Numbers of this way, which may be extended as far as is desired.

For here, it is as if one said, for example, that 111, or 7, is the sum of four, two, and one. And likewise 13, or rather 1101 here, is the sum of eight, four, and one.

					0
				1	1
			1	0	2
			1	1	3
		1	0	0	4
		1	0	1	5
		1	1	0	6
		1	1	1	7
	1	0	0	0	8
	1	0	0	1	9
	1	0	1	0	10
	1	0	1	1	11
	1	1	0	0	12
	1	1	0	1	13
	1	1	1	0	14
	1	1	1	1	15
1	0	0	0	0	16

100	4	1000	8
10	2	100	4
1	1	1	1
111	7	1101	13

Multiplication is also easy in this calculus; for example, five multiplied by three gives fifteen, which one gets commonly by the Pythagorean table, that is by one times one is one, etc. But here there is no need to learn anything by heart, because everything is found and is proved[11] from the source, for example:

101	5
11	3
101	
101	
1111	15

And all the other operations are done with the greatest ease in the world when these digits are used.

It is true that this way of calculating, which takes everything from first principles, would be inconvenient for common practice, to which we are accustomed and in which numbers are expressed with fewer digits. But in return this method is fundamental for the science, and yields new discoveries, which are useful even for the practice of numbers, and especially for geometry. The reason for this is that, numbers having been reduced to the simplest principles, 0 and 1, a marvelous order appears everywhere. For putting numbers in order, periods are found to recur, as in the first column 010101 etc., in the second

Column 4	Column 3	Column 2	Column 1	
0	0	0	0	0
0	0	0	1	1
0	0	1	0	2
0	0	1	1	3
0	1	0	0	4
0	1	0	1	5
0	1	1	0	6
0	1	1	1	7
1	0	0	0	8
1	0	0	1	9
1	0	1	0	10
1	0	1	1	11
1	1	0	0	12
1	1	0	1	13
1	1	1	0	14
1	1	1	1	15

---

10. Leibniz wrote both titles at the top of the page but did not strike out either. Transcription Z1 incorrectly states that Leibniz struck out the second of the two titles.

11. proved ▷ from itself ◁ *deleted*.

0011, 0011, in the third 0000 1111, 0000 1111, and so on forever. And it is found that also the square numbers, cubes, etc., likewise triangular, pyramidal, and others have similar periods, of a kind that makes it possible to write the tables immediately, without calculation.

But what is surprising is to find that this arithmetic of 0 and 1 contains the secret of the[12] lines of an ancient king and philosopher who lived more than four thousand years ago, named Fuxi, whom the Chinese see as the founder of their empire and of their sciences. The Chinese lost the significance of the gua or linear figures of Fuxi several thousand years ago, and made commentaries about that, where they sought I know not what[13] stretched meaning. And it has happened that the real explanation has come to them today from Europe. For when about three years ago I sent my new method of counting to Reverend Father Bouvet, the[14] renowned French Jesuit living in Peking, capital of China, he first recognized that it's the key to the figures of Fuxi, and writing me on 4 November 1701, he sent me that great figure of Fuxi, which goes to 64 and makes that evident.[15] But here we content ourselves with the 8 gua, which pass among the Chinese as fundamental. And to understand these, one need only remark that a whole line — means unity, or 1, and secondly, that a broken line - - means zero, or 0. Here they are with the explanation.

	000	001	010	011	100	101	110	111
		1	10	11	100	101	110	111
	0	1	2	3	4	5	6	7

This agreement gave me a high opinion of the depth of Fuxi's meditations, since what is easy to us was not so during these distant times; binary arithmetic is indeed very easy today for those few who think about it, because our ordinary method helps it a great deal. But ordinary arithmetic by ten, as we have it today, apparently did not yet exist in these olden times, since even the Greeks and Romans were unaware of it and were deprived of its benefits. And as Fuxi is believed in China to be the author of the ordinary Chinese characters (albeit much altered by the passage of time), something considerable might even be found in these characters with regard to numbers and ideas, if one could discover the foundation of Chinese writing, all the more since it is believed in China that he had consideration for numbers when establishing them.

---

12. the ▷ whole and broken lines ◁ *deleted.*
13. what ▷ mysteries ◁ *deleted.*
14. the ▷ French Jesuit who is my correspondent ◁ *deleted.*
15. That is, makes evident that Leibniz's new method is the key to understanding the figures of Fuxi.

[E]

Explanation of Binary Arithmetic, Which Uses Only the Digits 0 and 1, with Some Remarks on Its Usefulness, and on the Light It Throws on the Ancient Chinese Figures of Fuxi

The ordinary reckoning of arithmetic is done according to the progression of tens. Ten digits are used, which are 0, 1, 2, 3, 4, 5, 6, 7, 8, 9, which signify zero, one, and the successive numbers up to nine inclusively. And then, when reaching ten, one starts again, writing ten by 10, ten times ten, or a hundred, by 100, ten times a hundred, or a thousand, by 1000, ten times a thousand by 10000, and so on.

But instead of the progression of tens, I have for many years used the simplest progression of all, which proceeds by twos, having found it useful for the perfection of the science of numbers.[17] Thus I use no other digits in it bar 0 and 1, and when reaching two, I start again. This is why *two* is here expressed by 10, and two times two, or *four*, by 100, two times four, or *eight*, by 1000, two times eight, or *sixteen*, by 10000, and so on. Here is the *Table of Numbers* of this way, which may be extended as far as is desired.

Table of Numbers[16]

0	0	0	0	0	0	0
0	0	0	0	0	1	1
0	0	0	0	1	0	2
0	0	0	0	1	1	3
0	0	0	1	0	0	4
0	0	0	1	0	1	5
0	0	0	1	1	0	6
0	0	0	1	1	1	7
0	0	1	0	0	0	8
0	0	1	0	0	1	9
0	0	1	0	1	0	10
0	0	1	0	1	1	11
0	0	1	1	0	0	12
0	0	1	1	0	1	13
0	0	1	1	1	0	14
0	0	1	1	1	1	15
0	1	0	0	0	0	16
0	1	0	0	0	1	17
0	1	0	0	1	0	18
0	1	0	0	1	1	19
0	1	0	1	0	0	20
0	1	0	1	0	1	21
0	1	0	1	1	0	22
0	1	0	1	1	1	23
0	1	1	1	0	0	24
0	1	1	0	0	1	25
0	1	1	0	1	0	26
0	1	1	0	1	1	27
0	1	1	1	0	0	28
0	1	1	1	0	1	29
0	1	1	1	1	0	30
0	1	1	1	1	1	31
1	0	0	0	0	0	32

etc.

Here, one glance makes evident the reason for a *celebrated property of the double geometric progression* in whole numbers, which holds that if one has only one of these numbers of each degree, one can compose from them all the other whole numbers below the double of the highest degree. For here, it

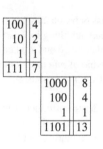

is as if one said, for example, that 111, or 7, is the sum of four, two, and one, and that 1101, or 13, is the sum of eight, four, and one. This property enables assayers to weigh all sorts of masses with few weights and could serve in coinage to give several values with few coins.

Establishing this expression of numbers enables us to perform all sorts of operations very easily.[18]

---

16. In M2, the table goes only up to 16.

17. In M2, the end of this sentence reads: "...useful for the perfection of science."

18. M2 here includes the following sentence: "*The dots* signify that the 1 must be reserved for a subsequent column in which the dot is marked." The second and third examples of subtraction are not in M2. In M2 near the multiplication examples, Leibniz wrote, "In this calculus, one can also multiply without difficulty." The third multiplication example is not in M2.

		110	6	101	5	1110	14
For *addition*,[19]		111	7	1011	11	10001	17
for example ☽		. .		. . . .			
		1101	13	10000	16	11111	31

	1101	13	10000	16	11111	31
For *subtraction*	111	7	1011	11	10001	17
	110	6	101	5	1110	14

	11	3	101	5	101	5
For *multiplic-*	11	3	11	3	101	5
ation ⊙	11		101		101	
	11		101		1010	
	. .					
	1001	9	1111	15	11001	25

For *division*   15 ‖ ̷1̷1̷11 ⎱ 101 ‖ 5
3 ‖ ̷1̷1̷̷1 1
̷1 1

And all these operations are so easy that there would never be any need to guess or try out anything, as has to be done in ordinary division. There would no longer be any need to learn anything by heart here, as has to be done in ordinary reckoning, where one has to know, for example, that 6 and 7 taken together make 13, and that 5 multiplied by 3 gives 15, in accordance with the table of *one times one is one*, which is called Pythagorean. But here, all of that is found and proved from the source, as is clear in the preceding examples under the signs ☽ and ⊙.

However I am not in any way recommending this method of counting in order to introduce it in place of the ordinary practice of counting by ten. For, aside from our being accustomed to that, there is no need to seek out what has already been learned by heart. Thus, the practice of counting by ten is shorter and the numbers not as long. And if we were accustomed to proceed by twelves or sixteens, there would be even more benefit. But reckoning by twos, that is, by 0 and 1, as compensation for its length, is the most fundamental way of reckoning for science, and offers up new discoveries, which are then found to be useful, even for the practice of numbers and especially for geometry. The reason for this is that, as numbers are reduced to the simplest principles, like 0 and 1, a wonderful order is apparent throughout. For example, in the *Table of Numbers* itself, we see each column is ruled by periods which always begin over again. In the first column this is 01, in the second 0011, in the third 0000 1111, in the fourth 0000 0000 1111 1111, and so on. And little zeros have been put into the table to fill the gap at the beginning of the column, and to emphasize these periods better. Also, lines have been drawn within the table, showing that what is contained within the lines always occurs again underneath them. And it even turns out that square numbers, cubic numbers, and other powers, likewise triangular numbers, pyramidal numbers, and other figurate numbers, have similar periods, so that tables of them can be written immediately, without any calculation. And this one drawn-out task in the beginning, which then gives the means to make reckoning economical and to proceed to infinity by rule, is infinitely advantageous.

What[20] is surprising in this calculus is that this arithmetic by 0 and 1 is found to contain the mystery of the lines of an ancient King and philosopher called Fuxi, who is believed to have lived more than 4000 years ago, and whom the Chinese regard as the founder of their empire and their sciences. There are several linear figures attributed to him, all of which come back to this arithmetic,

---

19. In M2, but not in E, the rightmost sum is erroneously presented as

1110	16
10001	17
11111	32

.

20. Manuscript M3 begins with this sentence.

but it is sufficient to give here the *Figure of Eight Gua*, as it is called, which is said to be fundamental, and to join to it the explanation which is obvious, provided one notices, firstly, that a whole line — means unity, or 1, and secondly, that a broken line - - means zero, or 0.

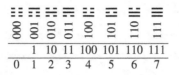

000	001	010	011	100	101	110	111
	1	10	11	100	101	110	111
0	1	2	3	4	5	6	7

    The Chinese lost the meaning of the *Gua* or Lineations of Fuxi, perhaps more than a thousand years ago, and they have written commentaries on the subject in which they have sought I know not what far out meanings, so that their true explanation now has to come from Europeans. Here is how: It was scarcely more than two years ago that I sent to Reverend Father Bouvet, the renowned French Jesuit who lives in Peking, my method of counting by 0 and 1, and nothing more was required to make him recognize that this was the key to the figures of Fuxi. Writing to me on 14 November 1701,[21] he sent me this philosophical prince's grand figure, which goes up to 64, and leaves no further room to doubt the truth of our interpretation, such that it can be said that this Father has deciphered the enigma of Fuxi with the help of what I had communicated to him. And as these figures are perhaps the most ancient monument of science which exists in the world, this restitution of their meaning, after such a great interval of time, will seem all the more extraordinary.

    The agreement between the figures of Fuxi and my Table of Numbers is more obvious when the initial zeros are provided in the Table; they seem superfluous, but are useful to better show the periods of the column, as I have in effect provided them with little circles, to distinguish them from the necessary zeros. And this agreement leaves me with a high opinion of the depth of Fuxi's meditations, since what seems easy to us now was not so at all in those far-off times. Binary or dyadic arithmetic is, in effect, very easy today, with little thought required, since it is greatly assisted by our method of counting, from which, it seems, only the excess is removed. But this ordinary arithmetic by tens does not seem very ancient; at least the Greeks and Romans were unaware of it, and were deprived of its benefits. It seems that Europe owes its introduction to Gerbert, who became Pope under the name of Sylvester II, who got it from the Moors of Spain.[22]

    Now, as it is believed in China that Fuxi is also the author of ordinary Chinese characters, although they were greatly altered in subsequent times, his specimen of arithmetic leads us to conclude that something considerable might even be found in these characters with regard to numbers and ideas, if one could discover the foundation of Chinese writing, all the more since it is believed in China that he had consideration for numbers when establishing them. Reverend Father Bouvet is strongly inclined to press this point, and very capable of succeeding in it in various ways. However, I do not know if there was ever an advantage in this Chinese writing similar to the one that there necessarily must be in the Characteristic I intend, which is that every reasoning derivable from notions could be derived from these notions' characters by a way of reckoning, which would be one of the more important means of assisting the human mind.

---

21. Bouvet's letter was in fact dated 4 November 1701; the mistake here is Leibniz's. In M2, Leibniz correctly wrote 4 November 1701.

22. Sylvester II (c. 946–1003), originally known as Gerbert of Aurillac, was pope from 999 to 1003. In his "Discourse on the Natural Theology of the Chinese" (1716), Leibniz repeated his claim that Gerbert introduced the decimal system to Christian Europe. See WOC 135. Leibniz's claim is mistaken; although Gerbert is traditionally believed to have introduced Arabic numerals to Christian Europe, he did not introduce the decimal system.

les uns font — — devant, et les autres — font vis à vis l'un de l'autre. Les figures Chinoises qui doivent servir d'explication pretendue, y sont jointes aussi. M le Comte de Schwartzbourg a dans son excellent Cabinet de Medailles une Chinoise, qui a les huit notes avec leur figures Chinoises

Je trouve que les Notes de Fohi pourroient encor servir à marquer les dichotomies continuées par des sous-divisions. En voicy le Tableau

Les deux notes marqueroient les membres de la premiere division, les quatre notes

marqueroient

Manuscripts:
M1: LBr 146 Bl. 5–6. Draft. French.
M2: LH 35, 4, 20 Bl. 1–6. Fair copy, dispatched. French.

Transcription:
Z: Zacher 1973, 345–352 (following M2).

Despite (or perhaps because of) Leibniz's uncertainty as to whether his "Explanation of Binary Arithmetic" had been published (see chapter 30), he continued to promote his invention through his network of correspondents, raising it in letters to mathematicians or those with some knowledge of the discipline, often in the hope that they would assist him in developing it. These promotional efforts occasionally earned Leibniz new correspondents. One such was César Caze, a French Huguenot émigré who was at the time living in Amsterdam manufacturing and inventing scientific instruments, among them a calculating machine.[1] The correspondence between Leibniz and Caze was initiated by a letter Leibniz wrote on 2 March 1704 to his long-time correspondent Nicolaas Witsen (1641–1717), a renowned mapmaker and naval architect, and mayor of Amsterdam. Leibniz's letter contained a brief outline of his "new kind of arithmetic," an enthusiastic report of Bouvet's claim that the "ancient characters of the renowned Fuxi … are precisely this arithmetic," and concluded on a note of resignation, with Leibniz acknowledging that Witsen's schedule probably did not leave him much time to delve into such curiosities (A I 23, 141). This proved to be correct: in his reply of 6 June 1704 (A I 23, 405–406), Witsen advised that as the matter required more time and attention than he had to offer, he had passed Leibniz's letter on to Caze. Enclosed with Witsen's letter was a short paper Caze had written for Witsen in May 1704 (see Zacher 1973, 305–306), in which he claimed that while Leibniz's binary system and the hexagrams of the Yijing were similar in that they both made use of two symbols (digits in the first case, lines in the second), the hexagrams of the Yijing were likely designed for something other than binary arithmetic. Caze also claimed that he had developed and completed the binary system five years earlier (i.e., in 1699). This prompted Leibniz to write a short response to Caze, of which there survives a draft, dated 24 June 1704 (see Zacher 1973, 309–310). Leibniz's opening remark of his response—"I think it was more than 20 years ago that I spoke and

---

1. Curiously, the first mention of Caze's arithmetic machine appeared in the May 1707 issue of the journal *Nouvelles de la république des lettres*, in the midst of a review of the contents of the 1703 edition of the *Histoire de l'Académie Royale des Sciences*, the very volume that contained Leibniz's "Explanation of Binary Arithmetic" (see chapter 30). In his review, the editor of the *Nouvelles de la république des lettres*, Jaques Bernard, wrote: "With regard to arithmetic, we are told of the new binary arithmetic invented by Mr. Leibniz, and which can be of great use. On this matter [i.e., arithmetic rather than binary arithmetic], Mr. Caze has invented a very curious machine, which I have seen, and which he must he implored to give to the public." *Nouvelles de la république des lettres* (May 1707), 574.

wrote to friends about my binary arithmetic" (Zacher 1973, 309)[2]—was clearly intended to establish priority for the invention. And while he insisted, *pace* Caze, that the hexagrams of the Yijing were invented with binary arithmetic in mind, he conceded that this may not have been their sole aim ("I also do not deny that besides arithmetic this Fuxi had perhaps yet another considerable purpose"; Zacher 1973, 309). Leibniz's response was enclosed with his letter to Witsen of 26 June 1704 (see A I 23, 469 and 471) and prompted Caze to write a much longer paper (see Zacher 1973, 311–338), which was sent to Leibniz on 14 October 1704 along with a short covering letter (see Zacher 1973, 339).

Caze's lengthy paper begins with a history of binary arithmetic. While he repeats his earlier claim that he had thought of binary arithmetic in 1699 (while in prison for nonpayment of debt), he does not assert priority for the invention, claiming instead that his research had revealed that it could be found in the *Arithmeticae localis liber* [*Book of Location Arithmetic*] at the end of John Napier's *Rabdologiæ* (Napier 1617),[3] and in a chapter on binary arithmetic in the *Mathesis biceps* (1670) of Caramuel, although Caze concedes that Caramuel's treatment of binary was thin and underdeveloped. Caze then proceeds to elaborate on his earlier claim that the hexagrams of the Yijing were probably not intended as binary arithmetic, as Bouvet and Leibniz had supposed. Specifically, Caze observes that the hexagrams thought to represent the numbers 0–31 in binary all begin with a broken line, taken by Bouvet and Leibniz to be equivalent to 0 in binary, which meant that if the hexagrams were intended to represent numbers in binary, then those representing decimal 0–31 would all contain apparently superfluous leading 0s; for example, the hexagram Bouvet and Leibniz took to represent the number 8, namely, ䷆, would be equivalent (in Leibniz's notation) to 00 1000, where the first two 0s serve no purpose. With no obvious explanation as to why the Chinese would include such "useless" digits in a binary numbering system, Caze concludes that the hexagrams were not intended to represent binary strings at all. Caze also asks why, if Fuxi had intended his hexagrams to express binary strings, he would have stopped at 63. Nevertheless, Caze conjectures that the hexagrams may have been based on a six-row abacus (*suanpan*) that may have been used by the ancient Chinese instead of the eight-row version known to be in use in the early eighteenth century. He also provides some analysis of the laws governing the columns of binary numbers. Leibniz takes up all of these matters (suggesting a potential function for the leading 0s Caze saw as useless) in the following letter, which is his somewhat delayed response to Caze's lengthy paper of October 1704,[4] and the last in his brief correspondence with Caze.

[M2]

To Mr. Caze   Hanover 23 June 1705

Sir

I am almost ashamed to write, having been so long without discharging my duty. You have pre-occupied me, Sir, most obligingly; you have sent me some excellent and interesting thoughts, but a long absence, illnesses, very great distractions, the loss of the Queen of Prussia,[5] who had limitless kindness towards me, and many other reasons discomposed me in a strange way. I am beginning to feel like myself again and I begin almost with you, Sir, in fulfilling what I owe to people who do me the honor of writing to me. I read with admiration your thoughts about binary arithmetic. You rightly note that it might be useful even for practice, in numbers of considerable magnitude, where the difference in length of binary and decimal expressions is hardly important in proportion to the

---

2. The claim is correct. See introduction, this volume.

3. For an English translation, see Napier (2017).

4. Indeed, such was the delay that Caze wrote to Leibniz again in January 1705 expressing concern that his lengthy paper might not have found its way to its recipient (see LBr 146 Bl. 2; English translation: Caze 1705).

5. Queen Sophie Charlotte died on 1 February 1705.

ease that binaries offer; but with all that I share your doubts that one could ever succeed in changing the ordinary practice of calculations, and I don't believe one should even think of doing so. I
have carefully read your doubts, Sir, as to whether Fuxi actually thought of binary arithmetic, and I
find, indeed, that it cannot be proved entirely convincingly. It is true that the combinations consist in
an excessively abstract speculation, and that if we think about them we could arrive at completely
different arrangements. For example, the eight trigrams could be arranged like this:

a^3	aab	baa	aba
0	1	4	2
b^3	bba	abb	bab
7	6	3	5

by putting together those which swap around a and b (or else 0 and 1) in order to observe the rules
of equity, and firstly doing the transpositions without divulsion,[6] for example, changing aab into baa
rather than aba. And we will not easily or always notice this order, which yields the numbers:

$$a^3 \quad aab \quad aba \quad abb \quad baa \quad bab \quad bba \quad b^3$$

Father Bouvet sent me from Peking a printed figure which represents the 64 characters twice,
namely in a centrally-placed square, which retains the order of the numbers perfectly, and then in a
circle around it, where the characters, which differ only because some start with -- and others —,
are opposite each other.[7] Chinese figures which are to serve as an alleged explanation are added. In
his excellent cabinet of medals, the Count of Schwarzburg has a Chinese one, which has the eight
trigrams along with their Chinese figures.[8]

I find that Fuxi's glyphs could even serve to indicate the dichotomies continued by sub-divisions.
Here is the table:

The two lines would indicate the members of the first division, the four glyphs would indicate the
members of the second division, the eight those of the third, and so on. In this way, — and ==
and == etc would not indicate the same thing. But there is more likelihood of an arithmetic among
the ancient Chinese than a logical method. However, I have taken it into consideration, likewise the
combinations, when I have spoken of some other uses that could be given to Fuxi's characters, although I have more penchant for arithmetic, since it is the most popular of our sciences. Supposing
that arithmetic was the goal of this ancient figure, we must believe that he would have been pleased

---

6. Leibniz used the term "divulsion," literally a tearing apart, to refer to breaking up a natural number into
partitions, that is, combinations of integers that add up to that number. For example, the number 3 has three
partitions: 3, 1 + 2, and 1 + 1 + 1.

7. See the figures at the end of chapter 28.

8. A reproduction of Schwarzburg's medal can be found in Tentzel (1705), immediately before the title page. In
the same work, the medal is described in detail in the middle of an essay on the binary system by Tentzel, who
also states that Leibniz had promised to send a rubber imprint of the medal to Bouvet; see Tentzel (1705, 93).

by the regularity, which is preserved by an equal number of strokes; and in fact I believe that without this regularity his figures would not have delighted posterity, nor given occasion to so many commentaries. He wouldn't have found it necessary to go beyond 64: this moderately high number is sufficient for perfect instruction. But your thought here offers a more precise and more ingenious reason for it, by supposing that it was arrived at through a small machine with six rows of skewered beads,[9] similar to the one the Romans used, and which the Chinese and Russians still use. I once saw Napier's book in Latin without considering then its relation to binary arithmetic, which is also noticeable in the divisions of weights among assayers without any thought being given to it.

I would like to deserve, by sending on something worthy of you, Sir, the trouble you have taken to share with me so many clever thoughts, represented so ingeniously even in figures, and in doing so I would like to make up for responding so late. But as I have not been in a position to pursue my meditations, I will give you only the demonstration, which I have constructed to show that any sequence of whole numbers in which the value of each term, like y, can be expressed generally by x, a variable natural number (that is, 0, or 1, or 2, or 3, or 4 etc.) without this variable quantity entering the denominator or the exponent; that any sequence, I say,[10] of this nature, yields periodic columns, that is, where the first digits always return endlessly after some interval. The same demonstration holds good not just in binaries but also in ternaries, quaternaries, and others, and especially in our decimal progression. But it is easier to find the rules of periods in binary, where the digits are the simplest. It follows that not only the powers, but also what can be formed through them, have their periods. And the demonstration even opens the way to determine the periods, as it is usual for demonstrations of possibility, taken *a priori*, to lead also to the means of putting the possibility into actuality.

The main use of binary arithmetic would be to perfect geometry in relation to the expression of infinitely determined sequences. For those we usually use are only indefinite, for example, where the arc of a circle is a, the tangent t, and the radius 1, I had found in my youth that a is equal to

$$\frac{t}{1} - \frac{t^3}{3} + \frac{t^5}{5} - \frac{t^7}{7} + \frac{t^9}{9} \text{ etc.}$$

But when it's a matter of finding an arc of a determinate circle, or some other determinate irrational quantity, in order to finish this matter one would have to be able to express it by a sequence of whole numbers, similar to our decimals, whose progression to infinity was known. We know that the value of a rational fraction can always be expressed by whole numbers in decimals to infinity, so that the digits recur periodically, for example, $\frac{1}{7} = 0.142857142857142857$ etc. But when the quantity is irrational, the digits cannot recur in this way. However, the law of progression to infinity is always determinate, albeit often difficult to know. If Ludolph, in giving the value of the circumference 3.14159 etc. had found the way of continuing these digits to infinity by a law of progression, he would have done in whole numbers what I did in fractions, by showing that if the square is 1, the circle is $1 - \frac{1}{3} + \frac{1}{5} - \frac{1}{7} + \frac{1}{9} - \frac{1}{11}$ etc. This determination of the law of progression in whole numbers would be the height of arithmetic applied to geometry, but I do not think we can get there so soon except through binaries.

The computation of binaries has this wonderful feature, that being applied to algebra and the digits 0 or 1 being taken for unknowns (as they are in effect, when the magnitude is not yet expressed by means of them), these unknowns never rise to any power and always remain in their first degree, which makes them easy to find. For supposing that the unknown x signifies only 0 or 1 (as is the nature of binary digits), it follows that x and xx and x^3 and x^4 etc. are nothing but the same thing, because $0 = 0^2 = 0^3 = 0^4$ etc., and $1 = 1^2 = 1^3 = 1^4$ etc. By this principle one can arrive at solutions which seem unachievable by any other route. However, over time it will be good to pass from binary expressions to others, in order to find the same rules for variations. This arithmetic would be the highest point of the knowledge of numbers.

---

9. Namely, the *suanpan*, or Chinese abacus.
10. say, ▷ expressed by binary digits ◁ [M1] *deleted*.

It would be a shame, Sir, if your fine research on this subject remained buried. So I urge you on behalf of the public to keep thinking about it as much as more necessary or more pleasing occupations allow you. I have no doubt you will easily determine the laws of the periods of powers, if you take the trouble to apply yourself to them. For myself, pleased as I am to have given some insights, I fear I will never do much without being assisted. For aside from the fact that I am beginning to no longer adequately carry through the work of meditations, and especially of calculations, I cannot say how overwhelmed I find myself in other respects.

Before finishing, since you speak of arithmetic machines, and among others of your own, different from those of Mr. Pascal,[11] Chevalier Cotterel,[12] and Mr. Grillet,[13] to which I add that of the late Chevalier Samuel Morland,[14] inventor of the speaking trumpet, prior to that of Mr. Grillet, but coming down to more or less the same thing; I will tell you, Sir, that the one I had demonstrated in Paris over 30 years ago, albeit in miniature, is of quite a different kind. It is absolutely independent of rabdology,[15] consisting of wheels, which lead one another and which produce the sum straightaway, without any presupposed multiplication and without any auxiliary addition. From then on it was seen in the royal societies of Paris and London, and the late Mr. Matthion,[16] who also saw it, spoke of it with praise in a plaque engraved in copperplate. In their letters, Mr. Huygens and Mr. Arnauld[17] urged me to push it forward, all the while I have been hindered for lack of workers. But I finally made it a point of conscience to let it go to ruin. A large version is now being constructed, to multiply 8 digits by 8 digits. With this machine, a large number is multiplied as quickly as a small number, and division is as easy as multiplication. Mr. Pascal's machine is really only the addition machine. Mr. Morland and Mr. Grillet have added rabdology to it, set in wheels; in this way, rabdology offers multiplications and Pascal's machine auxiliary additions, but I find little advantage in it. This saves us having to use what is called the Pythagorean table, which begins with 1 times 1 is 1, but the multitude of small additions makes this way of calculating almost as long and prone to errors as ordinary calculation. Whereas in my machine the mind does almost nothing, and a child, as Mr. Matthion said (not to mention a monkey), could make very long multiplications. However, I will be happy to learn of what you have also done on this matter, and I am sincerely

Sir   Your very humble and very obedient servant

                                                                                    Leibniz

---

11. Pascal's calculating machine, known as the Pascaline, was designed in 1642 and unveiled in 1645. See Pascal (1645).

12. That is, the courtier Sir Charles Cotterel (1615–1701), who in 1667 devised the "arithmetical compendium," a calculating instrument that utilized Napier's bones.

13. That is, René Grillet de Roven (dates unknown), watchmaker of Louis XIV. A lengthy description of his portable arithmetic machine can be found in Grillet de Roven (1673).

14. That is, the inventor Samuel Morland (1625–1695). For details of his calculating machine, see Morland (1673).

15. That is, the method of practicing arithmetic operations using Napier's bones (for an explanation of which, see note 10 of chapter 19).

16. That is, the mathematician Oded Louis Matthion (1620–1700).

17. That is, the theologian, philosopher, and mathematician Antoine Arnauld (1612–1694). Leibniz met him in Paris and corresponded with him thereafter; see Leibniz (2016b).

De Dyadicis

(§.1) Definitio:

Numerus dyadice expressus
est ... d, c, b, a ...
si idem significet quod ...
... ...
simul sumti {
    a
    b o
    c o o
    d o o o
}

et bo idem sit quod bis b, et coo
idem quod quater c, et dooo idem quod bis quaterna d seu
octies d, et ita porro.

(§.2) itaq 10 est 2, et 100 est 4,
et 1000 est 8, et 10000 est 16, et ita
porro. patet ex praecedenti si pro
a, b, c, d, ponantur 1, 1, 1, 1, et
... ... Tabula ...
et generaliter Numerus ...
progressionis geometricae ...
à binario incipientis ...
exprimitur dyadice per unitatem tot
nullitatibus postfixam, quot sunt unitates
in progressionis Geometricae exponente
seu $2^e = 10^e$. Tabulaq ita stabit

10	2	2^1
100	4	2^2
1000	8	2^3
10000	16	2^4
100000	32	2^5
1000000	64	2^6
10000000	128	2^7
100000000	256	2^8
1000000000	512	2^9
10000000000	1024	2^{10}

... et ... = ... seu $\frac{e}{10}$
per praecedentem

(§.4) 3 = 11 Nam 3 = 10 + 1 (per §.3)
fiet per ... $\frac{10}{11}$

(§.5) 5 = 101 Nam 5 = ... 4 + 1 = 100 + 1 per §.2
unde ... 100 idem quod $\frac{101}{101}$ per §.1

32 On Binary (late June 1705)

Manuscript:
M: LH 35, 3 B 1 Bl. 1–4. Draft. Latin.

Transcription:
GM: GM VII, 228–234.

While Leibniz made many attempts to find someone willing to help him investigate the binary system more thoroughly than his other preoccupations allowed him to do himself, most of them proved unfruitful. From 1698, he had many relatively short-lived exchanges about binary with noted mathematicians of the day, such as Johann Christian Schulenburg (see A II 3, 427, 432–433, 449, 524), Philippe Naudé (see chapter 21), and Johann Bernoulli (see A III 8, 602, 638–639, 670, 684), but none of them resulted in the sort of collaboration he had envisaged. This pattern was to be repeated in 1704–1705 with Jakob Hermann (1678–1733), a student of Jacob Bernoulli. The following paper was written in the hope of enlisting Hermann's help.

The correspondence between Leibniz and Hermann began on 15 October 1704 with a letter from Hermann thanking Leibniz for recommending him for the chair in mathematics at Padua (see GM IV, 259–261). In his reply of 24 November 1704, Leibniz added a lengthy postscript offering a brief (and unsolicited) outline of his binary system (GM IV, 263–264). Responding on 21 January 1705, Hermann expressed his admiration for Leibniz's innovation (see LBr 396 Bl. 3r), which would normally have been Leibniz's cue to offer more information along with a request for assistance, but other concerns at this time (mentioned at the start of his letter to Caze of 23 June 1705; see chapter 31) meant that, in his next two letters, he was unable to offer more than this tantalizing remark: "In the binary expression of numbers, many things lie hidden, as anyone may easily suspect" (GM IV, 269). His interest piqued even further, Hermann responded on 4 April 1705 that he eagerly awaited the appearance of a book by Charles René Reyneau (1656–1728), which he believed— mistakenly as it happens—would treat of the binary system (GM IV, 270).[1] Receiving nothing further, Hermann wrote again on 13 June 1705 to say that he thought he had worked out the four arithmetic operations in binary and would like Leibniz to pass on his own observations (see GM IV, 276–277).[2] Suspecting he may have found a willing collaborator, Leibniz decided to provide more information about his research on binary, a purpose which the following paper, "On Binary," was intended to serve, being a systematic treatment of the binary system, covering the four basic arithmetic operations before touching upon the periodicity of summatrix columns. Ultimately, however, the paper was left

---

1. The book was Reyneau (1708), and Hermann's mistake in thinking it would include a treatment of binary appears to be due to his misreading of Leibniz's letter of 10 March 1705, in which Leibniz had indicated, albeit rather obliquely, that Reyneau's book would include some of Leibniz's work on the differential calculus (see GM IV, 269).

2. Note that in GM IV, this letter is erroneously dated 13 July 1705.

unfinished and never sent. Instead, Leibniz treated the column periodicity of arithmetic progressions in his letter to Hermann of 26 June 1705 (see GM IV, 272–275), and the technique for binary addition in a letter of 2 July 1705 (see GM IV, 280–281), the latter reusing parts of §7 of the now-abandoned paper "On Binary." The overlap between "On Binary" and Leibniz's letter of 2 July 1705 suggests the former was written first, probably at the end of June 1705.

As already noted, Leibniz's efforts to elicit Hermann's assistance ultimately proved unsuccessful. On 21 September 1705, Leibniz expressed his hope that Hermann would make headway in his research on binary, "especially [regarding] the periods of a geometric progression exhibited in columns" (GM IV, 285). Responding on 28 October 1705, Hermann conceded that while he had examined the periodicity of columns of geometric progressions, he had not discovered anything worth sharing with Leibniz (see GM IV, 288). There was no further discussion of binary in their correspondence, which lasted until Leibniz's death in 1716.

On Binary

(§1.) Definition: $dcba$ is expressed as a binary number if it signifies the same as these taken together: $\left\{\begin{array}{l} a \\ b0 \\ c00 \\ d000 \end{array}\right.$

and $b0$ is the same as twice b, $c00$ is the same as twice the double of c, or four times c, and $d000$ is the same as twice four times d, or eight times d, and so on.

(§2.) And so, 10 is 2, and 100 is 4, and 1000 is 8, and 10000 is 16, and so on. It is evident from the preceding that, if 1, 1, 1, 1 is put down for a, b, c, d, and [since] generally a number of a geometric progression beginning from 2 is expressed in binary by a 1 prefixed to as many 0s as there are 1s in the exponent of the geometric progression, i.e. $2^e = 10^e$, then the table will be determined thus:

1	1	$= 2^0$
10	2	2^1
100	4	2^2
1000	8	2^3
1 0000	16	2^4
10 0000	32	2^5
100 0000	64	2^6
1000 0000	128	2^7
1 0000 0000	256	2^8
10 0000 0000	512	2^9
100 0000 0000	1024	2^{10}

[³]

---

3. 2^{10} ▷ ¶ (§3) $1 + 1 = 10$, that is, $\dfrac{\begin{array}{c}1\\1\end{array}}{10}$

by the preceding.

(§4) $3 = 11$, for $3 = 10 + 1$ (by §3), which will become, by §1, $\dfrac{\begin{array}{c}1\\10\end{array}}{11}$

(§5) $5 = 101$, for $5 = 4 + 1 = 100 + 1$ by §2, from which $\dfrac{1}{100}$ is the same as $\dfrac{\begin{array}{c}1\\00\\100\end{array}}{101}$, by §1.

(§3.) Every number can be expressed in binary using no other digits than 0 and 1. For since every number is formed by the continuous addition of 1s,[4] and 1 added to 1 makes 10, so that 0 is written in the last position and 1 in the penultimate, i.e. is added to the penultimate position, then 1 will be written there if a 0 is there, that is, if it is empty. But if a 1 already occupies that position, it will in turn change to 0 and cause a 1 to be added in the antepenultimate, as before in the penultimate, and so on from position to position. From which it is clear that, once we begin from 1, as long as it is done properly, there cannot appear any digits other than 0 and 1, however far along the position.[5]

(§5.)[6] Whenever a 1 is to be shifted or added to the subsequent position from the preceding one, then a dot is marked in the subsequent position as a reminder.[7]

$$\begin{array}{c} 1\ 1 \\ \cdot\ \cdot\ 1 \\ \hline 1\ 0\ 0 \end{array}$$

For example, if 11 and 1 (that is, 3 and 1) are to be added, evidently 1 and 1 is 10 in the last position, so 0 is written and a 1 marked by a dot is put in the next position. Again, in the penultimate position, 1 and 1 (because of course the dot there signifies 1) will become 10, so 0 is written and 1 is marked in the antepenultimate position. Now in the antepenultimate position there is only the transferred dot or unity, which signifies 1, and the result will be 100.

(§6.) The generation of numbers by the continuous additions of 1s is shown here from 1 to 16:

(§6) 6 = 110, for 6 = 5 + 1 = $\dfrac{\begin{array}{r}101\\1\end{array}}{110}$ ◁ *deleted*. Leibniz's deletions are not recorded in transcription GM.

4. 1s, ▷ and 1 added to 1 in the last position makes 10, that is, the 1 just shifts to the penultimate position ◁ *deleted*.

5. position. ▷ Let us show this in detail

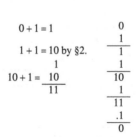

$$\begin{array}{ll} 0 + 1 = 1 & 0 \\ & 1 \\ 1 + 1 = 10 \text{ by §2.} & \overline{1} \\ & 1 \\ 10 + 1 = \dfrac{10}{11} & \overline{10} \\ & 1 \\ & \overline{11} \\ & .1 \\ & \overline{0} \end{array}$$

(§4.) The method of addition ◁ *deleted*. For the addition example on the right, the final sum should be 100; Leibniz crossed it all out before completing it.

6. The omission of §4 is a mistake on Leibniz's part.

7. reminder. ▷ For example, $\dfrac{\begin{array}{r}11\\ \cdot\ \cdot\ 1\end{array}}{100}$ In the last position, 1 and 1 is 10 so a 0 is written and the 1 in the subsequent position is marked by a dot. ◁ *deleted*.

```
    0 | 0
    1 | 1
 _____
    1
    1
 _____
   10 | 2
    1
 _____
   11 | 3
  · 1
 _____
  100 | 4
    1
 _____
  101 | 5
  · 1
 _____
  110 | 6
    1
 _____
  111 | 7
 ·· 1
 _____
 1000 | 8
    1
 _____
 1001 | 9
  · 1
 _____
 1010 | 10
    1
 _____
 1011 | 11
 ·· 1
 _____
 1100 | 12
    1
 _____
 1101 | 13
  · 1
 _____
 1110 | 14
    1
 _____
 1111 | 15
    1
 _____
10000 | 16
```

(§7.) General method of addition. However many 1s are expressed in one position by the binary number *edcba*, for each of these digits signifying 1—*a* or *b*, or *c*, or *d*, or *e* etc.—mark a dot in the position as many places to the left of the present position as the position of the digit is away from the last position, and they [i.e., the 1s] are marked in the present position by common numerals added to the last 1 of each double progression. I shall also observe that a dot in any column should be put in the first place if it is from the immediately preceding column, in the second place if it is from the one before that, in the third if it is from the one before that, and so on, for in this way one can always discern where the dot comes from, which is useful for checking; also, more than one dot cannot be inscribed in the same place. And I prefer the first place from the bottommost working upwards. In place of numerals one can also use ╱ or ╱╱.[8] Here is an example:

___

8. ╱╱. ▷ In the last column, the greatest number of 1s of the double geometric progression is 4, therefore because of these 1s I ascribe a 2 to the last [column], which indicates a dot to be marked under the second column and a 1 to the second [column] after this, because on account of the four 1s a dot ought to be marked under the second column from the present one, and on account of the two remaining 1s a dot should be marked under the first column from the present one, and because nothing remains, under the penultimate column itself I write 0. I proceed to the antepenultimate column; there again we shall add a 2 to the fourth 1, and to the second 1 after this, namely to the second dot, we add a 1, and mark dots in the first and second columns after the present one. ◁ *deleted*.

It is better to write 2
or 3 in place of ∥ or
⫻ but I prefer ∕ to
1, so that the 1
signifying the count
is not confused with
the 1 in the column.

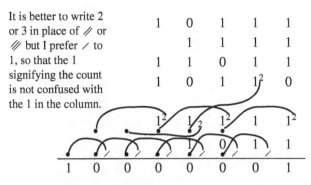

Here I have drawn lines
from the dots to the
numbers from which they
originate in order to show
the grounds for the
process, and not because
there is a need for these
lines in practice.[9]

In the *last column*, the 1s are first added, up to the highest number of the double progression which is present, namely, 4, and a 2 (on account of $4 = 2^2$) is added to the fourth or last 1, which indicates that a dot should be placed two columns further to the left (namely the antepenultimate), and indeed in the second place, that is, between its second and third mark. There remains no further number of the double progression except 1, and so 1 should be written under the column below the line drawn under all the columns. In the *penultimate column*, a 2 is again added to the fourth 1 and a 1 is added to the next one after that, since $4 = 2^2$ and $2 = 2^1$. And because of the 2, a dot is marked in the second place of the second column from here, and because of the 1 a dot is marked in the first place of the first column from here,[10] and a 0 is written under the column because nothing remains. The same happens in the *antepenultimate* or third [column]; the same again in preantepenultimate or fourth, and the same again in the antepreantepenultimate, or fifth. In the sixth there are two 1s, so I add a 1 to the second [of these] and mark a dot in the first place of the first column from here, and under the sixth column I write 0. Exactly the same will happen in the seventh column. But in the eighth, nothing remains except a 1, which is written below, and the result is 1000 0001.[11]

(§8.) The method of subtraction is this: if several [values] are to be subtracted, they may be subtracted individually or first combined into a single [value] and then this sum is subtracted. Either way, all that is required is the subtraction of one number. After that, a digit is subtracted from a digit of the same position, and if the digit to be subtracted is 1, but the digit from which it should be subtracted is 0, a remaining 1 will be written but a dot is added to the next nearest digit of the subtrahend, and if this is already a 1 then again to the next digit, and so on. Therefore, when two 1s occur in the subtrahend, they are considered as 0, and the subtraction will become only 0 from 1 or 1 from 1, from which 1 remains in the first case and 0 in the second. For example:

$$\begin{array}{c} 1\ 1\ 0\ 1\ 0\ 0\ 1\ 1\ 0 \\ \underline{1\ 1\ 0\ 1\ 1\ 0\ 1\ 1} \\ 1\ 1\ 0\ 0\ 1\ 0\ 1\ 1 \end{array}$$

(§9.) *The method of multiplication*. This is very easy, because the digits to be multiplied are 0 and 1. Now a 0 in the multiplicand makes 0, and a 1 leaves [the number] such as it was.[12] But since a 1 in the second position represents 2, it doubles the multiplicand, that is, moves it into the next nearest positions. And since a 1 in the third position represents 4, it quadruples the multiplicand, that is, moves [this number] into the third positions, and so on.

---

9. Leibniz redrew this diagram to improve its clarity. Our rendition approximates the second version, in which one of the lines is drawn underneath dots of the second row to reduce the number of line-crossings.

10. that ▷ . In the antepenultimate column, a 2 should be added to the fourth 1 (which is here designated by a dot) and a 1 to the second 1 from here, and in like manner marks will be made in subsequent [columns] ◁ *deleted*.

11. Leibniz mistakenly added a 0 in the sum, making it 1 0000 0001.

12. was. ▷ Nor, then, is there a need for the table called Pythagorean, which begins thus: once one is one, twice two is four, twice three is six, etc., and which has to be learned by heart. Indeed, we need only the start of it, that once one is one. ◁ *deleted*.

$$
\begin{array}{r}
e\ d\ c\ b\ a \\
1\ 1\ 0\ 1 \\
\hline
e\ d\ c\ b\ a \\
e\ d\ c\ b\ a\ 0 \\
e\ d\ c\ b\ a
\end{array}
$$

$$
\begin{array}{r|l}
1\ 1\ 0\ 1 & 13 \\
1\ 0\ 1 & 5 \\
\hline
1\ 1\ 0\ 1 & \\
1\ 1\ 0\ 1\ 0 & \\
\hline
1\ 0\ 0\ 0\ 0\ 0\ 1 & 65
\end{array}
$$

For 100 0000 is 64 and 100 0001 is $64 + 1$.

Hence in binary multiplication there is no need for the table commonly called Pythagorean, which has to be learned by heart. And we use only the start of it, namely that once one is one.

(§10.) *The method of division* is no less easy, because the quotient cannot be greater than 1. Therefore, only an example is needed, where for the sake of greater clarity I proceed with the common method.

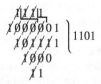

(§11.) The sequence of natural numbers from 1 to 64 [see end of this chapter].[13] The 0s not preceded by 1 can be omitted, but we have added them so that the progression of the periods is more apparent.

(§12.) Comparison of our notation with that left by Fuxi, most ancient king of China. What we designate by 0, he designated by $--$, a broken or interrupted line, and what we designate by 1, he designated by $—$, an unbroken line.

Following the same reasoning, Fuxi continued the tables of his own characters up to 64 and arranged them sometimes in a circle, sometimes in the form of a square.[14]

And when I explained my binary digits in a letter to Reverend Father Bouvet, a missionary of the Society of Jesus in China, he immediately noticed the agreement with the characters of Fuxi and reported it to me in detail. At the same time, he sent a table containing the 64 characters of Fuxi both in a square and again in a circumscribed circle, the only difference being that they follow the order of our [digits] in the square, while in the circle, one semicircle descending from the top on the left hand side contains the characters from 0 to 31, but the other semicircle, this time descending from the top on the right hand side, contains the characters from 32 to 63. For in this way, those characters which differ only in that the first digits of the left-hand ones is 0, or $\vdots$, and of the right-hand ones 1, or $|$, are placed directly opposite each other at the same height. For example, 0 and 32, 1 and 33, 2 and 34

$$
\begin{array}{r|l l|l}
0 & 00\ 0000 & 10\ 0000 & 32 \\
1 & 00\ 0001 & 10\ 0001 & 33 \\
2 & 00\ 0010 & 10\ 0010 & 34 \\
& \text{etc.} & \text{etc.} &
\end{array}
$$

---

13. In fact, Leibniz's table runs from 0 to 63, not 1 to 64, as he states here.

14. square. ▷ From this, the agreement of my binaries with Fuxi's characters is immediately ◁ *deleted.*

So something remarkable has happened, that things known three [thousand][15] or more years ago in the far east of our continent have now been rediscovered in its far west, albeit under better auspices, as I hope. For it does not seem that the use of this characteristic for augmenting the science of numbers was known before this. Indeed, the Chinese themselves, not even understanding arithmetical method, imagined who knows what mystical meanings for purely numeral characters.[16]

I shall describe the rest elsewhere, and show in general that the sums of a periodic sequence yield a periodic sequence. And that the summatrix columns of another column have a period, in such a way that the period of the first summatrix column is at least twice the length of the period of the summand, [the period] of the second [summatrix column] is quadruple the length, [the period] of the third [summatrix column] is eight times the length, etc. For at that point, a 0 of the summatrix column and of the preceding summatrices always falls at the same time as the end of a period, so everything proceeds over again as before. From this, the period of natural [numbers] is immediately evident. I shall show also that when adding two periodic columns into one, the result is periodic. Therefore, adding any number of periodic columns into one results in a periodic column. From this I demonstrate that the natural numbers, and those of an arithmetic progression, and all the polygonal numbers, as well as even quadratic and cubic numbers, yield periodic sequences etc. With regard to which, I have sketched out the foundation for a few things in a separate paper, in 4°.[17]

---

15. Leibniz omits "millia" [thousands], though the sense requires it.

16. characters. ▷ {1} (§13.) If the natural numbers are put down in order in binary, the period of the first column (from right to left) is 01. For the two first terms are 0 and 1, and subsequent numbers are formed by the continuous addition of 1s. Now if you add 1 to 1, a 0 returns, from which, by continuously adding a 1, everything will occur as from the beginning. {2} (§13.) The period of the second column is 0011 and of the third 0000 1111, and of the fourth 0000 0000 1111 1111, and so on, for there is a 0 at the beginning in the second column and a remaining 0 in the first period of the first [column]. Then, at the beginning of the second period of the first [column], it receives a second 1, and keeps it because nothing is changed in the remaining second period of the first [column]. But at the beginning of the third period of the first [column], a 1 is again received, whereby the 1 [already there] is changed to 0. Hence everything in each column returns as from the beginning. The process is the same in the third and subsequent columns. For each column has pure 0s, with the first one remaining from the preceding [column]; and it begins with 1 when the preceding ◁ deleted.

17. Leibniz is here referring to his paper "Demonstrations That Columns of Sequences Exhibiting Powers of Arithmetic Progressions, or Numbers Composed from These, Are Periodic," written in November 1701 (chapter 27).

00 0000	0
00 0001	1
00 0010	2
00 0011	3
00 0100	4
00 0101	5
00 0110	6
00 0111	7
00 1000	8
00 1001	9
00 1010	10
00 1011	11
00 1100	12
00 1101	13
00 1110	14
00 1111	15
01 0000	16
01 0001	17
01 0010	18
01 0011	19
01 0100	20
01 0101	21
01 0110	22
01 0111	23
01 1000	24
01 1001	25
01 1010	26
01 1011	27
01 1100	28
01 1101	29
01 1110	30
01 1111	31
10 0000	32
10 0001	33
10 0010	34
10 0011	35
10 0100	36
10 0101	37
10 0110	38
10 0111	39
10 1000	40
10 1001	41
10 1010	42
10 1011	43
10 1100	44
10 1101	45
10 1110	46
10 1111	47
11 0000	48
11 0001	49
11 0010	50
11 0011	51
11 0100	52
11 0101	53
11 0110	54
11 0111	55
11 1000	56
11 1001	57
11 1010	58
11 1011	59
11 1100	60
11 1101	61
11 1110	62
11 1111	63

0	0	¦
1	1	‖

00	0	¦¦
01	1	¦‖
10	2	‖¦
11	1	‖‖

000	0	¦¦¦
001	1	¦¦‖
010	2	¦‖¦
011	3	¦‖‖
100	4	‖¦¦
101	5	‖¦‖
110	6	‖‖¦
111	7	‖‖‖

0000	0	¦¦¦¦
0001	1	¦¦¦‖
0010	2	¦¦‖¦
0011	3	¦¦‖‖
0100	4	¦‖¦¦
0101	5	¦‖¦‖
0110	6	¦‖‖¦
0111	7	¦‖‖‖
1000	8	‖¦¦¦
1001	9	‖¦¦‖
1010	10	‖¦‖¦
1011	11	‖¦‖‖
1100	12	‖‖¦¦
1101	13	‖‖¦‖
1110	14	‖‖‖¦
1111	15	‖‖‖‖

Bibliography

Adamson, Donald. 1995. *Blaise Pascal. Mathematician, Physicist and Thinker about God.* Basingstoke: Palgrave.

Aiken, Howard, A. G. Oettinger, and T. C. Bartee. 1964. "Proposed Automatic Calculating Machine." *IEEE Spectrum* 1 (8): 62–69.

Aiton, E. J. 1981. "An Unpublished Letter of Leibniz to Sloane." *Annals of Science* 38 (1): 103–107.

Aiton, E. J. 1985. *Leibniz: A Biography.* Bristol: Adam Hilger.

Alexandre, Noel. 1700. *Conformite des ceremonies chinoises avec l'idolatrie grecque et romaine. Pour servir de confirmation à l'Apologie des Dominicains Missionnaires de la Chine.* Cologne: Corneille d'Egmond.

Andrés, Giovanni. 1790. *Dell'origine, progressi e stato attuale d'ogni letteratura.* Parma: Dalla Stamperia reale.

Anon. 1682. "Machines rares et surprenantes." *Journal des sçavans:* January 19, 16–18.

Anon. 1700. *Historia cultus Sinensium: seu varia scripta de cultibus Sinarum, inter vicarios apostolicos gallos aliosque missionarios, & patres societatis Jesu controversis, oblata Innocentio XII.* Cologne: n.p.

Anon. 1857. *Nuova enciclopedia popolare italiana, ovvero Dizionario Generale di Scienze, Lettere, Arti, Storia, Geografia, Ecc. Ecc.* 25 vols. Torino: Dalla Societa l'Unione Tipografico-Editrice.

Antognazza, Maria Rosa. 2009. *Leibniz: An Intellectual Biography.* Cambridge: Cambridge University Press.

Antognazza, Maria Rosa. 2018. "Leibniz as Historian." In *The Oxford Handbook of Leibniz,* edited by Maria Rosa Antognazza, 591–608. Oxford: Oxford University Press.

Ares, J., J. Lara, D. Lizcano, and M. A. Martinez. 2018. "Who Discovered the Binary System and Arithmetic? Did Leibniz Plagiarize Caramuel?" *Science and Engineering Ethics* 24:173–188.

Aristotle. 1984. *The Complete Works of Aristotle.* Edited by Jonathan Barnes. 2 vols. Princeton, NJ: Princeton University Press.

Astin, J. 1984. "Binary Expansion of Reciprocals." *International Journal of Mathematical Education in Science and Technology* 15 (4): 423–428.

Augustine, Saint. 1952. *The City of God, Books VIII–XVI.* Translated by Gerald G. Walsh and Grace Monahan. Washington, DC: Catholic University of America Press.

Bacon, Francis. 1620. *Novum Organum, sive indicia vera de interpretatione naturae.* London: Billium.

Bacon, Francis. 2000. *The New Organon.* Edited by Lisa Jardine and Michael Silverthorne. Cambridge: Cambridge University Press.

Barlow, Peter. 1814. *A New Mathematical and Philosophical Dictionary: Comprising an Explanation of the Terms and Principles of Pure and Mixed Mathematics, and Such Branches of Natural Philosophy as Are Susceptible of Mathematical Investigation.* London: Whittingham/Rowland.

Bauer, Friedrich L. 2010. *Origins and Foundations of Computing.* Dordrecht: Springer.

Bayer, Gottlieb Siegfried. 1730. *Museum sinicum in quo sinicae linguae et literaturae ratio explicatur. Tomus primus.* St. Petersburg: Academia imperatoria.

Beeley, Philip. 2020. "Leibniz and the Royal Society Revisited." In *Leibniz's Legacy and Impact,* edited by Julia Weckend and Lloyd Strickland, 23–52. New York: Routledge.

Belaval, Yvon. 2005. *Leibniz: Initiation à sa philosophie.* Paris: Vrin.

Bell, A. H. 1895. "The 'Cattle Problem' by Archimedes 251 B.C." *The American Mathematical Monthly* 2:140–141.

Bilfinger, Georg Bernhard. 1724. *Specimen doctrinae veterum sinarum moralis et politicae.* Frankfurt: Andreas and Hort.

Boerhaave, Herman. 1715. *Sermo academicus de comparando certo in physici.* Leiden: Vander.

Boole, George. 1854. *An Investigation of the Laws of Thought, on Which Are Founded the Mathematical Theories of Logic and Probabilities.* London: Walton/Maberly.

Bouvet, Joachim. 1697. *Portrait historique de l'Empereur de la Chine.* Paris: Michallet.

Bouvet, Joachim. 1704. "Extrait d'une lettre écrite à Mr. Leibnits par le P. Bouvet de la Compagnie de Jesus. A Pekin le 4. Novembre 1701." *Memoires pour l'histoire des sciences et des beaux arts:* January, 128–165.

Brancato, Mattia. 2021. "Leibniz's Binary Algebra and its Role in the Expression and Classification of Numbers." *Philosophia Scientiae* 25:71–94.

Brander, Georg Friedrich. 1775. *Arithmetica Binaria Sive Dyadica das ist Die Kunst nur mit zwey Zahlen in allen vorkommenden Fällen sicher und leicht zu rechnen.* Augsburg: Klett.

Brewster, David. 1832. *The Edinburgh Encyclopaedia.* 18 vols. Philadelphia: Joseph and Edward Parker.

Buffon, Comte de, Georges-Louis Leclerc. 1741. "Formule sur les échelles arithmetiques." *Histoire de l'Académie Royale des Sciences. Année MDCCXLI* (Paris): 219–221.

Cajori, Florian. 1916. "Leibniz's 'Image of Creation'." *The Monist* 26 (4): 557–565.

Cantor, Moritz. 1863. *Mathematische Beiträge zum Kulturleben der Volker.* Halle: Schmidt.

Caramuel, Juan. 1670. *Mathesis biceps, vetus et nova.* Campagna: Annison.

Carhart, Michael C. 2016. "Leibniz between Paris, Grand Tartary, and the Far East: Gerbillon's Intercepted Letter." In *China in the German Enlightenment,* edited by Bettina Brandt and Daniel Leonhard Purdy, 80–96. Toronto: University of Toronto Press.

Caze, César. 1705. *Letter to Gottfried Wilhelm Leibniz.* Translated by Lloyd Strickland. http://www.leibniz-translations.com/caze1705.htm.

Ceulen, Ludolph van. 1615. *Vanden circkel. Daer in gheleert werdt te vinden de naeste proportie des circkels-diameter tegen synen omloop daer door alle circkels (met alle figueren ofte landen met cromme linien besloten) recht ghemeten kunnen werde.* Delph: Andriesz.

Ching, Julia, and Willard G. Oxtoby. 1992. *Moral Enlightenment: Leibniz and Wolff on China.* New York: Routledge.

Clavius, Christopher. 1570. *In sphaeram Ioannis de Sacro Bosco commentarius.* Rome: Victorium Helianum.

Clüver, Philipp. 1663. *Germania antiqua, cum vindelicia et norico auctoris methodo, verbis & tabulis geographicis retentis contracta.* Wolfenbüttel: Buno.

Collani, Claudia von. 1988. "Tianxue benyi—Joachim Bouvets Forschungen zum Monotheismus." *China-Mission Studies (1550–1800)* Bulletin X:9–33.

Collani, Claudia von. 2007. "The First Encounter of the West with the Yijing: Introduction to and Edition of Letters and Latin Translations by French Jesuits from the 18th Century." *Momumenta Serica* 55:227–387.

Collignon, Eduard. 1897. "Note sur l'arithmétique binaire." *Journal de mathématiques élémentaires* 21:101–105.

Couturat, Louis. 1901. *La logique de Leibniz.* Paris: Félix Alcan.

Crippa, David. 2019. *The Impossibility of Squaring the Circle in the 17th Century. A Debate among Gregory, Huygens and Leibniz.* Cham: Birkhäuser.

Croker, Henry Temple, Thomas Williams, and Samuel Clark, eds. 1766. *The Complete Dictionary of Arts and Sciences. In Which the Whole Circle of Human Learning Is Explained, and the Difficulties Attending the Acquisition of Every Art, Whether Liberal or Mechanical, Are Removed, in the Most Easy and Familiar Manner.* London: J. Wilson & J. Fell.

Curione, Celio Secondo. 1617. *De amplitudine regni coelestis.* Frankfurt: Palthenius.

Dangicourt, Pierre. 1710. "De periodis columnarum in serie numerorum progressionis arithmeticae dyadice expressorum." *Miscellanea Berolinensia* 1:336–376.

Dascal, Marcelo. 2006. *Gottfried Wilhelm Leibniz: The Art of Controversies.* Dordrecht: Springer.

De Risi, Vincenzo. 2007. *Geometry and Monadology: Leibniz's Analysis Situs and Philosophy of Space.* Basel: Birkhäuser.

De Volder, Burcher. 1695. *Exercitationes academicae quibus Ren. Cartesii philosophia defenditur adversus Petri Danielis Huetii Episcopi Suessionensis Censuram philosophiae Cartesianae.* Amsterdam: Ravestein.

Descartes, René. 1898. *Oeuvres de Descartes: Correspondance II.* Edited by Charles Adam and Paul Tannery. Paris: Cerf.

Diaz, Don Francisco. 1640–1643. *Vocabulario de letra China con la explicacion Castellana hecho con gran propriedad y abundancia de palabras por el Padre Don Francisco Diaz de la Orden de Predicadores ministro incansable en esto Reyno de China.* Berlin. https://digital.staatsbibliothek-berlin.de/werkansicht/?PPN=PPN335 8707539.

Diderot, Denis, and Jean le Rond D'Alembert, eds. 1751. *Encyclopédie: ou dictionnaire raisonné des sciences, des arts et des métiers. Tome second. B–CEZ.* Paris: Briasson, David, Le Breton, and Durand.

Dotzler, Bernhard J. 2010. "Erschaffung aller Dinge aus Nichts: Leibniz und die Muster der Binarität." In *Die Enzyklopädik der Esoterik: Allwissenheitsmythen und universalwissenschaftliche Modelle in der Esoterik der Neuzeit,* edited by Andreas B. Kilcher and Philipp Theisohn, 21–34. Paderborn: Wilhelm Fink.

Eglash, Ron. 2005. *African Fractals: Modern Computing and Indigenous Design.* New Brunswick, NJ: Rutgers University Press.

Figuier, Louis. 1870. *Vies des savants illustrés du XVIIIe siècle: avec l'appréciation sommaire de leurs travaux.* Brussels: Lacroix, Verboeckhoven & co.

Fontenelle, Bernard le Bovier de. 1705a. *New Binary Arithmetic.* Translated by Lloyd Strickland. http://www.leibniz-translations.com/newbinary.htm.

Fontenelle, Bernard le Bovier de. 1705b. "Nouvelle arithmetique binaire." In *Histoire de l'Académie Royale des Sciences. Année MDCCIII. Avec les memoires de mathematique & de physique, pour la même année,* 58–63. Paris: Boudot.

Fontenelle, Bernard le Bovier de. 1705c. "Sur les tangentes et les secantes des angles." In *Histoire de l'Académie Royale des Sciences. Année MDCCIII. Avec les memoires de mathematique & de physique, pour la même année,* 64–65. Paris: Boudot.

Fontenelle, Bernard le Bovier de. 1718. "Eloge de M. Leibnitz." In *Histoire de l'Academie Royale des Sciences. Année M.DCCXVI. Avec les memoires de mathematique & de phisique, pour la même année,* 94–128. Paris: L'imprimerie Royale.

Gage, Thomas. 1696. *Histoire de l'Empire mexicain représentée par figures. Relation du Mexique ou de la Nouvelle Espagne, par Thomas Gages, traduite par Melchisédec Thévenot.* Paris: Moette.

Galilei, Galileo. 1638. *Discorsi E Dimostrazioni Matematiche.* Leiden: Elzevier.

Galilei, Galileo. 1914. *Dialogue Concerning Two New Sciences.* Translated by Henry Crew and Alfonso de Salvio. New York: Macmillan.

Garfinkel, Simson L., and Rachel H. Grunspan. 2018. *The Computer Book: From the Abacus to Artificial Intelligence, 250 Milestones in the History of Computer Science.* New York: Sterling.

Gimmestad, Beverly J. 1991. "The Russian Peasant Multiplication Algorithm: A Generalization." *The Mathematical Gazette* 75 (472): 169–171.

Glaser, Anton. 1981. *History of Binary and Other Nondecimal Numeration.* 2nd ed. Los Angeles: Tomash.

Golius, Jacob. 1655. "De regno catayo additamentum." In *Novus atlas sinensis,* edited by Martino Martini. Amsterdam: J. Blaeu.

Golius, Jacob. 1696. *Catalogus Insignium in omni facultate, linguisque, Arabica, Persica, Turcica, Chinensi &c. Librorum M.Ss.* Leiden: Vivie.

Grillet de Roven, René. 1673. *Curiositez mathematiques de l'invention du Sr Grillet horlogeur à Paris.* Paris: Coignard.

Guericke, Otto von. 1672. *Experimenta nova (ut vocantur) Magdeburgica de spatio vacuo.* Amsterdam: Waesberge.

Harriot, Thomas. 1590. *A briefe and true report of the new found land of Virginia: of the commodities and of the nature and manners of the naturall inhabitants. Discouered by the English colony there seated by Sir Richard Greinuile Knight in the yeere 1585. Which remained vnder the gouernement of twelue monethes; at the speciall charge and direction of the Honourable Sir Walter Raleigh Knight lord Warden of the stanneries who therein hath beene fauoured and authorised by her Maiestie and her letters patents.* Frankfurt: Johann Wechel, Theodor de Bry, and Sigmund Feyerabend.

Harsdörffer, Georg Philipp. 1651. *Delitiae mathematicae et physicae der Mathematischen und Philosophischen Erquickstunden. Zweiter Theil.* Nuremberg: Dümler.

Harsdörffer, Georg Philipp. 1653. *Delitiae mathematicae et physicae der Mathematischen und Philosophischen Erquickstunden. Dritter Theil.* Nuremberg: Endters.

Heinekamp, Albert. 1978. "Kochański als Leibniz-Korrespondent." *Organon* 14:73–106.

Heyse, Johann-Christian-August. 1859. *Allgemeines verdeutschendes und erklärendes Fremdwörterbuch.* 12th ed. Hanover: Hofbuchhandlung.

Hill, Thomas Wright. 1860. *Selections from the Papers of the Late Thomas Wright Hill, Esq. F.R.A.S.* London: John W. Parker / Son.

Hochstetter, Erich. 1966. *Herrn von Leibniz' Rechnung mit Null und Eins.* Berlin: Siemens.

Hoffmann, Friedrich. 1700. *Demonstrationes physicae curiosae, experimentis et observationibus mechanicis ac chymicis illustratae.* Halle: Zeitlerum.

Huet, Pierre-Daniel. 1690. *Censura philosophiae Cartesianae.* Frankfurt and Leipzig: Grenz.

Huet, Pierre-Daniel. 2003. *Against Cartesian Philosophy.* Edited and translated by Thomas M. Lennon. Amherst, NY: Humanity Books.

Hutton, Charles. 1795. *A Mathematical and Philosophical Dictionary: Containing an Explanation of the Terms, and an Account of the Several Subjects, Comprized under the Heads Mathematics, Astronomy, and Philosophy both Natural and Experimental.* 2 vols. London: J. Johnson and G. G. and J. Robinson.

Huygens, Christiaan. 1693. "De problemate Bernouilliano." *Acta eruditorum:* October, 475–476.

Huygens, Christiaan. 1698. Κοσμοθεωρός, *sive de terris coelestibus earumque ornatu conjecturae.* The Hague: Adriaen Moetjens.

Ifrah, Georges. 1998. *The Universal History of Numbers.* Translated by David Bellos, E. F. Harding, Sophie Wood, and Ian Monk. New York: John Wiley & Sons.

Ineichen, R. 2008. "Leibniz, Caramuel, Harriot und das Dualsystem." *Mitteilungen der deutschen Mathematiker-Vereinigung* 16 (1): 12–15.

Ingaliso, Luigi. 2017. "Leibniz e la lettura binaria dell'I Ching." In *Alterità e Cosmopolitismo nel Pensiero Moderno e Contemporaneo,* edited by Giancarlo Magnano San Lio and Luigi Ingaliso, 109–132. Soveria Mannelli: Rubbettino.

Intorcetta, Prospero, Christian Wolfgang Herdtrick, François de Rougemont, and Philippe Couplet, eds. and trans. 1687. *Confucius Sinarum philosophus.* Paris: Daniel Horthemels.

Irson, Claude. 1692. *Arithmetique pratique et raisonnée.* Paris: Chardon.

Jeake, Samuel. 1696. Λογιστολογία, *or Arithmetick Surveighed and Reviewed.* London: J. R / J. D.

Jones, Matthew L. 2018. "Calculating Machine." In *The Oxford Handbook of Leibniz,* edited by Maria Rosa Antognazza, 509–525. Oxford: Oxford University Press.

Jordan, G. J. 1927. *The Reunion of the Churches: A Study of G. W. Leibnitz and His Great Attempt.* London: Constable & Co.

Justinian. 1998. *The Digest of Justinian.* Translated by Alan Watson. 4 vols. Philadelphia, PA: University of Pennsylvania Press.

Kircher, Athanasius. 1650. *Obeliscus Pamphilius.* Rome: Grignani.

Kircher, Athanasius. 1652–1654. *Oedipus Aegyptiacus.* 3 vols. Rome: Mascardi.

Kircher, Athanasius. 1667. *China monumentis, qua sacris qua profanis, nec non variis naturæ & artis spectaculis, aliarumque rerum memorabilium argumentis illsutrata.* Amsterdam: Jansson.

Knobloch, Eberhard. 2004. "Leibniz and the Use of Manuscripts: Text as Process." In *History of Science, History of Text,* edited by K. Chemla, 51–79. Dordrecht: Springer.

Knobloch, Eberhard. 2015. "Analyticité, équipollence et théorie des courbes chez Leibniz." In *G. W. Leibniz, Interrelations between Mathematics and Philosophy,* edited by Norma B. Goethe, Philip Beeley, and David Rabouin, 89–110. New Studies in the History and Philosophy of Science and Technology 41. Dordrecht: Springer Netherlands.

Knobloch, Eberhard. 2018. "Determinant Theory, Symmetric Functions, and Dyadic." In *The Oxford Handbook of Leibniz,* edited by Maria Rosa Antognazza, 225–246. Oxford: Oxford University Press.

Knuth, Donald E. 1997. *The Art of Computer Programming, Volume 2: Seminumerical Algorithms.* 3rd ed. Upper Saddle River, NJ: Addison-Wesley Professional.

Kolpas, Sid. 2019. "Napier's Binary Chessboard Calculator." *Math Horizons* 27:25–27.

Kopp, Franz Otto, and Erwin Stein. 2008. "Es funktioniert doch—mit zwei Korrekturen: Zehnerüberträge der Leibniz-Maschine." *Unimagazain Hannover* 3/4:56–59.

Krist, Joseph. 1859. "Über Zahlensysteme und deren Geschichte." In *Vierter Jahres-Bericht der Kaiserlich-Königlichen Ober-Realschule der königlich freien Haupstadt Ofen*, 32–73. Ofen: Bagoschen.

L'Hospital, Guillaume-François-Antoine. 1696. *Analyse des infiniment petits, pour l'intelligence des lignes courbes*. Paris: L'imprimerie royale.

La Hire, Philippe de. 1695. *Traité de mécanique, ou l'on explique tout ce qui est nécessaire dans la pratique des Arts, & les propriétés des corps pesants lesquelles ont un plus grand usage dans la Physique*. Paris: Anisson.

Lagny, Thomas Fantet de. 1725. "Second memoire sur la goniometrie purement analytique, ou methode nouvelle & generale, pour déterminer exactement, lorsqu'il est possible, ou indéfiniment près lorsque l'exactitude est impossible, la valeur des trois angles de tout triangle rectiligne, soit rectangle, soit obliquangue, dont les trois côtés sont donnés en nombre; & cela par le seul calcul analytique sans tables des sinus, tangentes & secantes." In *Histoire de l'Académie Royale des Sciences. Année M.DCCXXV. Avec les memoires de mathematique & de physique, pour la même année*, 282–323. Paris: L'imprimerie royale.

Lamprecht, Jacob Friedrich. 1740. *Leben des Freyherrn Gottfried Wilhelm von Leibniz*. Berlin: Haude.

Laplace, Pierre-Simon de. 1825. *Essai philosophique sur les probilités*. 5th ed. Paris: Courcier.

Legge, James. 1871. *The Chinese Classics*. 7 vols. London: Trübner & Co.

Legge, James. 1882. *The Sacred Books of The East. Vol. XXVII: The Texts of Confucianism*. Translated by James Legge. Oxford: Clarendon Press.

Leibniz, Gottfried Wilhelm. 1682. "De vera proportione circuli ad quadratum circumscriptum in numeris rationalibus expressa." *Acta Eruditorum:* January, 41–46.

Leibniz, Gottfried Wilhelm. 1683. *Essay on the Number of Men*. Translated by Lloyd Strickland. http://www.leibniz-translations.com/numbermen.htm.

Leibniz, Gottfried Wilhelm. 1684. "Nova methodus pro maximis et minimis." *Acta eruditorum:* October, 457–473.

Leibniz, Gottfried Wilhelm. 1686a. "Brevis demonstratio erroris memorabilis Cartesii." *Acta eruditorum:* March, 161–163.

Leibniz, Gottfried Wilhelm. 1686b. "De geometria recondita et analysi indivisibilium atque infinitorum, Addenda his quae dicta sunt in Actis a. 1684, Maji p. 233; Octob. p. 264; Decemb. p. 586." *Acta Eruditorum:* June, 292–300.

Leibniz, Gottfried Wilhelm. 1689. "Tentamen de motuum coelestium causis." *Acta eruditorum:* February, 82–96.

Leibniz, Gottfried Wilhelm. 1691a. "De legibus naturae et vera aestimatione virium motricium contra Cartesianos." *Acta eruditorum:* September, 439–447.

Leibniz, Gottfried Wilhelm. 1691b. "De linea in quam exile se pondere proprio curvat." *Acta eruditorum:* June, 277–281.

Leibniz, Gottfried Wilhelm. 1692a. "De la chainette, ou solution d'un Problème fameux proposé par Galilei." *Journal des sçavans:* March, 147–153.

Leibniz, Gottfried Wilhelm. 1692b. "Generalia de natura linearum, anguloque contactus & osculi, provolutionibis aliisque cognatis, & eorum usibus nonnullis." *Acta eruditorum:* September, 440–446.

Leibniz, Gottfried Wilhelm, ed. 1693. *Codex juris gentium diplomaticus*. Hanover: n.p.

Leibniz, Gottfried Wilhelm. 1694. *Binary Division*. Translated by Lloyd Strickland. http://www.leibniz-translations.com/division.htm.

Leibniz, Gottfried Wilhelm. 1695a. *My Arithmetic Machine*. Translated by Lloyd Strickland. http://www.leibniz-translations.com/machine.htm.

Leibniz, Gottfried Wilhelm. 1695b. "Specimen dynamicum." *Acta eruditorum:* April, 145–157.

Leibniz, Gottfried Wilhelm. 1696. *Sketch for a Medal Commemorating His Formula for the Quadrature of the Circle*. Translated by Lloyd Strickland. http://www.leibniz-translations.com/medal.htm.

Leibniz, Gottfried Wilhelm. 1697. *Novissima sinica*. n.p.: n.p.

Leibniz, Gottfried Wilhelm. 1699. *Reflections on Locke's Second Reply*. Translated by Lloyd Strickland. http://www.leibniz-translations.com/locke.htm.

Leibniz, Gottfried Wilhelm, ed. 1700. *Mantissa codicis juris gentium diplomatici*. Hanover: Freytag.

Leibniz, Gottfried Wilhelm. 1701. *Letter to Jean de Fontaney.* Translated by Lloyd Strickland. http://www.leibniz -translations.com/fontaney1701.htm.

Leibniz, Gottfried Wilhelm. 1703a. *Binary and the Hexagrams of Fuxi.* Translated by Lloyd Strickland. http://www.leibniz-translations.com/fuxi.htm.

Leibniz, Gottfried Wilhelm. 1703b. *Letter to Jean Fontaney.* Translated by Lloyd Strickland. http://www.leibniz -translations.com/fontaney1703.htm.

Leibniz, Gottfried Wilhelm. 1705a. "Explication de l'aritmètique binaire, qui se sert des seuls caracteres 0 & 1; avec des Remarques sur son utilité, & sur ce qu'elle donne le sens des anciennes figures Chinoises de Fohy." In *Memoires de l'Académie Royale des Sciences. Année MDCIII. Avec les Memoires de Mathematique & de Physique, pour la même Année,* 85–89. Paris: Boudot.

Leibniz, Gottfried Wilhelm. 1705b. *Letter to Ferdinand Orban.* Translated by Lloyd Strickland. http://www .leibniz-translations.com/orban.htm.

Leibniz, Gottfried Wilhelm. 1718. *Otium Hanoverarum, sive miscellanea.* Edited by Joachim Friedrich Feller. Leipzig: J. C. Martini.

Leibniz, Gottfried Wilhelm. 1720. *Lehr-Sätze über die Monadologie: ingleichen von Gott und seiner Existentz, seinen Eigenschaften und von der Seele des Menschen.* Edited by Heinrich Köhler. Frankfurt: Meyers.

Leibniz, Gottfried Wilhelm. 1739. *Leibnitii Tentamina Theodicaeae.* Edited by Johann Ulrich Steinhofer. 2 vols. Frankfurt: Berger.

Leibniz, Gottfried Wilhelm. 1740. *Kleinere Philosophische Schriften.* Edited by Caspar Jacob Huth. Jena: Mayer.

Leibniz, Gottfried Wilhelm. 1742. "The True Proportion of the Circle to the Circumscribed Square, in Rational Numbers, Trans. A Society of Gentlemen." *Acta Germanica* (London) 1:59–62.

Leibniz, Gottfried Wilhelm. 1744. *Theodicee, das ist, Versuch von der Güte Gottes, Freyheit des Menschen, und vom Ursprunge des Bösen.* Translated by Johann Christian Gottsched. Hanover: Försters/Erben.

Leibniz, Gottfried Wilhelm. 1768. *G. G. Leibnitii Opera Omnia.* Edited by Ludovic Dutens. 6 vols. Geneva: Fratres de tournes.

Leibniz, Gottfried Wilhelm. 1845. *Godofredi Guilielmi L. B. de Leibnitz Lipsiensis epistolae XLVI at Teuberum concionatorem aulae cizensis.* Leipzig: Staritz.

Leibniz, Gottfried Wilhelm. 1849–1863. *Leibniziens Mathematische Schriften.* Edited by C. I. Gerhardt. 7 vols. Halle: H. W. Schmidt.

Leibniz, Gottfried Wilhelm. 1857. *Nouvelles lettres et opuscules inédits de Leibniz.* Edited by A. Foucher de Careil. Paris: Durand.

Leibniz, Gottfried Wilhelm. 1875–1890. *Die Philosophischen Schriften von Leibniz, Gottfried Wilhelm.* Edited by C. I. Gerhardt. 7 vols. Berlin: Weidmann.

Leibniz, Gottfried Wilhelm. 1903. *Opuscules et fragments inédits de Leibniz.* Edited by Louis Couturat. Paris: Félix Alcan.

Leibniz, Gottfried Wilhelm. 1923–. *Sämtliche Schriften und Briefe.* Edited by Berlin-Brandenburgische Akademie der Wissenschaften. 8 series, each divided into multiple volumes. Berlin: Akademie Verlag.

Leibniz, Gottfried Wilhelm. 1969. *Leibniz, Gottfried Wilhelm: Philosophical Papers and Letters.* Edited and translated by Leroy E. Loemker. 2nd ed. Dordrecht: D. Reidel.

Leibniz, Gottfried Wilhelm. 1971. *Calcolo con zero e uno.* Translated by Lia Ruffino. Milan: Etas Kompass.

Leibniz, Gottfried Wilhelm. 1973. *Leibniz: Philosophical Writings.* Edited and translated by Mary Morris and G. H. R. Parkinson. London: Everyman.

Leibniz, Gottfried Wilhelm. 1977. *Leibniziens Mathematische Schriften.* Edited by C. J. Gerhardt. Hildesheim: Georg Olms.

Leibniz, Gottfried Wilhelm. 1992. *De Summa Serum: Metaphysical Papers 1675–1676.* Edited and translated by G. H. R. Parkinson. New Haven, CT: Yale University Press.

Leibniz, Gottfried Wilhelm. 1993. "An Essay on the Causes of Celestial Motions." In *Equivalence and Priority: Newton versus Leibniz,* edited by Domenico Bertoloni Meli, 126–142. Oxford: Clarendon Press.

Leibniz, Gottfried Wilhelm. 1994. *Writings on China.* Edited and translated by Daniel J. Cook and Henry Rosemont Jr. Chicago: Open Court.

Leibniz, Gottfried Wilhelm. 2001. "Two Papers on the Catenary Curve and Logarithmic Curve." Translated by Pierre Beaudry. *Fidelio* 10 (1): 55–58.

Leibniz, Gottfried Wilhelm. 2006a. *Der Briefwechsel mit den Jesuiten in China (1689–1714)*. Edited by Rita Widmaier. Hamburg: Felix Meiner Verlag.

Leibniz, Gottfried Wilhelm. 2006b. *Shorter Leibniz Texts*. Edited and translated by Lloyd Strickland. London: Continuum.

Leibniz, Gottfried Wilhelm. 2011a. *Gottfried Wilhelm Leibniz: Die mathematischen Zeitschriftenartikel*. Edited by H.-J. Hess and Malte-L. Babin. Hildesheim: Georg Olms Verlag.

Leibniz, Gottfried Wilhelm. 2011b. *Leibniz and the Two Sophies*. Edited and translated by Lloyd Strickland. Toronto: CRRS.

Leibniz, Gottfried Wilhelm. 2013. *The Leibniz–De Volder Correspondence*. Edited and translated by Paul Lodge. New Haven, CT: Yale University Press.

Leibniz, Gottfried Wilhelm. 2016a. *Leibniz on God and Religion*. Edited and translated by Lloyd Strickland. London: Bloomsbury.

Leibniz, Gottfried Wilhelm. 2016b. *The Leibniz-Arnauld Correspondence*. Edited and translated by Stephen Voss. New Haven, CT: Yale University Press.

Leibniz, Gottfried Wilhelm. 2020. *Dissertation on Combinatorial Art*. Edited by Massimo Mugnai, Han van Ruler, and Martin Wilson. Oxford: Oxford University Press.

Lewis, Harry R., ed. 2021. *Ideas That Created the Future*. Cambridge: MIT Press.

Leybourn, William. 1700. *Arithmetick, Vulgar, Decimal, Instrumental, Algebraical. In Four Parts*. 7th ed. London: J. Matthews.

Lieber, Francis, E. Wigglesworth, and T. G. Bradford, eds. 1854. *Encyclopædia Americana: A Popular Dictionary of Arts, Sciences, Literature, History, Politics and Biography*. Vol. IX. Boston: B. B. Mussey & Co.

Loosen, Renate, and Franz Vonessen. 1968. *Zwei Briefe über das binäre Zahlensystem und die chinesische Philosophie*. Stuttgart: Belser.

Lübsen, Heinrich B. 1869. *Ausführliches Lehrbuch der Arithmetik und Algebra: zum Selbstunterricht und mit Rücksicht auf die Zwecke des practischen Lebens*. Leipzig: Brandstetter.

Lucas, Édouard. 1891. *Récréations mathématiques*. Paris: Gauthier-Villars.

Ludovici, Carl Günther. 1737. *Ausführlicher Entwurff einer vollständigen Historie der Leibnitzischen Philosophie zum Gebrauch seiner Zuhörer*. 2 vols. Leipzig: Johann Georg Löwe.

Mackensen, Ludolf von. 1974. "Leibniz als Ahnherr der Kybernetik—ein bisher unbekannter Leibnizscher Vorschlag einer 'Machina arithmeticae dyadicae'." In *Akten des II Internationalen Leibniz-Kongresses*, II:255–268. Wiesbaden: Steiner Verlag.

Madrolle, Claudius. 1901. *Les premiers voyages français à la Chine: la Compagnie de la Chine, 1698–1719*. Paris: Challamel.

Malebranche, Nicolas. 1692. *Des loix de la communication des mouvements*. Paris: Pralard.

Marcellinus, Ammianus. 1935. *History*. Translated by John C. Rolfe. 3 vols. London: Heinemann.

Mariotte, Edme. 1673. *Traité de la percussion ou choc des corps, dans lequel des principales regles du mouvement, contraires à celles que Mr Descartes, & quelques autres modernes ont voulu establir, sont demonstrées par leur veritable causes*. Paris: Michallet.

Martin, Bruce A. 1968. "Letters to the Editor: On Binary Notation." *Communications of the ACM* 11 (10): 658.

Martini, Martino. 1655. "Novus Atlas Sinensis." In *Theatrum orbis terrarum, sive Atlas Novus, Pars VI*, edited by Willem Blaeu and Joan Blaeu. Amsterdam: Blaeu.

Martini, Martino. 1658. *Sinicae historiae decas prima*. Munich: Lucas Straub.

McClatchie, Thomas. 1876. *A Translation of the Confucian Yi-king*. Shanghai: American Presbyterian Mission Press.

Merzbach, Uta C., and Carl B. Boyer. 2011. *A History of Mathematics*. 3rd ed. Hoboken, NJ: Wiley.

Mohl, Julius, ed. 1834, 1839. *Y-King antiquissimus Sinarum liber quem ex latina interpretatione P. Regis aliorumque ex Soc Jesu P.P.* 2 vols. Stuttgart: Cotta.

Montucla, Jean Étienne. 1799. *Histoire des mathématiques. Tome premier*. Paris: Agasse.

Morar, Florin-Stefan. 2015. "Reinventing Machines: The Transmission History of the Leibniz Calculator." *British Journal for the History of Science* 48 (1): 123–146.

Morland, Samuel. 1673. *The Description and Use of Two Arithmetick Instruments*. London: Moses Pitt.

Morrison, Theron V. 1952. "A Short Method of Long Division." *Journal of Applied Social Psychology* 52 (9): 684–686.

Müller, Andreas. 1671. *Disquisitio geographica & historica de Chataja.* Berlin: Rungian.

Müller, Andreas. 1674. *Inventum Brandenburgicum sive Andreae Mulleri Greiffenhagii, Praepositi Berlinensis, Proposito super Clave sua Sinica.* n.p.: n.p.

Mungello, David E. 1977. *Leibniz and Confucianism: The Search for Accord.* Honolulu: The University Press of Hawaii.

Napier, John. 1617. *Rabdologiæ.* Edinburgh: Andreas Hart.

Napier, John. 1990. *Rabdology.* Translated by William F. Richardson. Cambridge: MIT Press.

Napier, John. 2017. *The Life and Works of John Napier.* Edited and translated by Brian Rice, Enrique González-Velasco, and Alexander Corrigan. Cham: Springer.

Neumann, J. von. 1993. "First Draft of a Report on the EDVAC." *IEEE Annals of the History of Computing* 15 (4): 27–75.

Newman, James R., ed. 1956. *The World of Mathematics.* New York: Simon and Schuster.

Newton, Isaac. 1671/2. "A letter of Mr. Isaac Newton, Professor of the Mathematicks in the University of Cambridge; containing his new theory about light and colors: sent by the author to the publisher from Cambridge, Febr. 6. 1671/72; in order to be communicated to the R. Society." *Philosophical Transactions* 80 (February): 3075–3087.

Nilous, Horapollo. 1840. *The Hieroglyphics of Horapollo Nilous.* Translated by Alexander Turner Cory. London: William Pickering.

Nolte, Rudolph August. 1734. *Gottfried Wilhelms Baron von Leibnitz Mathematischer Beweis der Erschaffung und Ordnung der Welt.* Leipzig: Langenheim.

Nystrom, John W. 1862. *Project of a New System of Arithmetic, Weight, Measure and Coins, Proposed to be Called the Tonal System, with Sixteen to the Base.* Philadelphia, PA: J. B. Lippincott & Co.

Nystrom, John W. 1863. "On a New System of Arithmetic and Metrology, Called the Tonal System." *Journal of the Franklin Institute* 76:263–275, 337–348, 402–407.

Ozanam, Jacques. 1790. *Récréations mathématiques et physiques. Nouvelle édition, totalement refondue et considerablement augmentée par M. de M***. Tome premier.* Paris: Firmin Didot.

Papin, Denis. 1695. *Fasciculus dissertationum de novis quibusdam machinis atque aliis argumentis philosophicis.* Marburg: Jacob Estienne.

Pappas, Theoni. 1991. *More Joy of Mathematics.* San Carlos, CA: Wide World Publishing.

Parent, Antoine. 1700. *Élémens de méchanique et de physique.* Paris: Florentin/Laulne.

Pascal, Blaise. 1645. *Lettre dédicatoire à Monseigneur le Chancelier sur le sujet de la machine nouvellement inventée par le Sieur B.P. pour faire toutes sortes d'opérations d'arithmétique, par un mouvement reglé, sans plume ny jettons. Avec un advis necessaire à ceux qui auront curiosité de voir ladite machine, & de s'en servir.* n.p.: n.p.

Pascal, Blaise. 1665. *Traité du triangle arithmetique, avec quelques autres petits traitez sur la mesme materiere.* Paris: Guillaume Desprez.

Pasini, Enrico. 2016. "Leibniz's Concept of Unity and Its Aristotelian Origins." In *Beiträge zu Leibniz' Rezeption der Aristotelischen Logik und Metaphysik,* edited by J. A. Nicolas and N. Öffenberger, 325–332. Hildesheim: Olms.

Peano, Giuseppe. 1898. "La numerazione binaria applicata alla stenografia." *Accademia reale delle scienze di Torino* 34:4–11.

Peet, E. Eric, ed. 1923. *The Rhind Mathematical Papyrus, British Museum 10057 and 10058.* Translated by T. Eric Peet. London: Hodder/Stoughton.

Pelicanus, Wenceslaus Josephus. 1712. *Arithmeticus Perfectus Qui tria numerare nescit. Seu Arithmetica dualis, In qua numerando non proceditur, nisi ad duo, & tamen omnes quaestiones arithmeticae negotio facili enodari possunt.* Prague: Rosenmüller.

Petrocelli, Carla. 2019. "Konrad Zuse and His Plankalkül." *International Journal of Humanities and Arts Computing* 13 (1–2): 249–265.

Petty, William. 1682. *An Essay Concerning the Multiplication of Mankind: Together With Another Essay in Political Arithmetick, Concerning the Growth of the City of London.* London: Mark Pardoe.

Pignoria, Lorenzo. 1669. *Mensa Isiaca, qua sacrorum apud Aegyptios ratio & simulacra subjectis tabulis aenis simul exhibentur & explicantur.* Amsterdam: Frisi.

Privat-Deschanel, Augustin, and Adolphe-Jean Focillon. 1864. *Dictionnaire général des sciences théoriques et appliquées. Ire Partie.* Paris: Tandou and Garnier Brothers.

Probst, Siegmund. 2018. "The Calculus." In *The Oxford Handbook of Leibniz,* edited by Maria Rosa Antognazza, 211–224. Oxford: Oxford University Press.

Pythagorean Society. 1672. *Tetractyn tetracty Pythagoreae correspondentem, ut primum disceptationum suarum specimen, ulteriori curiosorum industriae exponit Societas Pythagorea in Alma Salana.* Jena: Meyer.

Qian, Sima. 1993. *Records of the Grand Historian: Han Dynasty II.* Rev. ed. Translated by Burton Watson. New York: Columbia University Press.

Rabouin, David. 2015. "The Difficulty of Being Simple: On Some Interactions Between Mathematics and Philosophy in Leibniz's Analysis of Notions." In *G. W. Leibniz, Interrelations between Mathematics and Philosophy,* edited by Norma B. Goethe, Philip Beeley, and David Rabouin, 49–74. New Studies in the History and Philosophy of Science and Technology 41. Dordrecht: Springer Netherlands.

Reyneau, Charles René. 1708. *Analyse démontrée, ou la méthode de résoudre les problèmes des mathématiques, et d'apprendre facilement ces sciences.* Paris: Quillau.

Riccioli, Giovanni Battista. 1672. *Geographiae et hydrographiae reformatae, nuper recognitae & auctae, libri duodecim.* Venice: LaNoù.

Rivest, R., A. Shamir, and L. Adleman. 1978. "A Method for Obtaining Digital Signatures and Public-Key Cryptosystems." *Communications of the ACM* 21 (2): 120–126.

Rudolph, Hartmut. 2018. "Scientific Organization and Learned Societies." In *The Oxford Handbook of Leibniz,* edited by Maria Rosa Antognazza, 543–562. Oxford: Oxford University Press.

Ryan, Michael T. 1991. "The Diffusion of Science and the Conversion of the Gentiles in the Seventeenth Century." In *In the Presence of the Past,* edited by Richard T. Bienvenu and Mordechai Feingold, 9–40. Dordrecht: Springer.

Schlömilch, Oskar, and B. Witzschel. 1858. *Zeitschrift für Mathematik und Physik.* Vol. 3. Leipzig: Teubner.

Schott, Gaspar. 1658. *Magia universalis naturae & artis operis quadripartiti. Tomus tertius & quartus.* Würzburg: Schönwetter.

Schwenter, Daniel. 1636. *Deliciae physico-mathematicae oder Mathematische und philosophische Erquickstunden.* Nuremberg: Dümmler.

Serra, Yves. 2010a. *Le manuscrit «De Progressione Dyadica» de Leibniz.* http://www.bibnum.education.fr/sites/default/files/leibniz-analyse.pdf.

Serra, Yves. 2010b. *Traduction du fac-similé du manuscrit de Leibniz du 15 mars 1679, «De Progressione Dyadica».* http://www.bibnum.education.fr/sites/default/files/leibniz-numeration-texte.pdf.

Shirley, John W. 1951. "Binary Numeration before Leibniz." *American Journal of Physics* 19:452–454.

Silva y Figueroa, García de. 1667. *L'ambassade de D. Garcia de Silva Figueroa en Perse, contenant la politique de ce grand Empire, les moeurs du Roy Schach Abbas, & une relation exacte de tous les lieux de Perse & des Indes, où cet ambassadeur a esté l'espace de huit années qu'il y a demeuré.* Translated by Abraham de Wicquefort. Paris: Billaine.

Stein, E., F. O. Kopp, K. Wiechmann, and G. Webe. 2006. "Calculemus! New Research Results and Functional Models Regarding Leibniz' Four Function Calculating Machine and Binary Calculating Machine." *Foundations of Civil and Environmental Engineering* 7:319–332.

Stein, Erwin. 2006. "Neue Forschungsergebnisse und Funktionsmodelle zur dezimalen Vier-Spezies-Rechenmaschine von und zur dyadischen Rechenmaschine nach Leibniz." In *Jahrbuch 2006 der Braunschweigischen Wissenschaftlichen Gesellschaf,* 56–57. Brunswick: J. Cramer Verlag.

Strabo. 1928. *The Geography of Strabo.* Translated by Horace Leonard Jones. 8 vols. London: Heinemann.

Strickland, Lloyd. 2005. "How Modern Was Leibniz's Biology?" *Studia Leibnitiana* 37 (2): 186–207.

Strickland, Lloyd. 2014. *Leibniz's Monadology: A New Translation and Guide.* Edinburgh: Edinburgh University Press.

Strickland, Lloyd. 2023. "Consilium Aegyptiacum." In *Companion to Leibniz's Political Thought and Activities,* edited by Luca Basso, Wenchao Li, Jaime de Salas, and Hartmut Rudolph. Berlin: De Gruyter.

Struik, Dirk Jan, ed. 1986. *A Source Book in Mathematics, 1200–1800.* Princeton, NJ: Princeton University Press.

Suetonius. 1914. *The Lives of the Caesars*. Translated by J. C. Rolfe. Vol. I. Cambridge, MA: Harvard University Press.

Swedenborg, Emanuel. 1754. "A Curious Memoir of M. Emanuel Swedenborg Concerning Charles XII of Sweden." *The Gentleman's Magazine, and Historical Chronicle* (London) XXIIII:423–424.

Swetz, Frank J. 2003. "Leibniz, the Yijing, and the Religious Conversion of the Chinese." *Mathematics Magazine* 76 (4): 276–291.

Swinton, John. 1789. *A Proposal for Uniformity of Weights and Measures in Scotland, By Execution of the Laws now in Force. With Tables of the English and Scotch Standards, and of the Customary Weights and Measures of the several Counties and Boroughs of Scotland;-Comparisons of the Standards with Each Other, and with the County-Measures;—Tables and Rules for their Reciprocal Conversion;-And some Tables of the Weight and Produce of Corn, &c*. 2nd ed. Edinburgh: Peter Hill.

Sypniewski, Bernard Paul. 2005. "China and Universals: Leibniz, Binary Mathematics, and the Yijing Hexagrams." *Monumenta Serica* 53:287–314.

Tachard, Guy. 1686. *Voyage de Siam, des peres Jesuites, envoyez par le Roy aux Indes & à la Chine*. Paris: Seneuze and Horthemels.

Tentzel, Wilhelm Ernst. 1696. *Epistola de sceleto elephantino Tonnae nuper effosso*. Gotha: Reyher.

Tentzel, Wilhelm Ernst. 1705. "Erklärung der Arithmeticae binariae, welche vor 3000 Jahren bey den Chinesern in Gebrauch gewesen und bißher bey ihnen selbst verlohren, neulich aber bey uns wieder funden worden." In *Curieuse Bibliothec oder Fortsetzung der Monatlichen Unterredungen*, edited by Wilhelm Ernst Tentzel, 81–112. Frankfurt: Stock.

Tentzel, Wilhelm Ernst. 2006. *Correspondence between Leibniz and Tentzel*. Edited and translated by Lloyd Strickland. http://www.leibniz-translations.com/tentzel.htm.

Terrentius, Joannes, and Johannes Kepler. 1630. *R.P.J. Terrentii e Societate Jesu Epistolium ex regno Sinarum ad mathematicos Europæos missum: cum Commentatiuncula Joannis Keppleri mathematici*. Żagań-Silesia: Cobius and Wiske.

Thévenot, Melchisédech. 1663. "Description des antiquitez de Persepolis, appellées maintenant Chimilnar, traduite de l'Anglois." In *Relations de divers voyages curieux, qui n'ont point esté publiées, ou qui ont esté traduites d'Hacluyt, de Purchas, & d'autres voyageurs Anglois, Hollandois, Portugais, Alemands, Espagnols; et de quelques Persans, Arabes, & autres autheurs orientaux. Premiere partie*. Paris: Langlois.

Thomas, Antoine, Filippo Grimaldi, Thomas Pereira, Jean-Francois Gerbillon, Joseph Suares, Joachim Bouvet, Kilian Stumpf, Jean-Baptiste Regis, Ludovicus Pernoti, and Dominique Parrenin. 1701. *Brevis relatio eorum, quae spectant ad declarationem Sinarum Imperatoris Kam Hi circa caeli, Cumfucii, et avorum cultum, datam anno 1700*. Peking: n.p.

Tropfke, Johannes. 1902. *Geschichte der Elementar mathematik in systematischer Darstellung. Erster Band: rechnen und Algebra*. Leipzig: Veit & Co.

Tropfke, Johannes. 1980. *Geschichte der Elementarmathematik. Band 1: Arithmetik und Algebra*. 4th ed. Berlin: De Gruyter.

Valeriano, Pierio. 1556. *Hieroglyphica, sive de sacris Aegyptiorum literis commentarii*. Basel: [Michael Isengrin].

Vieth, Gerhard Ulrich Anton. 1796. *Anfangsgründe der Mathematik: Erster Theil, Arithmetik und Geometrie*. Leipzig: Johann Ambrosius Barth.

Wakefield, Andre. 2010. "Leibniz and the Wind Machines." *Osiris* 25 (1): 171–188.

Weber, Heinric. 1922. *Enzyklopadie der Elementarmathematik: Ein Handbuch für Lehrer und Studierende. Erster Band. Arithmetik, Algebra und Analysis*. 4th ed. Leipzig: Teubner.

Weigel, Erhard. 1673. *Tetractys, Summum tum Arithmeticae tum Philosophiae discursivae compendium, artis magnae sciendi*. Jena: Meyer.

Weigel, Erhard. 1687a. "Aretologistica, die Tugendübende Rechen-Kunst." In *Wienerischer Tugend-Spiegel*. Nuremberg: Endtern.

Weigel, Erhard. 1687b. "Der Grund aller Tugenden, Nemlich eine Mathematische Demonstration wider alle Atheisten, Daraus man Rechenschafftlich erkennen kan dass ein Gott sey." In *Wienerischer Tugend-Spiegel*. Nuremberg: Endtern.

Weigel, Erhard. 1687c. "Kurzer Bericht auf eingenommenen Augenschein eines guten Grundes J. M. P. P. von dem Muster einer auf die Aretologistic gegründeten Tugend-Schul zu Jena." In *Wienerischer Tugend-Spiegel*, 100–112. Nuremberg: Endtern.

Weigel, Erhard. 1687d. "Logisticae virtutum genitricis, Das ist: Tugendübender Zahlen-Rechen-Kunst." In *Wienerischer Tugend-Spiegel,* own pagination. Nuremberg: Endtern.

Weigel, Erhard. 1687e. *Wienerischer Tugend-Spiegel.* Nuremberg: Endtern.

Whitney, William Dwight. 1895. *The Century Dictionary. An Encyclopedic Lexicon of the English Language. Volume VII.* New York: The Century Co.

Wideburg, Johann Bernhard. 1718. *Dissertatio mathematica, de praestantia arithmeticae binariae prae decimali.* Jena: Krebsian.

Wieleitner, Heinrich, and Anton von Braunmühl. 1911. *Geschichte der mathematik. T.2, Von Cartesius bis zur Wende des 18. Jahrhunderts; von Heinrich Wieleitner; bearbeitet unter Benutzung des Nachlasses von Anton von Braunmühl. Hälfte 1, Arithmetik, Algebra, Analysis.* Leipzig: G.J. Göschen.

Wiener, Norbert. 1961. *Cybernetics, or Control and Communication in the Animal and the Machine.* 2nd ed. Cambridge, MA: MIT Press.

Wilhelm, Richard, and Cary F. Baynes. 1950. *The I Ching or Book of Changes.* Princeton, NJ: Princeton U. Press.

Wilkins, John. 1668. *An Essay Towards a Real Character, and a Philosophical Language.* London: Gellibrand.

Will, Georg Andreas. 1778. *Bemerkungen über einige Gegenden des katholischen Deutschlands auf einer kleinen gelehrten Reise gemachet Nebst sechs noch ungedruckten Leibnitzischen Briefen.* Nuremberg: Lochnerischen.

Witsen, Nicolaas. 1687. *Nieuwe Lantkaarte van het Noorder en Oosterdeel van Asia en Europa, strekkende van Nova Zemla tot China.* [Amsterdam]: n.p.

Wolf, Mark J. P. 2000. *Abstracting Reality: Art, Communication, and Cognition in the Digital Age.* Lanham, NY: Oxford University Press of America.

Zacher, Hans J., ed. 1973. *Die Hauptschriften zur Dyadik von G. W. Leibniz.* Frankfurt: Vittorio Klostermann.

Zhang, Dainian. 2002. *Key Concepts in Chinese Philosophy.* Translated by Edmund Ryden. New Haven, CT: Yale University Press.

Zhonglian, Shi. 2000. "Leibniz's Binary System and Shao Yong's Xiantiantu." In *Das Neueste über China: G.W. Leibnizens Novissima Sinica von 1697,* edited by Wenzhao Li and Hans Poser, 165–169. Stuttgart: Franz Steiner Verlag.

Zuse, Konrad. 1980. "Some Remarks on the History of Computing in Germany." In *A History of Computing in the Twentieth Century,* edited by N. Metropolis, J. Howlett, and G. C. Rota, 611–627. Saint Louis: Elsevier Science & Technology.

Zuse, Konrad. 1987. *Computer Design: Past, Present, Future: Talk in Lund, Sweden.* https://history.computer.org/pioneers/zuse.html.

Zuse, Konrad. 1993. *The Computer—My Life.* Translated by Patricia McKenna and J. Andrew Ross. Berlin: Springer.

Index